Environmental and Energetics Series

Edited by A. K. Gupta and D. G. Lilley

Edward B. Magrab

Computer Integrated Experimentation

With 167 Figures

Springer-Verlag
Berlin Heidelberg NewYork
London Paris Tokyo Hong Kong Barcelona

Editors of the series:

Professor Ashwani K. Gupta
Dept. of Mechanical Engineering
University of Maryland
College Park, MD 20742
USA

Professor David G. Lilley
School of Mech. and Aero. Eng.
Oklahoma State University
Stillwater, OK 74078-0545
USA

Author:

Professor Edward B. Magrab
Dept. of Mechanical Engineering
College of Engineering
University of Maryland
College Park, MD 20742
USA

ISBN-13: 978-3-642-95640-9   e-ISBN-13: 978-3-642-95638-6
DOI: 10.1007/978-3-642-95638-6

© Springer-Verlag Berlin, Heidelberg 1991
Softcover reprint of the hardcover 1st edition 1991

61/3020-543210 – Printed on acid-free paper

To June
and to the memory of
Ramesh N. Vaishnav, friend and teacher

Preface

The rapidly growing and widespread integration of computers to control and manage experiments and processes marks the dramatic change that has taken place in the use of instrumentation for experimentation. In recognizing these changes this book attempts to provide a basis by which instruments and transducers can be selected, assembled and integrated with a computer to measure and control physical processes in an accurate and predictable manner. It has as its goals to encourage the reader (a) to be aware of the complexities of integrating the computer with the experiment and what can be done to minimize them, (b) to acquire a global outlook towards computer integration and take advantage of the attributes common to all instruments and transducers, (c) to understand the major sources and magnitudes of errors in a system's components, (d) to use standard performance metrics to specify and judge the quality of instruments and transducers, and (e) to realize that computers can be used for more than data acquisition, they can be used for control.

In order to discuss the integration of computers with the constitutive components of an experimentation system, one has to modify the traditional approach to experimentation by emphasizing the component's common characteristics and de-emphasizing their functional uniqueness. To accomplish this the book consists of two parts. The first part lays the theoretical foundation for the second part. It starts with some introductory remarks concerning accuracy and repeatability. Then the Fourier analysis of signals are summarized and illustrated with numerous examples emphasizing ideas that recur throughout the remaining chapters. The next three chapters introduce, from a systems point of view, the fundamental components of the great majority of instrumentation systems: filters, amplifiers and analog-to-digital converters. The sixth chapter introduces several common, but important, general purpose instruments.

The first chapter of the book's second part introduces the considerations that must be given to the timing of the computer's interaction with its instruments, transducers and actuators and suggests procedures that can be used to insure that computers will be correctly integrated with these devices. It also discusses several of the most common communication interfaces. This is followed by a chapter that summarizes the more widely used transducers (with electrical output) and actuators, and organizes them in the traditional manner: according to the type of physical quantity measured or controlled. It also illustrates the use the majority of these transducers and actuators with discussions of specific computer integrated experiments taken primarily from the author's most recent research and from a newly developed instructional laboratory in the Department of Mechanical Engineering at the University of Maryland. The last chapter presents the details of four relatively complex experiments in which the computer integration of each experiment is emphasized.

The author wishes to thank Dr. Alexander H. Slocum and Dr. Clarence E. Thomas, Jr. for reading the manuscript and providing numerous suggestions for its improvement. He also wishes to express his gratitude to his colleagues from the Department of Mechanical Engineering at the University of Maryland who made substantial contributions to several of the experiments described in Chapters 8 and 9: Dr. Ioannis E. Minis for Sections 8.5.4, 8.5.5 and 9.5; Dr. Mohamed K. Abdelhamid for Sections 8.5.5 and 8.6.2; Dr. Reinhard Radermacher for Section 8.8.3; and Dr. Ashwani K. Gupta for Section 8.8.4.

College Park, Maryland   Edward B. Magrab
July, 1990

Table of Contents

PART I

Instrumentation Fundamentals

1 Introduction

1.1 Introduction

The integration of the computer with electronic instrumentation and devices has greatly enhanced the usefulness and capabilities of the instruments themselves. It has made experiments easier to perform, reduced a large amount of the tedium associated with data collection, improved the reliability and repeatability of the data collected, provided decision-making capabilities and error checking, and has directly coupled the experimental investigation, or process control, with the analysis, display and reporting of results. These advantages and enhancements, however, require one's knowledge to go beyond that normally needed for just the selection of instruments and transducers. It now requires the user to additionally learn the manner in which this integration can be correctly and effectively carried out.

It is the objective of this book to provide a basis by which instruments and transducers can be selected, assembled, and integrated with a computer to measure and control physical processes in an accurate, predictable and economical manner. The book is comprised of two parts. The first part introduces both the language used to describe the performance and the characteristics of the basic instrumentation components and summarizes the theoretical foundation governing the performance of all instrumentation systems and their interconnections. It discusses in detail the major sources of errors of the many commonly used classes of instruments and what can be done to minimize or avoid these errors. For each class relationships between their input and output signals are developed and estimates of the time required to perform their respective functions within prescribed error limits are obtained. It also emphasizes the interactive effects that each component in an instrumentation chain may have on one another and the principles that must be followed to minimize these interactions. The second part of the book applies the material of the first part by introducing techniques, procedures and examples that illustrate how instruments, transducers and actuators can be placed under computer control.

In this chapter the definitions of the types of errors one uses to describe instrumentation and their mathematical formulation are introduced.

1.2 Measurement Errors

1.2.1 Accuracy and Repeatability

In a sense all measurements are in error; that is, there exists a difference between the value measured and its true value. However, in practice, it is the magnitude of this error that makes some measurements more acceptable than others. To quantify the concept of measurement error the term *accuracy* of a measured quantity is defined as its deviation from the true value. On the other hand the quantity that characterizes the ability of a measuring instrument to give the same value of the quantity being measured under the *same* conditions is its *repeatability*. The term "same conditions" means using the same method, observer, instruments, location and during an appropriately short period of time. Thus accuracy is a measure of the approach to the true value, whereas repeatability is merely a measure of consistency. Consequently, repeatability is no guarantee of good accuracy, although poor repeatability does indicate poor accuracy.

When a group of values are measured over a long period of time or by different persons or with different methods, the closeness of these measured values is called *reproducibility*. The *resolution* of an instrument or transducer is its ability to respond to a small change in the measured quantity. It does not relate to accuracy.

If a set of N measurements $x_i, i = 1, 2, \ldots, N$ are made under the same conditions, then the mean value of the measurements is, assuming that the x_i have a normal (gaussian) distribution,

$$\bar{x} = \frac{1}{N} \sum_{i=1}^{N} x_i \tag{1.1}$$

and its standard deviation

$$\sigma = \left[\frac{1}{N-1} \sum_{i=1}^{N} (x_i - \bar{x})^2 \right]^{1/2} = \left[\frac{1}{N-1} \left(\sum_{i=1}^{N} x_i^2 - N \bar{x}^2 \right) \right]^{1/2}. \tag{1.2}$$

The repeatability r of this measurement is defined as (Haywood [1977])

$$r = t_{N-1,(1-\alpha)/2} \, \sigma \tag{1.3}$$

where $t_{N-1,(1-\alpha)/2}$ is the student-t distribution parameter at a confidence level $\alpha(<1)$. Representative values of $t_{m,(1-\alpha)/2}$ are given in Table 1.1. Equation (1.3) indicates that $100\alpha\%$ of the time the measurements will be within

$$\bar{x} - r \le x_i \le \bar{x} + r. \tag{1.4}$$

The mean and standard deviation are descriptors of random errors, which are errors that cannot be predicted on an individual basis. On the other hand there is a class of errors called systematic errors (or bias) d_s, which are those errors that, from past knowledge or measurement, can be predicted or estimated. These

Table 1.1 Selected values for the t-distribution, $t_{m,(1-\alpha)/2}$

	$100\alpha\%\ [(1-\alpha)/2]$		
m	90% [.05]	95% [.025]	99% [.005]
5	2.015	2.571	4.032
10	1.812	2.228	3.169
12	1.782	2.179	3.055
15	1.753	2.131	2.947
17	1.740	2.110	2.898
20	1.725	2.086	2.845
30	1.697	2.042	2.750
∞	1.645	1.960	2.576

systematic errors are part of every measurement x_i such that if $\overline{x}_T$ is the true mean, then as shown in Figure 1.1,

$$\overline{x}_T = \overline{x} - d_s. \tag{1.5}$$

Although there is no theoretically justifiable way to relate random and systematic errors, it has been proposed that the accuracy of the measurement be given as (Haywood [1977] and Coleman and Steele [1989])

$$\text{accuracy} = r_{acc} = \sqrt{r^2 + d_s^2}. \tag{1.6}$$

Once a systematic error becomes known and quantified, the magnitude of d_s decreases and in the limit, when all the systematic uncertainties are known $d_s = 0$. From Eq.(1.6) it is seen that the accuracy is greater than or equal to the repeatability.

In its simplest form d_s may be caused by a mistake in the calibration of a transducer or instrument, which shifts each x_i by d_s. In other cases it may be caused by small mechanical or thermal changes in the experimental setup. Systematic errors can often be disclosed by repeated measurements under different conditions, or with a different apparatus, or where possible, by using an entirely different method. Whether or not one should pursue the determination of the

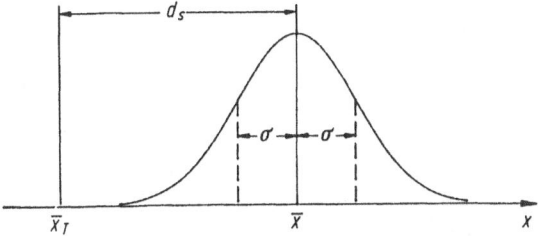

Figure 1.1 Systematic error d_s and its relationship to the measured mean $\overline{x}$ and the true mean $\overline{x}_T$

magnitude of d_s depends primarily on the purpose of the experiment and an *a priori* knowledge of the process itself.

Example 1: Determination of Accuracy. A series of 16 measurements are performed under the same conditions and it is found that the mean value is 5.017 and the standard deviation is 0.024. The suspected systematic error in the measurement is 0.008. For a confidence level of 95% the repeatability of the process is

$$r = t_{15, .025}(0.024) = (2.131)(0.024) = 0.0511 .$$

Hence, from Eq.(1.4) it is 95% certain that the next measurement will fall in the range $4.967 \leq x_i \leq 5.068$. The estimate of the accuracy of the measurement is, from Eq.(1.6),

$$\text{accuracy} = \sqrt{(0.0511)^2 + (0.008)^2} = 0.0517 .$$

Thus we are 95% certain that the true value is 5.017 ± 0.052.

1.2.2 Multiple Measured Quantities

When more than one quantity is measured and there exists the functional relationship between these quantities $A = f(x, y, \ldots)$, the repeatability of the process r_a can be estimated by (Coleman and Steele [1989])

$$r_a = \sqrt{\left(\frac{\partial f}{\partial x}\right)^2 r_x^2 + \left(\frac{\partial f}{\partial y}\right)^2 r_y^2 + \ldots} \tag{1.7}$$

where $x, y, \ldots$ are independently measured quantities and $r_x, r_y, \ldots$ are the respective repeatabilities of the measurement of each $x, y, \ldots$ Some special cases of Eq.(1.7) are:

1) If $A = x \pm y$ then

$$r_a = \sqrt{r_x^2 + r_y^2} ; \tag{1.8a}$$

2) If $A = kxy$ then

$$r_a = k\bar{x}\bar{y}\sqrt{\frac{r_x^2}{\bar{x}^2} + \frac{r_y^2}{\bar{y}^2}} ; \tag{1.8b}$$

3) If $A = kx/y$ then

$$r_a = \frac{k\bar{x}}{\bar{y}}\sqrt{\frac{r_x^2}{\bar{x}^2} + \frac{r_y^2}{\bar{y}^2}} ; \tag{1.8c}$$

where the bar denotes the average value. It is seen from Eq.(1.8a) that if $r_y = r_x/2$ then $r_a = 1.12 r_x$. Thus, to improve the overall repeatability in this case one should concentrate on improving the repeatability of the measurement of x.

Example 2: Repeatability of a Functionally Related Quantity. Consider the measurement of a quantity B that is related to the fuel flow F, the engine torque T and the engine speed N by

$$B = \frac{C_1 F}{TN}$$

where C_1 is a constant. Three sets of independent measurements at the set of nominal operating conditions $\overline{F}_o = 3.71$, $\overline{T}_o = 6.25$ and $\overline{N}_o = 2100$ yield the following respective repeatabilities at a 95% confidence level: $r_F = 0.28$, $r_T = 0.61$ and $r_N = 84.0$. Consequently, from Eq.(1.7),

$$r_B = \overline{B} \sqrt{\left(\frac{0.028}{3.71}\right)^2 + \left(\frac{0.61}{6.25}\right)^2 + \left(\frac{84}{2100}\right)^2} = 0.13\overline{B}$$

where

$$\overline{B} = \frac{C_1 \overline{F}_o}{\overline{T}_o \overline{N}_o}.$$

Thus the repeatability of the measurement of B at a 95% confidence level and at these nominal operating conditions is ±13%.

1.2.3 Linearity

Amplitude Linearity
The majority of the components of instrumentation systems are designed to provide an output signal that is linearly proportional to its input signal over their operating ranges. The deviation from linearity is a source of error that affects the overall accuracy of a measurement. If one were to determine the repeatability of a set of measurements over a range of inputs x_i there would be a corresponding repeatability at each output y_i over the selected range. A plot of the results of such a series of measurements would produce a collection of points representing the mean value at each group of measurements $\overline{Y}_i$ along with their corresponding repeatabilities r_i as shown in Figure 1.2a. It is customary, however, to assume that the statistics governing the mean values $\overline{Y}_i$ are the same; that is, the repeatabilities are the same, and that these mean values coincide with the true value of the straight line connecting them (Natrella [1962]). This is shown in Figure 1.2b. With this assumption then, it is sufficient to simply determine the statistics governing the

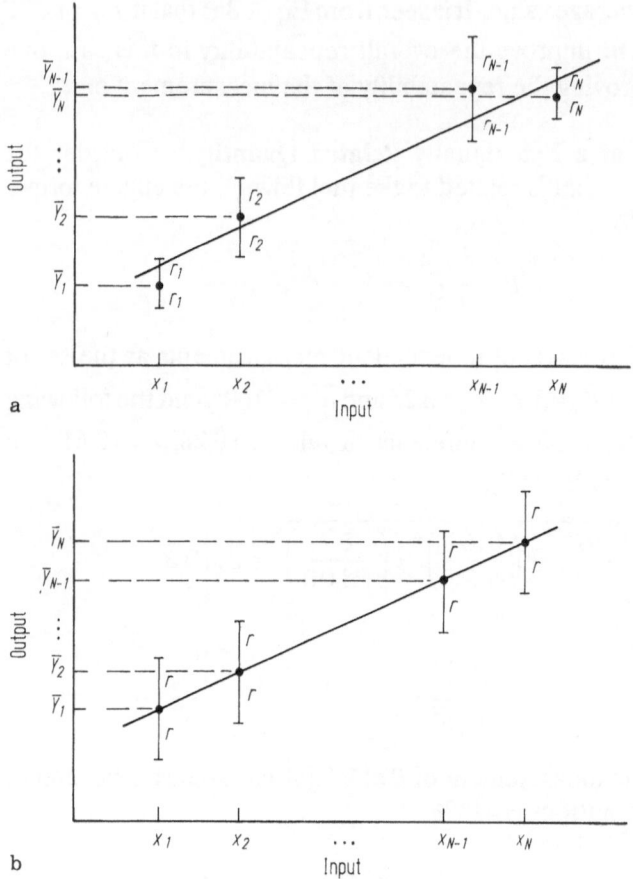

Figure 1.2 (a) Mean and repeatability of a general process over a range of input values (b) mean and repeatability of a process for which it is assumed that the mean values lie on a straight line and the repeatability is the same over the range

straight line fit and use them as an indicator of the repeatability of the process over the range of values, as well as at a specific value within the range. In determining linearity it is usually the case that the input quantity x_i is known to sufficient accuracy so that its error is small with respect to the variability (randomness) of y_i.

If the input is x_i and the corresponding output y_i, $i = 1, 2, ..., N$, then the equation for the best straight line through these data, as shown in Figure 1.3a, is (Bowker and Lieberman [1959])

$$y = mx + b \tag{1.9}$$

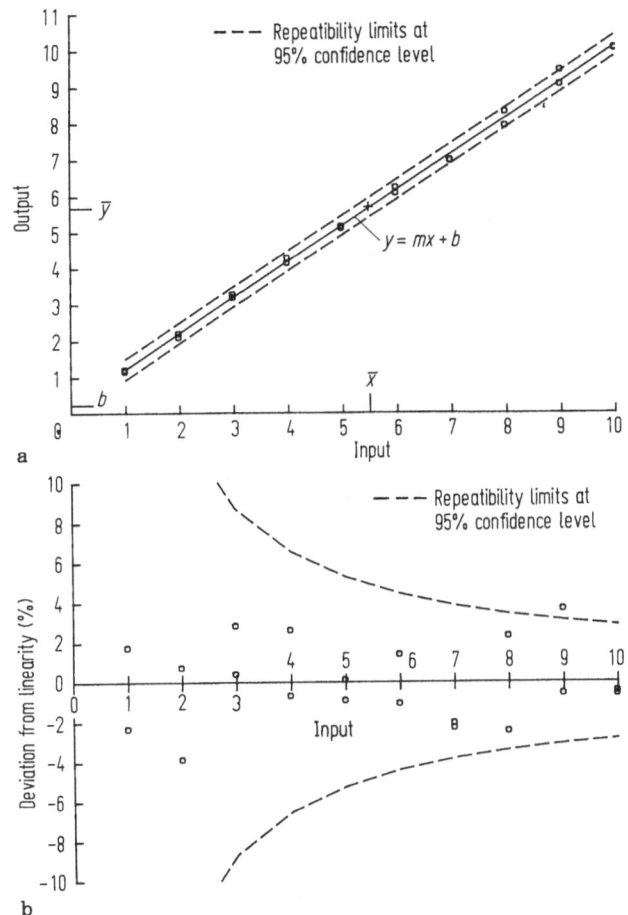

Figure 1.3 (a) Least squares straight line fit to data points and the corresponding confidence limits (b) the same data replotted to emphasize the percentage deviation from linearity

where

$$m = \frac{1}{(N-1)\sigma_x^2}\left(\sum_{i=1}^{N} x_i y_i - N\overline{x}\,\overline{y}\right)$$

$$\overline{x} = \frac{1}{N}\sum_{i=1}^{N} x_i, \qquad \overline{y} = \frac{1}{N}\sum_{i=1}^{N} y_i$$

$$b = \overline{y} - m\overline{x}, \qquad \sigma_x^2 = \frac{1}{N-1}\left(\sum_{i=1}^{N} x_i^2 - N\overline{x}^2\right).$$

The estimate of the variability of the data about the line $s_{y/x}$ is

$$s_{y/x} = \sqrt{\frac{N-1}{N-2}(\sigma_y^2 - m^2\sigma_x^2)} \qquad (1.10)$$

where

$$\sigma_y^2 = \frac{1}{N-1}\left(\sum_{i=1}^{N} y_i^2 - N\bar{y}^2\right).$$

The confidence intervals of the slope, intercept and the value of y itself are, respectively,

$$m \to m \pm t_{N-2,(1-\alpha)/2}\frac{s_{y/x}}{\sigma_x\sqrt{N-1}}$$

$$b \to b \pm t_{N-2,(1-\alpha)/2}s_{y/x}\sqrt{\frac{1}{N} + \frac{\bar{x}^2}{(N-1)\sigma_x^2}}$$

$$y' \to mx' + b \pm r_{yx}$$

where

$$r_{yx} = t_{N-2,(1-\alpha)/2}s_{y/x}\sqrt{1 + \frac{1}{N} + \frac{(x'-\bar{x})^2}{(N-1)\sigma_x^2}} \qquad (1.11)$$

is the repeatability and x' is any value of x within the range for which m and b were determined. Consequently r_{yx} is a measure of the deviation from linearity. It is noted that r_{yx} is not a constant over the range of x and has its minimum value at $x' = \bar{x}$ and its maxima at the outer limits of x.

The *range* of y is determined by the maximum and minimum values of x, x_{max} and x_{min}, respectively. Therefore $y_{max} = mx_{max} + b$ and $y_{min} = mx_{min} + b$. The *span* of the linearity relationship is $x_{max} - x_{min}$, which results in the output span $y_{max} - y_{min}$. The quantity y_{max} is often referred to as the full scale value. With these definitions Eq.(1.9) has the implied limits $x_{min} \leq x \leq x_{max}$.

In certain situations the system's components, or the process itself, is inherently nonlinear; that is, $y = f(x) \neq mx + b$ (see for example Eqs.(8.65) and (8.67)). For these situations the nonlinearity, that is, the deviation from a straight line, is often defined in terms of the span as

$$\text{percentage nonlinearity} = \frac{f(x) - mx - b}{y_{max} - y_{min}}100 \ \%$$

$$x_{min} \leq x \leq x_{max}. \qquad (1.12)$$

Another form of nonlinearity is *hysteresis*, which is the difference between the output y_i^+ obtained by approaching the corresponding input x_i from x_{min} and the output y_i^- obtained by approaching the corresponding input x_i from x_{max}. It is usually expressed as a function of the span as

$$\text{percentage hysteresis} = \left| \frac{y_i^+ - y_i^-}{y_{max} - y_{min}} \right| 100 \ \%. \qquad (1.13)$$

When determining the linearity of electronic components the input and output quantities are usually voltages. On the other hand when the component is a transducer or sensor (see Section 8.1 for their definitions) the input is a physical quantity, such as displacement, temperature, pressure, etc., and the output is again a voltage. In both cases r_{yx} is a measure of its linearity and b the amount of offset from zero. For electronic components the slope m has no special significance other than that already mentioned. However, for transducers and sensors the slope is the *sensitivity* of the device; that is, the conversion factor from physical units to electrical units. If the output of the device is S_T, in volts, and the input is a physical quantity S_P, in physical units, then

$$S_T = S_s S_P + b \quad V \qquad (1.14)$$

where S_s is the sensitivity of the device in (Volts)/(physical units). In general both S_s and b are functions of temperature and S_s is a function of frequency (see Eq.(3.4) ff).

Returning to Figure 1.3a it is seen that the deviation from a straight line is not emphasized in this type of representation. Therefore, an alternative representation is frequently used in which the individual outputs y_i (corresponding to x_i) are expressed as the percentage $100(y_i/(mx_i + b) - 1)\%$ and the repeatability as $100r_{yx}/(mx_i + b)\%$. This representation is shown in Figure 1.3b. Notice that the percentage repeatability (uncertainty) increases for decreasing x. To account for the variability of r_{yx} with the input quantity, the linearity of instruments and transducers are usually specified as some percentage of full scale or as some percentage of the data value or as a combination of both. Application of this representation is presented in Section 6.1.2.

Frequency Linearity
A large number of instruments and transducers are designed to give uniform amplitude response over a frequency range $FR = f_2 - f_1$. This linearity with frequency is usually expressed either as a percentage or in decibels (see Section 1.3) over FR, and is referenced to one frequency f_r, where $f_1 \leq f_r \leq f_2$. In order for this specification to be meaningful FR must be given. A typical instrument's or transducer's frequency response is illustrated in Figure 1.4. (This is further discussed in Section 3.2.) It is usually assumed that the frequency response is the

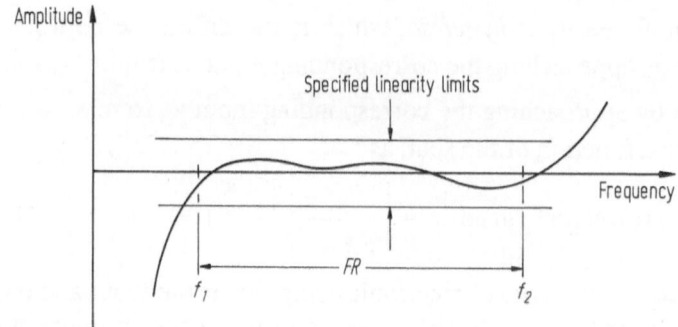

Figure 1.4 Frequency response linearity and its specification

same for all values of the input amplitude. Equations (1.9) and (1.11) also apply to this situation. It should be realized that the frequency response uncertainty (deviation with respect to frequency) must be added to the linearity uncertainty in the manner shown in Eq.(1.8a).

1.3 The Decibel (dB)

The decibel is defined as

$$dB = 10\log_{10}\left(\frac{P}{P_{ref}}\right) = 20\log_{10}\left(\frac{A}{A_{ref}}\right) \qquad (1.15)$$

where P is power or a power-like quantity, P_{ref} is a reference quantity having the same dimensions (units) as P, A is a linear quantity and A_{ref} is a reference quantity having the same dimensions (units) as A. Examples of P are electrical power in watts and acoustic intensity in watts/m^2; examples of A are voltage, displacement and acceleration. The designation dB does not pertain to a physical quantity and, as such, it is not a unit in the ordinary sense. It simply indicates the transformation of a ratio of two like quantities as shown in Eq.(1.15).

The main reason for the introduction and use of the decibel is to compress very large and very small numbers into more manageable ones and to provide a convenient manner in which to talk about them. From Eq.(1.15) it is seen that for linear quantities each factor of 10 increase with respect to the reference quantity corresponds to 20 dB, whereas a decrease by a factor of 10 corresponds to -20 dB. Thus 60 dB means that a linear quantity is 1000 times larger than either it originally was or its reference quantity and -60 dB 1000 times smaller. The following ratios are of special interest in Chapter 3: $A/A_{ref} = \sqrt{2}$ and $A/A_{ref} = 1/\sqrt{2}$, which correspond to 3 dB and -3 dB, respectively; and $P/P_{ref} = 2$ and $P/P_{ref} = 1/2$, which also correspond to 3 dB and -3 dB, respectively.

Frequently errors are expressed in dB. Consider the usual definition of percentage error ϵ of a linear quantity Q with respect to a reference quantity Q_{ref}:

$$\epsilon = 100\frac{Q-Q_{ref}}{Q_{ref}} = \left(\frac{Q}{Q_{ref}} - 1\right)100 \quad \% \tag{1.16a}$$

or

$$\frac{Q}{Q_{ref}} = 1 + \frac{\epsilon}{100}. \tag{1.16b}$$

If the ratio Q/Q_{ref} is expressed as Δ dB, then from Eq.(1.15)

$$\Delta = 20\log_{10}(Q/Q_{ref})$$

or

$$\frac{Q}{Q_{ref}} = 10^{\Delta/20}.$$

Therefore, from Eqs.(1.16)

or

$$\epsilon = (10^{\Delta/20} - 1)100 \quad \% \tag{1-17a}$$

$$\Delta = 20\log_{10}\left(1 + \frac{\epsilon}{100}\right) \quad dB. \tag{1-17b}$$

If we let $Q/Q_{ref} = x_i/\bar{x}$ and $r = r_{acc}$, then Eqs.(1.4) and (1.16a) give

$$\epsilon = \pm(r_{acc}/\bar{x})100 \quad \%$$

and Eq.(1.17b) becomes

$$\Delta = 20\log_{10}\left(1 \pm \frac{r_{acc}}{\bar{x}}\right) \quad dB.$$

Table 1.2 Relationship between an error expressed in dB to one expressed as a percentage

Δ (dB)	ϵ (%)
+0.01	+0.12
-0.01	-0.12
+0.1	+1.16
-0.1	-1.14
+0.5	+5.93
-0.5	-5.59
+1.0	+12.20
-1.0	-10.87
+1.5	+18.85
-1.5	-15.86
+2.0	+25.89
-2.0	-20.56
+3.0	+41.25
-3.0	-29.20

14

Typical equivalent values for ϵ and Δ are given in Table 1.2. Notice that $\pm \Delta$ dB does not, in general correspond to $\pm \epsilon \%$, although for $\Delta \leq 1$ dB it is very closely true.

References

1. Bowker, A. J., and Lieberman, G. J., *Engineering Statistics,* Prentice-Hall, Englewood Cliffs, New Jersey, 1959.
2. Coleman, H. W., and Steele, W. G., *Experimentation and Uncertainty Analysis for Engineers,* John Wiley and Sons, New York, 1989.
3. Cunningham, M. J., "Measurement Errors and Instrument Inaccuracies", in *Instrumentation: A Reader,* Loxton, R. and Pope, P., Eds., Open University Press, Philadelphia, 1986.
4. Haywood, A. T. J., *Repeatability and Accuracy,* Mechanical Engineering Publications Ltd, London, 1977.
5. Natrella, M. G., "Experimental Statistics", Section I, Ordnance Corp Pamphlet, ORDP 20-110, June 1962.
6. Sydenham, P. H., Ed., *Handbook of Measurement Science, Vol. 1, Theoretical Fundamentals,* John Wiley and Sons, New York, 1982.

Exercises

1. For the set of data given below determine the range of uncertainty at the 99% confidence level for the slope, offset and repeatability of the output.

Input	Output	Input	Output
-5.0	-5.482	-5.0	-4.525
-4.0	-4.130	-4.0	-4.048
-3.0	-3.321	-3.0	-2.934
-2.0	-2.129	-2.0	-1.991
-1.0	-0.969	-1.0	-1.026
0.0	-0.018	0.0	-0.034
1.0	0.936	1.0	0.925
2.0	2.093	2.0	2.160
3.0	2.916	3.0	2.913
4.0	3.898	4.0	4.257
5.0	4.862	5.0	4.636

2. The data below were obtained from a frequency response test from 50 to 1000 Hz. Over what frequency range will the response vary by ± 1 dB ($\pm 12\%$) at the 99% confidence level when referred to the average amplitude at 525 Hz.

Frequency (Hz)	Amplitude (V)	Frequency (Hz)	Amplitude (V)
50	6.908	550	6.384
100	6.637	600	6.098
150	6.826	650	6.551
200	6.774	700	6.408
250	6.370	750	6.483
300	6.344	800	6.613
350	6.232	850	6.206
400	6.698	900	6.184
450	6.383	950	6.838
500	6.393	1000	6.378

2   Analysis of Signals

2.1 Introduction

The purpose of this chapter is to (a) introduce and summarize the mathematics required to describe signals commonly encountered in engineering experimentation, (b) to elicit from the mathematics the physical interpretation and subsequent relevance to instrumentation systems and (c) to introduce the important concept of the equivalence of the frequency and time domain representations. We will discuss the form of their representation in both the time and frequency domains and the corresponding engineering units that are used to describe the signal in each representation.   To facilitate the physical understanding of the results numerous examples will be given, which have been chosen to simulate characteristics of signals often encountered in practice.

2.2 Classification of signals

Observed signals representing physical phenomena can be classified as being either deterministic or nondeterministic. Deterministic signals are those signals whose amplitude as a function of time can be predicted. Nondeterministic signals cannot, and are usually termed random signals. Random signals are, therefore, those signals whose qualities can only be described by statistical measures.

Deterministic signals can be further classified as being either periodic or aperiodic (often called transient). A signal $f(t)$ is periodic if, for all t, $f(t)$ repeats itself in its entirety every interval T. This interval is called the period. An aperiodic signal is every other type of deterministic signal.

Nondeterministic signals can be further classified as either stationary or nonstationary. However, these classifications require the introduction of several terms which themselves must be defined. Therefore, the discussion of these signals is postponed until Section 2.6.

As we will see in the subsequent sections of this chapter, each of these types of signals requires different analysis methods and, as it will turn out, different engineering units to represent the information in the frequency domain.

There is another classification of signals that can be superimposed on those defined above: they are either analog or discrete. A discrete signal is one in which the amplitude information assumes a finite number of values. Discrete signals are dealt with where appropriate and are independent of the signal classifications given above.

2.3 Periodic Signals

2.3.1 Fourier Series

A signal $f(t)$ is periodic on some interval T if $f(t) = f(t+T)$. The frequency at which $f(t)$ is periodic is $f_0 = 1/T$ Hz. This frequency is related to the angular frequency ω_0 by $\omega_0 = 2\pi f_0 = 2\pi/T$ rad/s. Physically realizable periodic waveforms can be represented by the following series, called the Fourier series expansion of $f(t)$:

$$f(t) = \sum_{n=-\infty}^{\infty} c_n e^{jn\omega_0 t} \tag{2.1}$$

where $j = \sqrt{-1}$ and

$$c_n = \frac{1}{T} \int_0^T f(t) e^{-jn\omega_0 t} dt$$

$$= \frac{1}{T} \int_{-T/2}^{T/2} f(t) e^{-jn\omega_0 t} dt \qquad n = 0, \pm 1, \pm 2, \dots \tag{2.2}$$

The form of Eq.(2.1) is chosen for mathematical convenience. Equation (2.2) can be written as

$$c_n = \frac{1}{T} \int_{-T/2}^{T/2} f(t) \cos(n\omega_0 t) dt - \frac{j}{T} \int_{-T/2}^{T/2} f(t) \sin(n\omega_0 t) dt$$

$$= R(n) - jX(n) \qquad n = 0, \pm 1, \pm 2, \dots \tag{2.3}$$

If $f(t)$ is real, which is the physically realizable case, then from the above equations it is seen that

$$c_{-n} = R(-n) - jX(-n) = R(n) + jX(n) = c_n^* \tag{2.4}$$

where the asterisk indicates the complex conjugate. Therefore, using Eq.(2.3) and Eq.(2.4) we see that c_n can be expressed as

$$c_n = |c_n| e^{-j\phi_n} \tag{2.5}$$

where

$$|c_n| = \sqrt{c_n c_n^*} = \sqrt{c_n c_{-n}} = \sqrt{R(n)^2 + X(n)^2} = |c_{-n}| \qquad (2.6a)$$

and

$$\tan \phi_n = X(n)/R(n). \qquad (2.6b)$$

Using Eq.(2.4) it is seen that

$$\phi_{-n} = -\phi_n.$$

Therefore Eq.(2.1) becomes

$$f(t) = \sum_{n=-\infty}^{\infty} |c_n| e^{j(n\omega_0 t - \phi_n)}$$

$$= \sum_{n=1}^{\infty} |c_{-n}| e^{j(-n\omega_0 t - \phi_{-n})} + |c_0| + \sum_{n=1}^{\infty} |c_n| e^{j(n\omega_0 t - \phi_n)}$$

$$= |c_0| + \sum_{n=1}^{\infty} |c_n| (e^{-j(n\omega_0 t - \phi_n)} + e^{j(n\omega_0 t - \phi_n)})$$

$$= |c_0| + 2 \sum_{n=1}^{\infty} |c_n| \cos(n\omega_0 t - \phi_n). \qquad (2.7)$$

The c_n are, in general, complex quantities that are independent of time and only a function of the fundamental frequency ω_0 and its harmonics $n\omega_0$, $n = 2, 3, \ldots$ It can be said, therefore, that $2|c_n|$ represents the amplitude contribution of each harmonic comprising the periodic time waveform $f(t)$. In other words Eq.(2.2) represents the transform from the time domain into the frequency domain. Furthermore this transformed information is *discrete*; that is, it only contains amplitude information at the frequencies nf_0, $n = 0, 1, 2, \ldots$ and is zero everywhere else. In addition the spacing between each harmonic is

$$\Delta f = (n+1)f_0 - nf_0 = f_0 = 1/T. \qquad (2.8)$$

Thus, as the period of the signal increases the smaller the frequency spacing between the harmonic amplitudes; that is, the amplitudes tend to crowd together. In Section 2.4.1 we will examine what happens when T approaches infinity. Lastly, although negative frequencies were introduced in the mathematically convenient form given by Eqs.(2.1) and (2.2), for signals of practical interest the results can always be expressed in terms of only the positive frequencies as shown in Eq.(2.7).

Several definitions are now introduced.

Bandwidth and Rise Time

Consider a periodic signal $f(t)$ whose highest frequency is $N_H f_o$. Equation (2.7) can then be written as

$$f(t) = 2 \sum_{n=N_L}^{N_H} |c_n| \cos(n\omega_0 t - \phi_n)$$

when $N_L \geq 1$, and

$$f(t) = |c_0| + 2 \sum_{n=1}^{N_H} |c_n| \cos(n\omega_0 t - \phi_n)$$

when $N_L = 0$. The *bandwidth* of a periodic signal is defined as the difference of the highest and lowest frequencies appearing in its Fourier series expansion; that is, $BW = (N_H - N_L) f_o$.

The *rise time* of a signal is (usually) defined as the time it takes for a signal to go from 10% to 90% of its final value when the signal undergoes an abrupt change in its amplitude. In Example 3 the significance of these definitions and their relationship to each other will be established.

Average Power

If $g(t)$ represents the voltage across a 1 ohm resistor, then the average power of a periodic signal is given by

$$P_{avg} = \frac{1}{T} \int_{-T/2}^{T/2} g^2(t) dt \quad W. \tag{2.9a}$$

Substituting Eq.(2.7) into Eq.(2.9a) and integrating gives

$$P_{avg} = \sum_{n=-\infty}^{\infty} |c_n|^2 = |c_0|^2 + 2 \sum_{n=1}^{\infty} |c_n|^2 \quad W. \tag{2.9b}$$

Equation (2.9b) is known as Parseval's theorem for periodic signals. It says that the sum of the squares of the magnitudes of the fundamental frequency, its harmonics and the zero frequency component, called the dc value, equals the total power in the signal. Notice that the average power of a periodic signal is independent of the phase relations of the harmonics, ϕ_n.

RMS Value

The root-mean-square (rms) value of a periodic signal is

$$\{f(t)\}_{rms} = \sqrt{P_{avg}} = \sqrt{\frac{1}{T} \int_{-T/2}^{T/2} g^2(t) dt}. \tag{2.10}$$

It is seen from Eqs.(2.9b) and (2.10) that to correctly measure the rms value of a periodic waveform one must be sure that the measuring device has sufficient bandwidth so that all the significant harmonic contributions are included in the measurement. (See Eq.(6.15) ff.)

Amplitude and Power Spectrum
From the preceding definitions it is seen that if we considered a sine wave of frequency $f_0 = \omega_0/(2\pi)$, then $\sqrt{2}\,|c_1|$ is the rms value of the amplitude at f_0 and $2\,|c_1|^2$ its average power. Therefore, the sequence of amplitude contributions represented by Eqs.(2.9b) and (2.10) are called the power spectrum and (rms) amplitude spectrum of the signal $g(t)$, respectively. As indicated in Eq.(1.14) a voltage $g(t)$ can be converted to a physical quantity by means of a sensitivity factor S_s V/(physical unit). Thus if $f(t) = g(t)/S_s$, whose amplitude now is expressed in physical units, then from Eq.(2.2) it is seen that the $|c_n|$ are also expressed in these same physical units.

The quantity ϕ_n is called the phase spectrum of the signal. It will be shown subsequently [see Eq.(3.12)] that the phase spectrum must be preserved if one wants to maintain the original waveform of the signal in the time domain.

Percentage Total Harmonic Distortion (%THD)
In many situations it is often necessary to obtain a measure of how much an instrumentation system influences the signal of interest. One indicator of signal modification is to determine the amplitude of all the harmonics in an output signal of a system when the input signal is a single-frequency sine wave. This leads to the following definition of %THD:

$$\%THD = \frac{100}{|c_1|}\left[\sum_{n=2}^{\infty}|c_n|^2\right]^{1/2} = 100\left[\sum_{n=2}^{\infty}\frac{|c_n|^2}{|c_1|^2}\right]^{1/2} \quad \%. \quad (2.11)$$

The significance of this measure is illustrated in Section 4.2.

Example 1: Fourier Series of a Square Wave. Consider the square wave of frequency $\omega_0 = 2\pi/T$ shown in Figure 2.1a. Using Eq.(2.2) gives

$$c_n = -\frac{A}{T}\int_{-T/2}^{-T/4} e^{-jn\omega_0 t}\,dt + \frac{A}{T}\int_{-T/4}^{T/4} e^{-jn\omega_0 t}\,dt - \frac{A}{T}\int_{T/4}^{T/2} e^{-jn\omega_0 t}\,dt$$

$$= \frac{A}{n\pi}[2\sin(n\omega_0 T/4) - \sin(n\omega_0 T/2)]$$

$$= \frac{2A}{n\pi}(-1)^{(n-1)/2} \qquad n=1,3,5,\ldots$$

and $c_0 = 0$. Although the c_n are real, because of its alternating sign

$$\phi_n = -(n-1)\pi/2 \qquad n=1,3,5,\ldots$$

and, therefore, Eq.(2.7) becomes

$$f(t) = \frac{4A}{\pi} \sum_{n=1,3,5}^{\infty} \frac{1}{n} \cos(n\omega_o t + (n-1)\pi/2). \qquad (2.12)$$

If we integrate Eq.(2.12) we obtain

$$g(t) = \int f(t)dt = \frac{4A}{\pi\omega_0} \sum_{n=1,3,5}^{\infty} \frac{1}{n^2} \sin(n\omega_o t + (n-1)\pi/2) \qquad (2.13)$$

which is the Fourier series expansion for a triangular wave shown in Figure 2.1b.

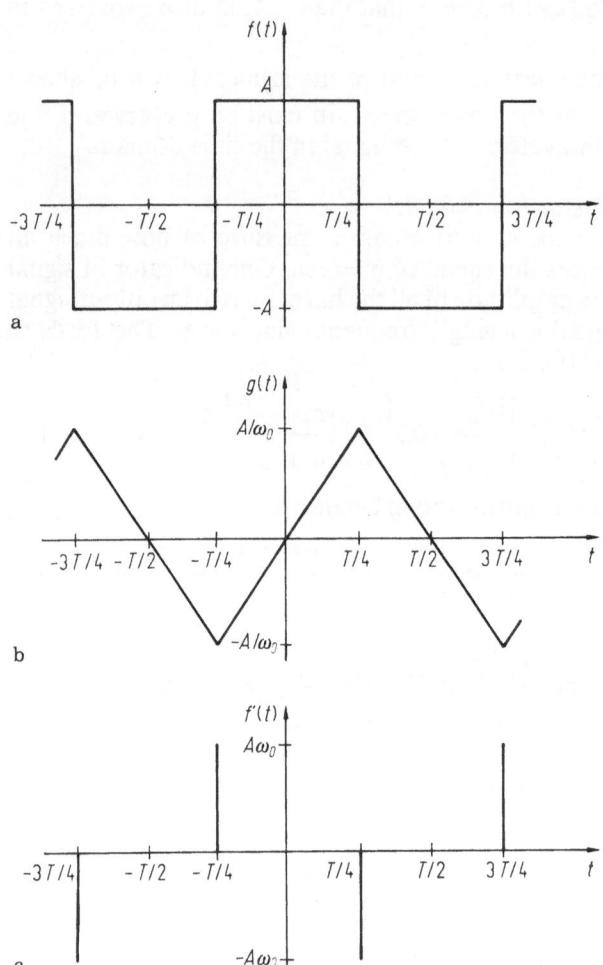

Figure 2.1 (a) Square wave   (b) integrated square wave   (c) differentiated square wave

On the other hand if we differentiate Eq.(2.12) we get

$$f'(t) = \frac{d}{dt} f(t) = \frac{4 A \omega_0}{\pi} \sum_{n=1,3,5}^{\infty} \sin(n \omega_0 t + (n+1)\pi/2) \qquad (2.14)$$

which is a series of periodic impulse-like quantities [see Section 2.3.2] shown in Figure 2.1c. Both Eqs.(2.13) and (2.14) will be used to verify the integrating and differentiating properties of RC filters discussed in Sections 3.3.2 and 3.3.3, respectively.

To determine the average power we use Eq.(2.9b) to obtain

$$P_{avg} = \frac{8 A^2}{\pi^2} \sum_{n=1,3,5}^{\infty} \frac{1}{n^2}.$$

However (Jolley [1961]),

$$\sum_{n=1,3,5}^{\infty} \frac{1}{n^2} = \frac{\pi^2}{8}$$

and, therefore, $P_{avg} = A^2$ and from Eq.(2.10),

$$\{f(t)\}_{rms} = \sqrt{P_{avg}} = A. \qquad (2.15)$$

Notice that the rms value of a square wave is equal to its peak value.

These results are plotted in Figure 2.2, which shows the power spectral plot of the square wave. It clearly demonstrates that its power and amplitude spectra are *discrete*, with each contribution to the total power in the signal appearing at integer multiples of the fundamental frequency.

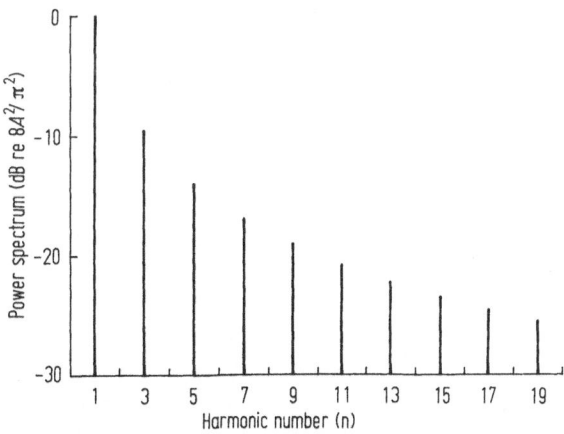

Figure 2.2 Normalized power spectrum of a square wave

Continuing the exercise we now determine the %THD of a square wave. Using Eq.(2.11) yields

$$\%THD = \frac{100}{|c_1|}\left[\sum_{n=3,5,7}^{\infty}|c_n|^2\right]^{1/2} = 100\left[\sum_{n=3,5,7}^{\infty}\frac{1}{n^2}\right]^{1/2}$$

$$= 100\left[\frac{\pi^2}{8} - 1\right]^{1/2} = 48.3\%.$$

Thus one could interpret a square wave as a sine wave with 48.3% total harmonic distortion.

Example 2: Fourier Series of a Pulse Train.   Consider the periodic train of pulses of duration $\tau_0 = 2t_0$ and frequency $\omega_o = 2\pi/T$ shown in Figure 2.3. Using Eq.(2.2) gives

$$c_n = \frac{A_0}{T}\int_{-t_0}^{t_0} e^{-jn\omega_o t}\,dt = \frac{\tau_0 A_0}{T}\frac{\sin(n\pi\tau_0/T)}{n\pi\tau_0/T} \qquad n=0,1,2,\dots \qquad (2.16)$$

Thus, from Eqs.(2.7) and (2.9b) we get, respectively,

$$f(t) = \frac{\tau_0 A_0}{T}\left[1 + 2\sum_{n=1}^{\infty}\frac{\sin(n\pi\tau_0/T)}{(n\pi\tau_0/T)}\cos(2n\pi t/T)\right] \qquad (2.17)$$

and

$$P_{avg} = \left[\frac{\tau_0 A_0}{T}\right]^2\left[1 + 2\sum_{n=1}^{\infty}\left[\frac{\sin(n\pi\tau_0/T)}{(n\pi\tau_0/T)}\right]^2\right]. \qquad (2.18)$$

When $t_0 = T/4$ Eq.(2.17) becomes

$$f(t)_{pulsetrain} = \frac{A_0}{2A}f(t)_{square} + \frac{A_0}{2}$$

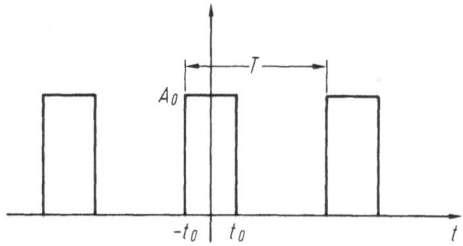

Figure 2.3   Pulse train

where $f(t)_{square}$ is given by Eq.(2.12). If $A_0 = 2A$ the pulse train is equal to a square wave plus a dc component equal to A, the peak amplitude of the square wave.

It is seen from Eq.(2.16) that for a constant value of τ_0 the spectral components $c_n = 0$, $n = 1, 2, 3 \ldots$, whenever

$$2n\pi t_0/T = m\pi \qquad m = 1, 2, 3 \ldots$$

or

$$n = \left[\frac{2\pi}{\tau_0} \right] \frac{m}{\omega_0} = \frac{m}{\tau_0 f_0} = m \frac{T}{\tau_0}$$

where n is the number of c_n that appear between m zeros of $|\sin(x)/x|^2$, where $x = n\pi\tau_0/T$. Using these results and referring to Figure 2.4 it is seen that the envelope of the curve is always the same, only the number of the spectral components within the envelope changes. This change is brought about by a change in T relative to τ_0. As τ_0/T decreases the number of c_n between minima increase and *vice versa*. Although the number and the normalized magnitude of the spectral lines increase with decreasing τ_0/T, for constant A_0 an evaluation of Eq.(2.18) reveals that the total power is proportional to τ_0/T; that is, when τ_0/T decreases by a factor of two P_{avg} decreases by a factor of two. Therefore, if for each decrease in τ_0/T the amplitude A_0 was increased by $\sqrt{T/\tau_0}$ the average power would remain constant. As τ_0/T approaches zero it would appear that the power spectrum becomes continuous. This is discussed in detail in Section 2.4.1.

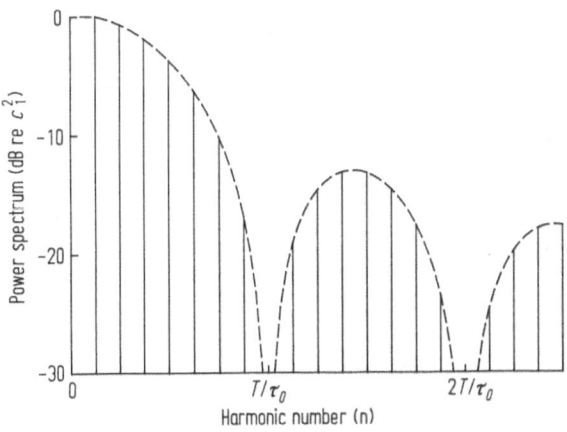

Figure 2.4 Normalized power spectrum of a pulse train

Example 3: Rise Time of a Pulse Train vs Bandwidth. The results of the previous example are now re-examined to determine the rise time of the leading edge of each pulse in the pulse train as a function of its bandwidth $BW = N_H f_0 = N_H / T$. Obviously when N_H is not infinite the periodic signal is not a pulse train. However, in anticipation of results to be obtained [see Eq.(3.33)] it is instructive to examine the case when N_H is finite, since the practical implementation of this waveform will always be with N_H finite. Equations (2.17) and (2.18) are rewritten as follows:

$$f(t) = \frac{\tau_0 A_0}{T}\left[1 + 2\sum_{n=1}^{N_H} \frac{\sin(n\pi\tau_0/T)}{(n\pi\tau_0/T)}\cos(2n\pi t/T) \right] \qquad (2.19)$$

and

$$P'_{avg} = \left[1 + 2\sum_{n=1}^{N_H}\left[\frac{\sin(n\pi\tau_0/T)}{(n\pi\tau_0/T)} \right]^2 \right] \qquad (2.20)$$

where

$$P'_{avg} = P_{avg}/(A_0^2\tau_0^2/T^2)$$

is the normalized average power in the signal as a function of N_H.

Numerically evaluating Eq.(2.19) to determine the rise time as a function of N_H yields the results shown in Figure 2.5 for $\tau_0/T = 1/\sqrt{19}$. (This value was chosen so that there exists no n for which $c_n = 0$; that is, we have a full spectrum.) Also

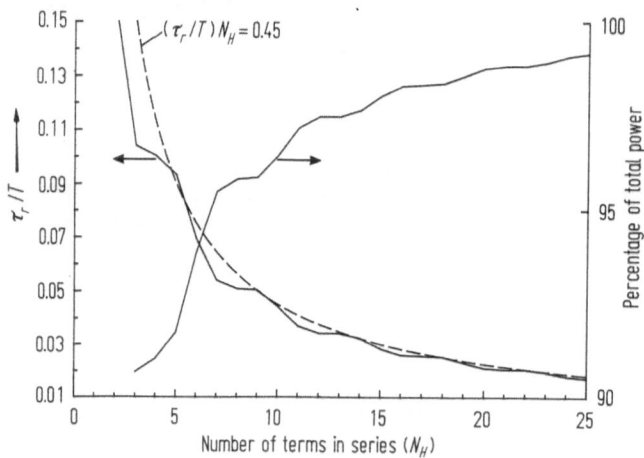

Figure 2.5 Rise time and percentage of the total power of a pulse train as a function of the number of harmonics in its Fourier series

plotted in this figure is the approximation to this curve

$$(\tau_r/T)N_H = \tau_r(N_H f_0) = 0.45 \qquad (2.21)$$

where, as stated above, $N_H f_0$ is the bandwidth of the signal. *Thus the product of the bandwidth and the rise time is equal to a constant*, which in this case is 0.45. From Eq.(2.21) we see that as $N_H \to \infty$, $\tau_r \to 0$.

Equation (2.20) is also plotted in Figure 2.5 and shows that relatively large increases in N_H result in only small increases in signal power.

Example 4: Fourier Series of a Sawtooth-like Wave. Consider the periodic waveform shown in Figure 2.6. Using Eq.(2.2) yields

$$c_n = \frac{1}{T} \int_0^T \left(1 - \frac{t}{T}\right) e^{-jn\omega_0 t} dt = \frac{-j}{2n\pi} \qquad n=1,2,3\ldots$$

and $c_0 = 1/2$. Since c_n is imaginary Eq.(2.6b) gives

$$\phi_n = \frac{\pi}{2} \qquad n=1,2,3,\ldots$$

Using the above results in Eq.(2.7) yields

$$f(t) = \frac{1}{2} + \frac{1}{\pi} \sum_{n=1}^{\infty} \frac{1}{n} \sin(n\omega_0 t). \qquad (2.22)$$

This result is needed in Section 2.3.2.

Example 5: Periodicity in the Frequency Domain. It is sometimes useful (see Section 2.5.1) to reverse the interpretation of the symbols appearing in Eqs.(2.1) and (2.2) by considering a function that is periodic in the frequency domain, instead of the time domain. In this case Eq.(2.1) becomes

$$f(\omega) = \sum_{n=-\infty}^{\infty} d_n e^{jn\omega T}, \qquad (2.23a)$$

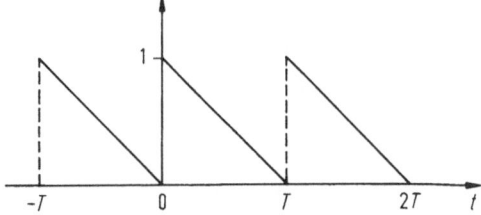

Figure 2.6 Sawtooth-like periodic waveform

and Eq.(2.2)

$$d_n = \frac{1}{\omega_c} \int_{-\omega_c/2}^{\omega_c/2} f(\omega)e^{-jn\omega T} \cdot d\omega \qquad (2.23b)$$

where the function is periodic on the interval ω_c and the d_n are equally spaced amplitudes appearing at each nT_s in the time domain.

2.3.2 Generalized Functions

We now define a useful mathematical function called the unit impulse function $\delta(t)$ (also called the dirac delta function) by the property that

$$\int_{-\infty}^{\infty} \delta(t-t_0)\phi(t)dt = \phi(t_0) \qquad (2.24)$$

provided that $\phi(t)$ is continuous at $t = t_0$. It is seen that the units of $\delta(t)$ are the reciprocal of its argument, that is, $1/t$.

In order to obtain an important formula for use in this, and subsequent, chapters we introduce the definition of the generalized derivative of a discontinuous function. If $f(t)$ is a piecewise continuous function having jump discontinuities of magnitude a_k at times t_k, then

$$f'(t) = g'(t) + \sum_k a_k \delta(t-t_k) \qquad (2.25)$$

where $g'(t)$ is the ordinary derivative of $f(t)$ where it exists.

We now use this definition to obtain the Fourier series expansion of a periodic train of unit impulse functions $\delta_T(t)$; that is,

$$\delta_T(t) = \sum_{n=-\infty}^{\infty} \delta(t-nT). \qquad (2.26)$$

Consider first Eq.(2.22) whose derivative is

$$f'(t) = \frac{2}{T} \sum_{n=1}^{\infty} \cos(2n\pi t/T).$$

Equation (2.22) is the Fourier expansion of the waveform

$$f(t) = 1 - \frac{t}{T} \qquad 0 \leq t \leq T$$

which is periodic on the interval T and from which we see that

$$g'(t) = -\frac{1}{T}.$$

In addition we note from Figure 2.6 that $f(t)$ has discontinuities occurring every $t_n = nT$, $n = 0, \pm 1, \ldots$ with magnitude $a_n = 1$. Thus Eq.(2.25) becomes

$$f'(t) = -\frac{1}{T} + \sum_{n=-\infty}^{\infty} \delta(t - nT).$$

Equating the two expressions for $f'(t)$ yields

$$\delta_T(t) = \sum_{n=-\infty}^{\infty} \delta(t - nT) = \frac{1}{T} + f'(t) = \frac{1}{T} + \frac{2}{T} \sum_{n=1}^{\infty} \cos(2n\pi t/T)$$

$$= \frac{1}{T} \sum_{n=-\infty}^{\infty} e^{jn\omega_0 t}. \tag{2.27}$$

We will make use of this result in subsequent sections.

2.3.3 Correlation

Correlation methods are used in statistics to determine the degree of linear dependence between two variables, say x and y. A normalized quantity called the correlation coefficient, which varies from -1 to +1, is the traditional indicator of this linear dependence. When the coefficient equals +1 there is complete dependency, that is, x is proportional to y. When the coefficient is -1 there is again complete dependency, but in the reverse sense. When the two variables are time dependent functions, say $x(t)$ and $y(t)$, the same interpretations hold, except that now an additional variable is introduced: a time delay τ between the two time dependent functions. In this section, and in Sections 2.4.2 and 2.6 for aperiodic and random signals, respectively, the traditional methods will be applied directly to quantities that are functions of time. For numerous examples of the application of correlation see Bendat and Piersol [1980] and Beck and Plaskowski [1987].

Consider two periodic signals $f_1(t)$ and $f_2(t)$, each with the period T. Since these signals are periodic each can be expressed as a Fourier series. Thus, from Eqs.(2.1) and (2.2)

$$f_1(t) = \sum_{n=-\infty}^{\infty} c_n e^{jn\omega_0 t} \tag{2.28a}$$

$$c_n = \frac{1}{T} \int_{-T/2}^{T/2} f_1(t) e^{-jn\omega_0 t} dt \tag{2.28b}$$

and

$$f_2(t) = \sum_{n=-\infty}^{\infty} c'_n e^{jn\omega_0 t} \tag{2.29a}$$

$$c'_n = \frac{1}{T} \int_{-T/2}^{T/2} f_2(t) e^{-jn\omega_0 t} dt. \tag{2.29b}$$

We now form the quantity

$$R_{12}(\tau) = \frac{1}{T} \int_{-T/2}^{T/2} f_2(t) f_1(t+\tau) dt \qquad (2.30)$$

where τ is a continuous time displacement over the range $(-\infty, \infty)$ and is independent of T.

Substituting Eq.(2.28a) into Eq.(2.30) yields

$$R_{12}(\tau) = \frac{1}{T} \int_{-T/2}^{T/2} \sum_{n=-\infty}^{\infty} c_n e^{jn\omega_0(t+\tau)} f_2(t) dt$$

$$= \sum_{n=-\infty}^{\infty} c_n e^{jn\omega_0\tau} \left[\frac{1}{T} \int_{-T/2}^{T/2} f_2(t) e^{jn\omega_0 t} dt \right]$$

$$= \sum_{n=-\infty}^{\infty} c_n c'_{-n} e^{jn\omega_0\tau} = \sum_{n=-\infty}^{\infty} c''_n e^{jn\omega_0\tau} \qquad (2.31)$$

where $c''_n = c_n c'_{-n}$ and we have used Eq.(2.29b). The function $R_{12}(\tau)$ is called the *cross-correlation* function. Since Eq.(2.31) is a Fourier series, the following must also hold:

$$c''_n = \frac{1}{T} \int_{-T/2}^{T/2} R_{12}(\tau) e^{-jn\omega_0\tau} d\tau. \qquad (2.32)$$

The cross-correlation function is most straightforwardly interpreted as a means of determining the time delay between $f_1(t)$ and $f_2(t)$, with the time delay usually being that value of τ corresponding to the maximum value of $R_{12}(\tau)$, either positive or negative, within each period.

If $f_1(t) = f_2(t)$ we obtain the *autocorrelation* function. Thus, using Eqs.(2.30), (2.31) and (2.6a) yields

$$R_{11}(\tau) = \frac{1}{T} \int_{-T/2}^{T/2} f_1(t) f_1(t+\tau) dt = \sum_{n=-\infty}^{\infty} |c_n|^2 e^{jn\omega_0\tau}$$

$$= |c_0|^2 + 2 \sum_{n=1}^{\infty} |c_n|^2 \cos(n\omega_0\tau) \qquad (2.33)$$

and, conversely, from Eq.(2.32)

$$|c_n|^2 = \frac{1}{T} \int_{-T/2}^{T/2} R_{11}(\tau) e^{-jn\omega_0\tau} d\tau. \qquad (2.34)$$

From Eq.(2.33) we see that the autocorrelation function is real and that when $\tau = 0$ we find from Eq.(2.9b) that

$$R_{11}(0) = P_{avg}. \qquad (2.35)$$

If the measured voltage $f(t)$ has been related to the physical quantities of interest by the sensitivity S_s, then the units of the correlation functions for periodic signals are (physical units)^2. The autocorrelation function is most directly interpreted as a measure of how well future values of $f(t)$ can be predicted from past observations.

Example 6: Autocorrelation of a Sine Wave. This problem will be solved two ways. The signal is

$$f_1(t) = f_2(t) = A \sin \omega_0 t.$$

We first use Eq.(2.30) to obtain

$$R_{11}(\tau) = \frac{A^2}{T} \int_{-T/2}^{T/2} \sin(\omega_0 t) \sin[\omega_0(t+\tau)] dt = \frac{A^2}{2} \cos(\omega_0 \tau). \qquad (2.36)$$

If instead Eq.(2.33) is used, we first must determine the c_n. Hence Eq.(2.28b) gives

$$c_n = \frac{A}{T} \int_{-T/2}^{T/2} \sin(\omega_0 t) e^{-jn\omega_0 t} dt$$

$$= -c_{-n} = -\frac{j}{2} A \qquad n = 1$$

$$= 0 \qquad n \neq 1. \qquad (2.37)$$

Substituting this result into Eq.(2.33) also yields Eq.(2.36).

Equation (2.36) shows that the autocorrelation of a sine wave of frequency ω_0 is another sine wave of frequency ω_0 shifted 90°, but of the correlation time displacement τ. Hence we see that Eq.(2.36) predicts exactly the future values of $f(t)$ from past observations. At $\tau = 0$ Eq.(2.36) gives the average power of the sine wave. Hence

$$R_{11}(0) = \frac{A^2}{2}$$

and, from Eqs.(2.35) and (2.10)

$$\{f(t)\}_{rms} = \sqrt{R_{11}(0)} = \frac{A}{\sqrt{2}} \qquad (2.38)$$

which is a well-known result (see also Eq.(6.13)).

Example 7: Cross-correlation of a Sine Wave and a Pulse Train. To obtain the cross-correlation of these two waveforms we use Eqs.(2.16), (2.37) and (2.31) to obtain

$$R_{12}(\tau) = c_{-1}c'_{1}e^{-j\omega_0\tau} + c_{1}c'_{-1}e^{j\omega_0\tau}$$

wherein $c_{\pm1}$ is given by Eq.(2.16) and $c'_{\pm1}$ by Eq.(2.37). Therefore,

$$R_{12}(\tau) = \frac{t_0 A A_0}{T}\frac{\sin\omega_0 t_0}{\omega_0 t_0}[e^{-j(\omega_0\tau+\pi/2)} + e^{j(\omega_0\tau+\pi/2)}]$$

$$= -\frac{2t_0 A A_0}{T}\frac{\sin\omega_0 t_0}{\omega_0 t_0}\sin\omega_0\tau. \tag{2.39}$$

Notice that $R_{12}(\tau)$ is a positive maximum at $\tau = -T/4$. This is because the two waveforms were initially 90° ($T/4$ s) out of phase at the start.

If $\omega_0 t_0 \ll 1$, then Eq.(2.39) becomes

$$R_{12}(\tau) \cong -\frac{2t_0 A A_0}{T}\sin\omega_0\tau.$$

Thus, the cross-correlation of a pulse train in which $t_0/T \ll 1$ can be used to recover the original sine wave. It should also be noted that this limiting case could have been obtained directly by using the c_n implied in Eq.(2.27) instead of those given by Eq.(2.16).

2.4 Aperiodic Signals

2.4.1 Fourier Transform

To introduce the Fourier transform we return to Eqs.(2.1) and (2.2) and recall from Eq.(2.8) that $\omega_0 = \Delta\omega = 2\pi/T$ is the spacing between the spectral lines. Thus

$$f(t) = \sum_{n=-\infty}^{\infty} c_n e^{j(n\Delta\omega)t} \tag{2.40}$$

and

$$c_n = \frac{\Delta\omega}{2\pi}\int_{-\pi/\Delta\omega}^{\pi/\Delta\omega} f(t)e^{-j(n\Delta\omega)t}dt. \tag{2.41}$$

Equation (2.41) shows that as $\Delta\omega \to 0 (T \to \infty)$, $c_n \to 0$. However, the ratio $c_n / \Delta\omega$ remains finite provided that

$$\int_{-\infty}^{\infty} |f(t)| dt \text{ is finite.}$$

Hence, if

$$\lim_{\Delta\omega \to 0} \frac{2\pi c_n}{\Delta\omega} = F(\omega) \tag{2.42}$$

then Eq.(2.41) becomes

$$F(\omega) = \int_{-\infty}^{\infty} f(t) e^{-j\omega t} dt \tag{2.43}$$

where we have selected n such that the product $n\Delta\omega$ remains finite as $\Delta\omega \to 0$ and we have denoted this finite value as ω.

Returning to Eq.(2.40) we see that it can be written as

$$f(t) = \frac{1}{2\pi} \sum_{n=-\infty}^{\infty} \frac{2\pi c_n}{\Delta\omega} e^{j(n\Delta\omega)t} \Delta\omega$$

which, after taking the limit as $\Delta\omega \to 0$ (recall the definition of an integral) and using Eq.(2.42), becomes

$$f(t) = \frac{1}{2\pi} \int_{-\infty}^{\infty} F(\omega) e^{j\omega t} d\omega . \tag{2.44}$$

Equations (2.43) and (2.44) comprise the Fourier transform pair. This pair is often expressed symbolically as

$$f(t) \leftrightarrow F(\omega)$$

which means that if $F(\omega)$ can be expressed by Eq.(2.43) then $f(t)$ is given by Eq.(2.44), and *vice versa*.

This heuristic approach very clearly shows two important differences between periodic and aperiodic signals. For the aperiodic signal the amplitude spectrum is *continuous* and if $f(t)$ is the voltage across a 1 ohm resistor, the units of the spectrum are now V-s or V/Hz.

If Eq.(2.43) is written as

$$F(\omega) = \int_{-\infty}^{\infty} f(t) \cos(\omega t) dt - j \int_{-\infty}^{\infty} f(t) \sin(\omega t) dt$$

$$= R(\omega) - jX(\omega) \tag{2.45}$$

then

$$F(\omega) = |F(\omega)|e^{-j\phi(\omega)} \qquad (2.46)$$

where

$$|F(\omega)| = \sqrt{R^2(\omega) + X^2(\omega)} \qquad (2.47a)$$

and

$$\phi(\omega) = \tan^{-1}[X(\omega)/R(\omega)]. \qquad (2.47b)$$

The quantity $|F(\omega)|$ is called the amplitude density spectrum (the amount of amplitude in a frequency band 1 Hz wide) and $\phi(\omega)$ the corresponding phase spectrum. It shall be seen in Section 3.2 that in certain situations the phase spectrum indicates the amount of delay of the signal at frequency ω. If the amplitude of $f(t)$ has been related to a physical quantity through the sensitivity S_s, then the amplitude density is expressed as (physical units)/Hz.

If $f(t)$ is real, which is the physically realizable case, Eq.(2.45) can be used to show that [recall Eq.(2.4)]

$$F(-\omega) = F^*(\omega) \qquad (2.48)$$

where the asterisk denotes the complex conjugate. Consequently Eqs.(2.48) and (2.47a) yield that

$$|F(-\omega)| = |F(\omega)|. \qquad (2.49)$$

2.4.2 Properties of the Fourier Transform

In this section we shall present several properties of the Fourier transform pair. These results will prove useful in both understanding aperiodic signals and in facilitating the derivation of several important relations.

Linearity
If a_1 and a_2 are constants, then

$$a_1 f_1(t) + a_2 f_2(t) \leftrightarrow a_1 F_1(\omega) + a_2 F_2(\omega). \qquad (2.50)$$

Equation (2.50) says that the complex frequency spectra of aperiodic signals can be added.

Time Shifting
If a function $f(t)$ is shifted in time an amount t_0, then

$$f(t-t_0) \leftrightarrow F(\omega)e^{-j\omega t_0} = |F(\omega)|e^{-j(\phi(\omega)+\omega t_0)} \qquad (2.51)$$

where we have used Eq.(2.46). Equation (2.51) states that a time shift of an aperiodic signal corresponds to a constant (linear) phase shift over all frequencies.

Frequency Shifting

If the time waveform $f(t)$ is multiplied by a harmonic signal of frequency ω_0, then

$$f(t)e^{j\omega_0 t} \leftrightarrow F(\omega - \omega_0).\qquad(2.52)$$

Thus, this multiplication in the time domain corresponds to a shift of the frequency spectrum by an amount ω_0.

Convolution

Convolution of two functions $f_1(t)$ and $f_2(t)$ is defined as

$$f(t) = f_1(t) * f_2(t) = \int_{-\infty}^{\infty} f_1(x) f_2(t-x) dx \qquad(2.53)$$

where the asterisk is a symbolic shorthand denoting the integral operation shown in the right-hand portion of Eq.(2.53).

For notational convenience we now introduce the unit step function $u(t)$, which is defined as

$$u(t) = 1 \qquad t>0$$

$$= 0 \qquad t<0 \qquad(2.54)$$

and $u(0) = 1/2$. Its derivative is (Papoulis [1962])

$$\frac{d}{dt}u(t) = \delta(t).\qquad(2.55)$$

Then if $f_1(t) = g_1(t)u(t)$ and $f_2(t) = g_2(t)u(t)$ Eq.(2.53) becomes

$$f(t) = [g_1(t)u(t)] * [g_2(t)u(t)]$$

$$= \int_{-\infty}^{\infty} g_1(x)u(x)f_2(t-x)u(t-x)dx$$

$$= \int_{0}^{t} g_1(x)g_2(t-x)dx \qquad(2.56)$$

where we have used the fact that

$$u(x)u(t-x) = 1 \qquad 0<x<t$$

$$= 0 \qquad \text{otherwise}.\qquad(2.57)$$

Substituting Eq.(2.53) into Eq.(2.43) yields the very useful Fourier transform pair:

$$f_1(t) * f_2(t) \leftrightarrow F_1(\omega)F_2(\omega)$$

or

$$\int_{-\infty}^{\infty} f_1(x)f_2(t-x)dx \Leftrightarrow F_1(\omega)F_2(\omega) \tag{2.58}$$

and its converse

$$(2\pi)f_1(t)f_2(t) \Leftrightarrow F_1(\omega)*F_2(\omega)$$

or

$$f_1(t)f_2(t) \Leftrightarrow \frac{1}{2\pi}\int_{-\infty}^{\infty} F_1(x)F_2(\omega-x)dx. \tag{2.59}$$

Equation (2.58) is the basis of the definition of a linear system introduced in Section 3.2. Equation (2.59) is used below to obtain the total energy in a signal.

Signal Energy
The total energy E in an aperiodic signal across a 1 ohm resistor is defined as

$$E = \int_{-\infty}^{\infty} f^2(t)dt. \tag{2.60}$$

To see how this relates to the frequency content of a signal we start with Eq.(2.59). Thus

$$\int_{-\infty}^{\infty} f_1(t)f_2(t)e^{-j\omega t}dt = \frac{1}{2\pi}\int_{-\infty}^{\infty} F_1(x)F_2(\omega-x)dx.$$

Setting $\omega = 0$ in the above equation yields

$$\int_{-\infty}^{\infty} f_1(t)f_2(t)dt = \frac{1}{2\pi}\int_{-\infty}^{\infty} F_1(x)F_2(-x)dx.$$

If we let $f(t) = f_1(t) = f_2(t)$ and use the above equation and Eq.(2.48), Eq.(2.60) becomes

$$E = \int_{-\infty}^{\infty} f^2(t)dt = \frac{1}{\pi}\int_{0}^{\infty} |F(\omega)|^2 d\omega. \tag{2.61}$$

The quantity $|F(\omega)|^2$ is called the energy density spectrum of $f(t)$. (Recall that $F(\omega)$ had the units V/Hz or V-s; therefore $|F(\omega)|^2$ has the units V^2-$s^2 = V^2$-s/(1/s) = Energy/Hz.) Equation (2.61) is called Parseval's theorem for aperiodic signals: it states that the total energy in an aperiodic signal is the integral of its energy spectral density. If again the voltage amplitude of $f(t)$ is related to the physical units by the sensitivity S_s, then the corresponding energy spectral density is (physical units)^2-s/Hz.

Correlation

The cross-correlation function for aperiodic signals is defined as

$$R_{12}(\tau) = \int_{-\infty}^{\infty} f_1(t+\tau)f_2(t)dt \qquad (2.62a)$$

$$R_{21}(\tau) = \int_{-\infty}^{\infty} f_1(t)f_2(t+\tau)dt \qquad (2.62b)$$

and the autocorrelation as

$$R_{11}(\tau) = \int_{-\infty}^{\infty} f_1(t)f_1(t+\tau)dt \qquad (2.63)$$

where τ is a continuous time displacement and, for physically realizable signals, $R_{ij}(\tau)$ is a real function.

Taking the Fourier transform of Eqs.(2.62) and (2.63) yields

$$S_{12}(\omega) = \int_{-\infty}^{\infty} R_{12}(\tau)e^{-j\omega\tau}d\tau = F_1(\omega)F_2^*(\omega) \qquad (2.64a)$$

$$S_{21}(\omega) = \int_{-\infty}^{\infty} R_{21}(\tau)e^{-j\omega\tau}d\tau = F_1^*(\omega)F_2(\omega) \qquad (2.64b)$$

$$S_{11}(\omega) = \int_{-\infty}^{\infty} R_{11}(\tau)e^{-j\omega\tau}d\tau = F_1(\omega)F_1^*(\omega) = |F_1(\omega)|^2 \qquad (2.64c)$$

where $S_{ij}(\omega)$ is called the cross-spectral energy density, $S_{ii}(\omega)$ the energy spectral density and we have used Eq.(2.48). Equations (2.64) imply that

$$R_{12}(\tau) = \frac{1}{2\pi}\int_{-\infty}^{\infty} S_{12}(\omega)e^{j\omega\tau}d\omega = \text{Re}\frac{1}{\pi}\int_{0}^{\infty} S_{12}(\omega)e^{j\omega\tau}d\omega \qquad (2.65a)$$

$$R_{21}(\tau) = \frac{1}{2\pi}\int_{-\infty}^{\infty} S_{21}(\omega)e^{j\omega\tau}d\omega = \text{Re}\frac{1}{\pi}\int_{0}^{\infty} S_{21}(\omega)e^{j\omega\tau}d\omega \qquad (2.65b)$$

$$R_{11}(\tau) = \frac{1}{2\pi}\int_{-\infty}^{\infty} S_{11}(\omega)e^{j\omega\tau}d\omega = \frac{1}{\pi}\int_{0}^{\infty} |F_1(\omega)|^2 e^{j\omega\tau}d\omega. \qquad (2.65c)$$

Equation (2.65c) is known as the Weiner-Khintchine theorem. Note that when $\tau = 0$ in Eq.(2.65c)

$$R_{11}(0) = \int_{-\infty}^{\infty} f_1^2(t)dt = \frac{1}{\pi} \int_0^{\infty} |F_1(\omega)|^2 d\omega = E$$

where we have used Eqs.(2.61) and (2.63). Thus the autocorrelation function of an aperiodic signal at $\tau = 0$ equals the total energy in the signal. If $f(t)$ has been converted by S_s to represent a physical quantity, then the correlation functions for aperiodic signals have the units of (physical units)²-s or (physical units)²/Hz. (Recall the analogous result, Eq.(2.35), for periodic signals.)

An application of the cross-correlation of two aperiodic signals is given in Section 8.3.2.

We now illustrate the results of this section with several examples.

Example 8: Amplitude Spectral Density of a Rectangular Pulse.   Consider a pulse of duration $\tau_d = 2t_0$ shown in Figure 2.7. From Eq.(2.43) we obtain

$$F(\omega) = A_p \int_{-t_0}^{t_0} e^{-j\omega t} dt = A_p \tau_d \frac{\sin(\omega t_0)}{\omega t_0}. \qquad (2.66)$$

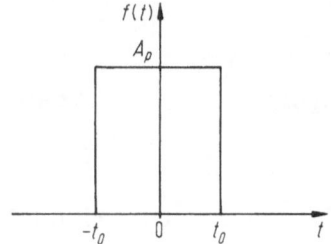

Figure 2.7   Rectangular pulse of duration $2t_0$

Equation (2.66) is plotted in Figure 2.8. When Figure 2.8 is compared to Figure 2.4, which was obtained for a periodic pulse train, it is seen that the *continuous* spectral curve shown in Figure 2.8 is exactly the same as the *envelope* of the discrete amplitude spectrum of Figure 2.4. However, the vertical scales in these figures are different: (physical units)/Hz for the aperiodic signal *versus* physical units for the periodic signal. This result is examined further in Example 9.

Example 9: Rise Time of a Rectangular Pulse versus Bandwidth.   Consider the pulse given in Example 8. If it is assumed that the bandwidth of a system is $\omega_c (= 2\pi f_c)$, then using Eq.(2.66) in Eq.(2.44) yields

$$f(t) = \frac{A_p \tau_d}{2\pi} \int_{-\omega_c}^{\omega_c} \frac{\sin(\omega t_0)}{(\omega t_0)} e^{j\omega t} d\omega = \frac{2A_p}{\pi} \int_0^{\omega_c t_0} \frac{\sin x}{x} \cos(xt/t_0)dx. \qquad (2.67)$$

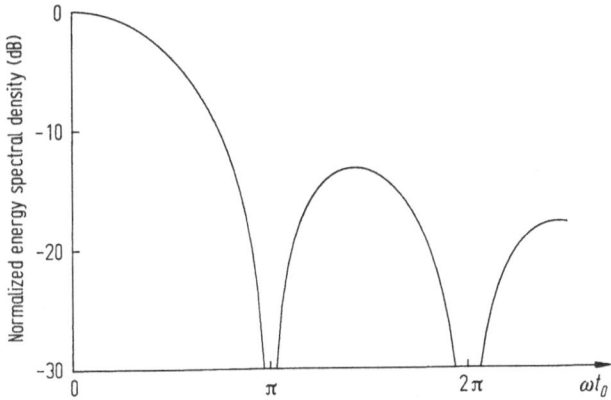

Figure 2.8 Normalized energy spectral density of a rectangular pulse of duration $2t_o$

Using the definition of rise time given in Section 2.3.1 and evaluating Eq.(2.67) numerically results in Figure 2.9. The total energy in the signal as a function of its bandwidth ω_c is obtained by substituting Eq.(2.66) in Eq.(2.61). Thus

$$E = \frac{2E_T}{\pi} \int_0^{\omega_c t_o} \left(\frac{\sin x}{x}\right)^2 dx$$

where $E_T = A_p^2 \tau_d$ is the total energy of the original signal. This result is also numerically evaluated and plotted in Figure 2.9. The rise time curve in Figure 2.9 can be expressed as

$$\left(\frac{2\tau_r}{\tau_d}\right)(\tau_d f_c) = 2\tau_r f_c = 0.9$$

or

$$\tau_r f_c = 0.45 \tag{2.68}$$

which is exactly what was obtained in Example 3 for a pulse train. Examining the relation between bandwidth and percentage total signal energy we again see that relatively small increases in signal energy result from relatively large increases in signal bandwidth.

Example 10: Energy Spectral Density of a Tone Burst. A tone burst is a sine wave whose duration lasts N periods. Thus

$$f(t) = A_0 \sin(\omega_1 t) \qquad 0 \le t \le 2N\pi/\omega_1 \qquad N = 1,2,\ldots$$

$$= 0 \qquad \text{otherwise.} \tag{2.69}$$

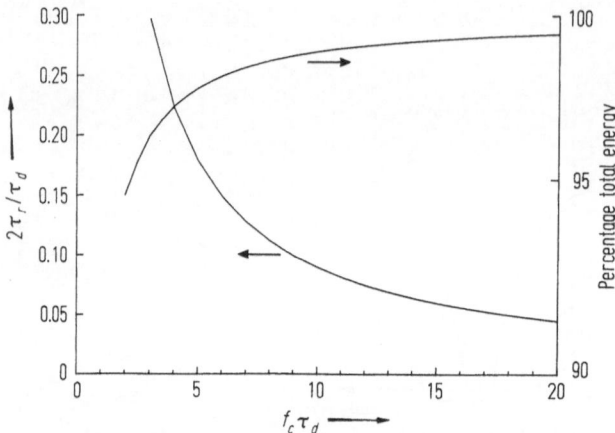

Figure 2.9 Normalized rise time and percentage of the total energy of a rectangular pulse as a function of its normalized bandwidth $f_c \tau_d$

Substituting Eq.(2.69) into Eq.(2.43) yields

$$F(\omega) = A_0 \int_0^{2N\pi/\omega_1} \sin(\omega_1 t) e^{-j\omega t} dt$$

$$= \frac{A_0 \omega_1}{\omega_1^2 - \omega^2} [1 - \cos(2N\pi\omega/\omega_1) + j\sin(2N\pi\omega/\omega_1)] \qquad \omega \neq \omega_1$$

$$= -j \frac{A_0}{\omega_1} N\pi \qquad \omega = \omega_1.$$

Consequently, from Eq.(2.64c) the energy spectral density is

$$|F(\omega)|^2 = \left[\frac{2A_0}{\omega_1(1 - \omega^2/\omega_1^2)} \sin(N\pi\omega/\omega_1) \right]^2 \qquad \omega \neq \omega_1 \quad (2.70a)$$

$$= \left(\frac{2A_0}{\omega_1} \right)^2 \left(\frac{N\pi}{2} \right)^2 \qquad \omega = \omega_1. \qquad (2.70b)$$

Equations (2.70) are plotted in Figure 2.10. It is seen that as the number of periods of the tone burst increases the more the spectral energy is concentrated at $\omega = \omega_1$. This is to be expected, for as $N \to \infty$ the tone burst approaches a periodic sine wave, which we know from Eq.(2.9b) has all of its power concentrated at $\omega = \omega_1$. In order to actually go to the limit the unit impulse function has to be used [see Eq.(2.74)].

An application of the tone burst is given in Section 8.3.2.

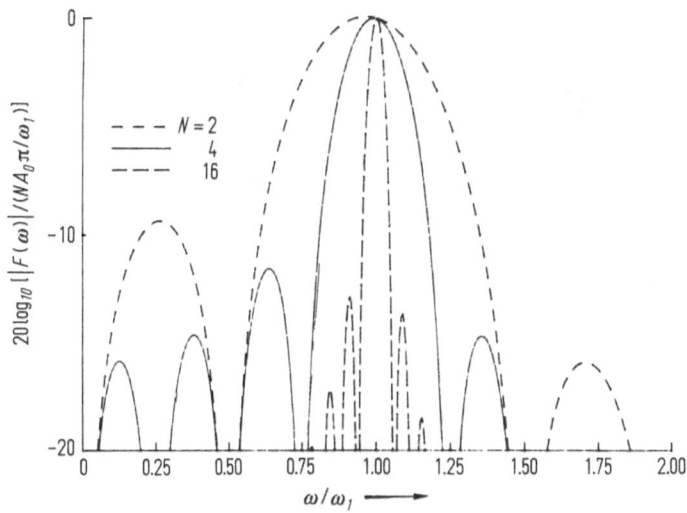

Figure 2.10 Normalized spectral energy density of a tone-burst as a function of its number of periods N

2.4.3 Fourier Transform and Generalized Functions

In this section several useful Fourier transform relationships involving generalized functions are introduced.

Unit Impulse Function
Using Eq.(2.43) and Eq.(2.24) gives

$$F(\omega) = \int_{-\infty}^{\infty} \delta(t) e^{-j\omega t} dt = 1. \qquad (2.71a)$$

From the inverse relation given by Eq.(2.44), Eq.(2.71a) implies that

$$\delta(t) = \frac{1}{2\pi} \int_{-\infty}^{\infty} e^{j\omega t} d\omega. \qquad (2.71b)$$

Equation (2.71a) states that the amplitude density spectrum of a unit impulse function has constant amplitude and infinite bandwidth.

Constant (dc) Value
If $f(t) = 1$, then Eq.(2.43) gives

$$F(\omega) = \int_{-\infty}^{\infty} e^{-j\omega t} dt \qquad (2.72a)$$

which, upon using Eq.(2.44), has as its inverse

$$1 = \frac{1}{2\pi} \int_{-\infty}^{\infty} F(\omega)e^{j\omega t} d\omega . \qquad (2.72b)$$

Equation (2.72b) implies that $F(\omega) = 2\pi\delta(\omega)$ and, therefore,

$$1 \leftrightarrow 2\pi\delta(\omega). \qquad (2.73)$$

The form of this result is as expected since a constant (dc) value is, by definition, the value at zero frequency.

Harmonic Waveforms
If

$$f(t) = e^{j\omega_0 t}$$

then from Eqs.(2.73) and (2.52) it is found that

$$e^{j\omega_0 t} \leftrightarrow 2\pi\delta(\omega - \omega_0). \qquad (2.74)$$

Periodic Function
The Fourier series expansion of a periodic waveform is given by Eq.(2.1). Substituting Eq.(2.1) into Eq.(2.43) yields

$$F(\omega) = \int_{-\infty}^{\infty} \left[\sum_{n=-\infty}^{\infty} c_n e^{jn\omega_0 t} \right] e^{-j\omega t} dt = \sum_{n=-\infty}^{\infty} c_n \int_{-\infty}^{\infty} e^{jn\omega_0 t} e^{-j\omega t} dt$$

$$= 2\pi \sum_{n=-\infty}^{\infty} c_n \delta(\omega - n\omega_0). \qquad (2.75)$$

where $\omega_0 = 2\pi/T$ and we have used Eq.(2.74). Equation (2.75) is what we expect, a series of equally spaced amplitudes of magnitude $2\pi |c_n|$. Notice that the c_n, after suitable conversion to represent physical units, still have their magnitudes expressed in physical units. Since the unit impulse function has the units of the reciprocal of its argument the product is (physical units)/Hz.

Periodic Train of Unit Impulses
Recall from Eqs.(2.26) and (2.27) that the Fourier series expansion for the periodic pulse train is given by

$$\delta_T(t) = \sum_{n=-\infty}^{\infty} \delta(t - nT) = \frac{1}{T} \sum_{-\infty}^{\infty} e^{jn\omega_0 t} \qquad (2.76)$$

where $\omega_0 = 2\pi/T$. However, it is seen that the Fourier transform of the right hand side of Eq.(2.76) is given by Eq.(2.75) with $c_n = 1$. Therefore

$$\delta_T(t) \leftrightarrow \Delta_T(\omega) = \omega_0 \sum_{n=-\infty}^{\infty} \delta(\omega - n\omega_0). \qquad (2.77)$$

This result will be used in Section 2.5.1 to develop the sampling theorem in the time domain.

2.4.4 Some Useful Fourier Transform Pairs

In anticipation of several results to be developed in Sections 3.3.1 to 3.3.3 several Fourier transform pairs shall be derived using the results of Section 2.4.3. Consider first the transform of

$$f(t) = e^{-\alpha t} u(t) \qquad \alpha > 0.$$

Substituting this expression in Eq.(2.43) yields

$$F(\omega) = \int_{-\infty}^{\infty} e^{-\alpha t} u(t) e^{-j\omega t} dt = \int_{0}^{\infty} e^{-(j\omega + \alpha)t} dt = \frac{1}{\alpha + j\omega}.$$

Therefore

$$e^{-\alpha t} u(t) \Leftrightarrow \frac{1}{\alpha + j\omega}. \tag{2.78}$$

Equation (2.78) will be used to determine the response of a low pass RC filter in Section 3.3.2.

Next consider the derivative of a function $f(t)$. Its Fourier transform is

$$\frac{d}{dt} f(t) \Leftrightarrow \int_{-\infty}^{\infty} \frac{d}{dt} f(t) e^{-j\omega t} dt. \tag{2.79}$$

Integrating Eq.(2.79) by parts and assuming that

$$\lim_{t \to \pm \infty} f(t) \to 0$$

gives

$$\frac{d}{dt} f(t) \Leftrightarrow j\omega \int_{-\infty}^{\infty} f(t) e^{-j\omega t} dt$$

or

$$\frac{d}{dt} f(t) \Leftrightarrow j\omega F(\omega) \tag{2.80}$$

where $f(t) \Leftrightarrow F(\omega)$.

To illustrate this Fourier transform pair let

$$G(\omega) = j\omega F(\omega) = \frac{j\omega}{\alpha + j\omega}.$$

We see that the inverse transform of $F(\omega)$ is given by Eq.(2.78). Thus from Eqs.(2.78), (2.80) and (2.55) we obtain

$$[\delta(t) - \alpha u(t)]e^{-\alpha t} \leftrightarrow \frac{j\omega}{\alpha + j\omega}. \tag{2.81}$$

Equation (2.81) will be used to analyze the response of a high pass RC filter in Section 3.3.3.

Now consider the expression

$$e^{-\alpha t}\sin(\omega_1 t)u(t).$$

Using the frequency shifting property given by Eq.(2.52) it is straightforward to show that

$$f_0(t)\sin\omega_1 t \leftrightarrow \frac{1}{2j}[F_0(\omega - \omega_1) - F_0(\omega + \omega_1)] \tag{2.82}$$

where $f_0(t) \leftrightarrow F_0(\omega)$ and

$$f_0(t) = e^{-\alpha t}u(t).$$

But from Eq.(2.78)

$$F_0(\omega) = \frac{1}{\alpha + j\omega}. \tag{2.83}$$

Therefore, substituting Eq.(2.83) into Eq.(2.82) gives

$$e^{-\alpha t}\sin(\omega_1 t)u(t) \leftrightarrow \frac{\omega_1}{(\alpha + j\omega)^2 + \omega_1^2}. \tag{2.84}$$

Equation (2.84) will be used in Section 3.3.1 to determine the response of a second-order linear system.

Lastly we state without proof the following pair (Papoulis [1962]):

$$u(t) \leftrightarrow \pi\delta(\omega) + \frac{1}{j\omega} \tag{2.85}$$

where $u(t)$ is the unit step function defined in Eq.(2.54). The right-hand side of Eq.(2.85) is interpreted as follows: the first term is valid only at $\omega = 0$ and the second term everywhere except at $\omega = 0$. This transform pair will be useful in developing certain relations in Chapter 3.

2.5 Sampling Theorem

2.5.1 Sampling Theorem in the Time Domain

Consider a bandlimited signal $f(t)$; that is, one whose Fourier transform is zero above a certain frequency ω_c. Thus

$$F(\omega) = 0 \qquad |\omega| > \omega_c. \qquad (2.86)$$

The sampling theorem in the time domain states that $f(t)$ can be uniquely determined from its values $f(nT_s)$ at the equally spaced times nT_s provided that $T_s = \pi/\omega_c$. Referring to Figure 2.11 consider a sampled waveform $f_s(t)$, which is the product of $f(t)$ and the unit impulse train $\delta_T(t)$. Thus

$$f_s(t) = f(t)\delta_T(t) = f(t) \sum_{n=-\infty}^{\infty} \delta(t - nT_s) \qquad (2.87)$$

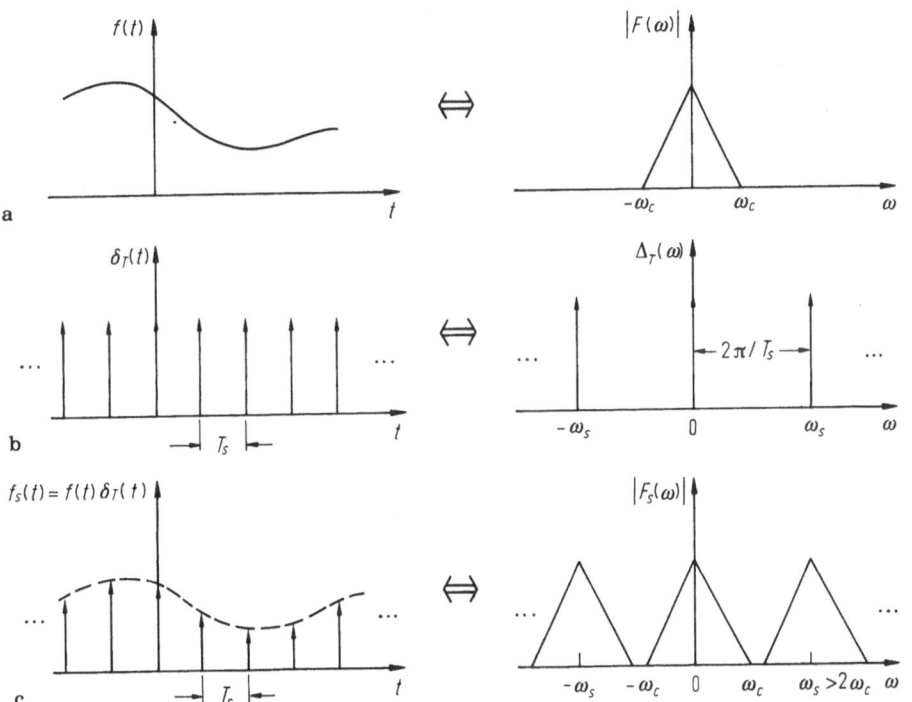

Figure 2.11 Fourier transform pairs of (a) a bandlimited signal $f(t)$ (b) a unit impulse train (c) the product of (a) and (b), which is the sampled waveform

where $T_s = 2\pi / \omega_s$ and ω_s is the sampling frequency in rad/s. If

$$f_s(t) \leftrightarrow F_s(\omega)$$

then using Eqs.(2.59) and (2.77) we obtain

$$F_s(\omega) = \frac{1}{2\pi} F(\omega) * \Delta_T(\omega) = \frac{1}{T_s} F(\omega) * \sum_{n=-\infty}^{\infty} \delta(\omega - n\omega_s)$$

$$= \frac{1}{T_s} \sum_{n=-\infty}^{\infty} F(\omega - n\omega_s). \tag{2.88}$$

Examining Eq.(2.88) it is seen that the Fourier transform of $f_s(t)$ is *periodic* on the interval ω_s and *without overlap* provided that

$$\omega_s > 2\omega_c \tag{2.89a}$$

or

$$T_s < \frac{1}{2f_c}. \tag{2.89b}$$

The maximum value of T_s is called the Nyquist interval. Conversely, the minimum value of $f_s = 1/T_s$ is called the Nyquist frequency. If Eq.(2.89a) is observed, then the Fourier transform of $f_s(t)$ will be a periodic replica of $F(\omega)$ and will, therefore, contain all the information about $f(t)$. These points are emphasized in Figure 2.11.

Since $F_s(\omega)$ is periodic we can use Eqs.(2.23) to express $F(\omega)$ as the following Fourier series:

$$F(\omega) = \sum_{n=-\infty}^{\infty} d_n e^{jn\omega T_s} \tag{2.90a}$$

and

$$d_n = \frac{1}{2\omega_c} \int_{-\omega_c}^{\omega_c} F(\omega)^{-jn\omega T_s} d\omega. \tag{2.90b}$$

Now consider the inverse Fourier transform of $F(\omega)$. Using Eq.(2.86) in Eq.(2.44) gives

$$f(t) = \frac{1}{2\pi} \int_{-\omega_c}^{\omega_c} F(\omega) e^{j\omega t} d\omega. \tag{2.91}$$

If $t = -nT_s$ in Eq.(2.91), then

$$f(-nT_s) = \frac{1}{2\pi} \int_{-\omega_c}^{\omega_c} F(\omega) e^{-jn\omega T_s} d\omega. \tag{2.92}$$

Comparing Eq.(2.90b) with Eq.(2.92) it is seen that they are equal when

$$d_n = \frac{\pi}{\omega_c} f(-nT_s).$$
(2.93)

Substituting Eq.(2.93) into Eq.(2.90a) and the result into Eq.(2.91) yields, upon reversing the order of summation and integration,

$$f(t) = \sum_{n=-\infty}^{\infty} \frac{1}{2\omega_c} f(-nT_s) \int_{-\omega_c}^{\omega_c} e^{j\omega(t+nT_s)} d\omega$$

$$= \sum_{n=-\infty}^{\infty} f(nT_s) \frac{\sin[\omega_c(t-nT_s)]}{\omega_c(t-nT_s)} \qquad T_s < \pi/\omega_c \quad (2.94)$$

which is what we set out to show.

In many applications the sampled signal does not have a finite frequency spectrum. In this case the choice of f_s becomes a problem, for the cutoff frequency f_c is not clearly defined. This is illustrated graphically in Figure 2.12, where it is seen that a portion of the periodic spectra centered around nf_s, $n = \pm 1, \pm 2, \ldots$, now have some overlap with each other. The overlapped portion, shown as a shaded area in Figure 2.12, is now a region subjected to a distortion called *aliasing*. This aliasing error occurs because the information in this region is sampled at an insufficient rate; that is, $f_s < 2f$ for $f > f_c$.

The frequency about which the spectra overlap is called the *folding frequency*; that is, the frequency about which the spectrum "folds" into the true spectrum, in this case f_c. This folding happens in such a way that the folded portion of the spectrum becomes indistinguishable from the true spectrum; that is, it's as if it were the original data. Hence the term aliasing. Once there, these folded data are irremovable from the true data.

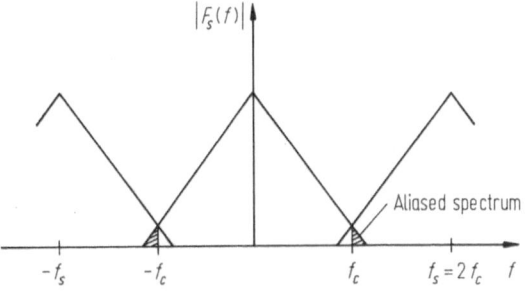

Figure 2.12 Aliased portion of a spectrum when the periodic spectra overlap because of insufficient sampling rate

To show this aliasing phenomenon we examine the following trigonometric reality. Let $0 < f < f_c$ and $t = nt_s = n/f_s = n/2f_c$ be the times at which the signal is sampled. Then

$$\cos(2\pi ft) = \cos(2\pi fnt_s) = \cos(2n\pi f/f_s). \qquad (2.95)$$

Now consider the case where $f = f_0 > f_c$, but $t = nt_s = n/f_s = n/2f_c$ remains the same. If we choose f_0 such that

$$f_0 = 2mf_c \pm f = mf_s \pm f \qquad m = 1, 2, 3 \ldots \qquad (2.96)$$

then Eq.(2.95) becomes

$$\cos(2\pi f_0 t) = \cos[2\pi n(2mf_c \pm f)/f_s]$$

$$= \cos(2n\pi f/f_s). \qquad (2.97)$$

Thus Eq.(2.95) and Eq.(2.97) are equal and, therefore, f_0 is indistinguishable from f. For example, if f_c is 100 Hz, then signals of frequencies 170, 230, 370, and so on, would be aliased with a signal having a frequency of 30 Hz. This is shown pictorially in Figure 2.13 where we have a sine wave of frequency f_i that is sampled at a frequency $f_s = 8f_i/7$. From Eq.(2.97) it is seen that the original signal at f_i now appears as the aliased signal at $f_i/7$. Further implications of aliasing errors are discussed in Section 5.5.

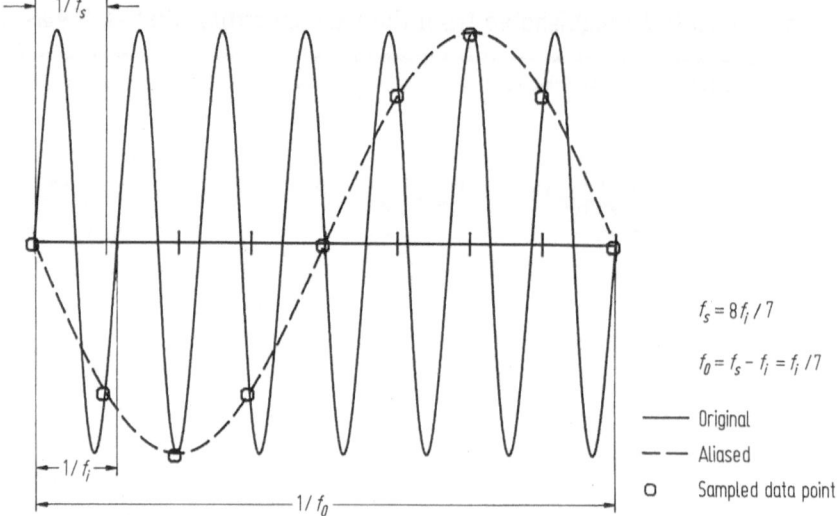

$f_s = 8f_i/7$

$f_0 = f_s - f_i = f_i/7$

—— Original

— — Aliased

○ Sampled data point

Figure 2.13 Aliasing of a sine wave of frequency f_i when the sampling frequency is $8f_i/7$

Finally, in a manner similar to that used to obtain Eq.(2.94) we can generate the sampling theorem in the frequency domain (Brigham [1974]):

$$F(\omega) = \sum_{n=-\infty}^{\infty} F(n\pi/T) \frac{\sin(\omega T - n\pi)}{(\omega T - n\pi)} \qquad (2.98)$$

provided that $f(t) = 0, |t| > T$.

2.5.2 Discrete Fourier Transform (DFT)

To obtain the discrete Fourier transform pair one must make both the original time domain signal and the corresponding Fourier transformed function periodic; that is, N time samples and N frequency samples represent one period of the time and frequency waveforms, respectively.

Let the time sampling interval be T and the frequency sampling interval be $1/T_o$, where $T_o = NT$ is the total time to take N samples. Note that because of aliasing the frequency range is limited to $f < f_c = 1/(2T)$. Then, if the given signal $h(t)$ is sampled every T seconds, Eq.(2.26) gives

$$h(t)\delta_T(t) = \sum_{k=-\infty}^{\infty} h(t)\delta(t - kT). \qquad (2.99)$$

If we sample this function over the interval T_o as shown in Figure 2.14, then Eq.(2.99) becomes

$$h(t)\delta_T(t)x(t) = \sum_{k=0}^{N-1} h(t)\delta(t - kT) \qquad (2.100)$$

where $x(t)$ is the rectangular windowing function of duration T_o. We now modify the original Fourier transform pair by uniformly sampling the Fourier transform of $h(t)$, which is denoted $H(\omega)$. The choice of T_o should now be apparent; since we want to eliminate aliasing in the time domain $x(t)$ was shifted the small amount $T/2$ shown in Figure 2.14. If this shift had not been done the Nth point of one period would have coincided with (and added to) the first point of the next period.

To obtain the Fourier transform of Eq.(2.100) we first substitute it into Eq.(2.2) to obtain

$$c_n = \frac{1}{T_o} \int_{-T/2}^{T_o-T/2} \sum_{k=0}^{N-1} h(t)\delta(t - kT)e^{-jn\omega_o t} dt$$

$$= \frac{1}{NT} \sum_{k=0}^{N-1} h(kT)e^{-j2n\pi k/N} \qquad (2.101)$$

since $\omega_o = 2\pi/T_o = 2\pi/NT$. Substituting Eq.(2.101) into Eq.(2.75) gives

$$H(n\omega_o) = H\left(\frac{2n\pi}{NT}\right) = T \sum_{k=0}^{N-1} h(kT)e^{-j2n\pi k/N} \quad n = 0, 1, \ldots, N-1 \qquad (2.102)$$

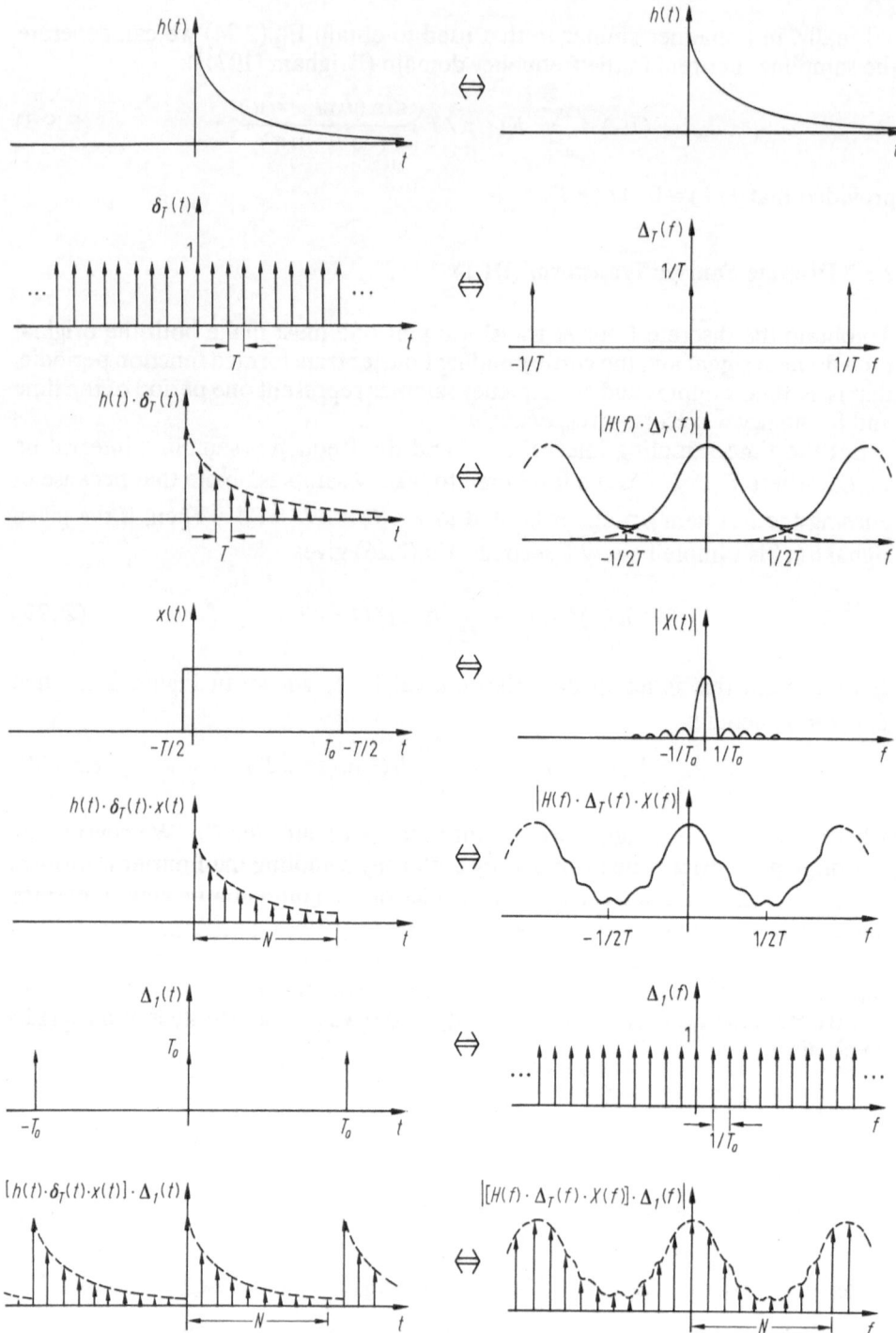

Figure 2.14   Graphical derivation of the discrete Fourier transform pair

where we have "adjusted" the constant multiplying the series so that the resulting physical units are the same as those for the Fourier integral transform. In a similar manner

$$h(kT) = \frac{1}{NT} \sum_{k=0}^{N-1} H\left(\frac{2n\pi}{NT}\right) e^{j2n\pi k/N} \quad k=0,1,\ldots,N-1 \quad (2.103)$$

which can be verified by directly substituting Eq.(2.103) into Eq.(2.102).

The truncation of the original signal $h(t)$ by the rectangular window function $x(t)$ results in a rippling effect in the Fourier transform of $h(t)\delta(t)x(t)$ as shown in Figure 2.14. This is brought about by the side lobes of $X(\omega)$, the magnitude of which can be determined from either Eq.(2.66) or Figure 2.8. To reduce the magnitude of these side lobes it is necessary to employ a time domain truncation function in which the side lobes are considerably less than those of the rectangular truncation function. One such function is the Hanning function given by (Brigham [1974])

$$x_H(t) = \frac{1}{2} - \frac{1}{2} \cos \frac{2\pi t}{T_o} \quad 0 < t < T_o$$

where T_o is the truncation interval. The magnitude of the Fourier transform of $x_H(t)$ is

$$|X_H(f)| = 0.5Q(f) + 0.25[Q(f+1/T_o) + Q(f-1/T_o)]$$

where

$$Q(f) = \frac{\sin \pi T_o f}{\pi f}.$$

The peak value of the second lobe of $|X_H(f)|$ is found to be 0.02 of its maximum value, which is 1. On the other hand the peak value of the second lobe of the $\sin f/f$ function is $1/(1.5\pi)$. Thus the Hanning function produces a second side lobe that is a factor of 10 smaller than that produced by the rectangular window. The Hanning window is frequently used when analyzing continuous signals. It is not needed for aperiodic (transient) signals if the duration of the signal is $\leq T_o$. For a discussion of other types of windowing functions see Otnes and Enochson [1978].

The previous discussion was for signals in general. If the signal is a bandlimited periodic signal, however, two special cases exist: the truncation interval is either equal to or is not equal to the signal's period. As detailed in Brigham [1974] when the truncation interval equals the signal's period (or an integer multiple of the period) the effect in the time domain is to recreate the original signal $h(t)$ with the final quantity $h(t)\delta_T(t)x(t)\Delta_1(t)$. On the other hand when the truncation interval is not equal to the signal's period the effect is to create a different periodic signal in the time domain, one with abrupt discontinuities at nT_1, $n = \pm 1, \pm 2, \ldots$, where $T_1 (\neq T_o)$ is the truncation interval. These abrupt discontinuities in the time domain result in additional frequency components in the frequency domain. This effect is termed leakage and can be minimized by either using a Hanning (or equivalent) function or by limiting the magnitude of the discontinuity. In Sections 8.6.2 and 9.5 are applications that consider these aspects.

2.5.3 Fast Fourier Transform (FFT)

The fast Fourier transform is an efficient means of calculating the discrete Fourier transform (DFT) given by Eq.(2.102), which can be rewritten as

$$X(n) = \sum_{k=0}^{N-1} x(k)W^{nk} \tag{2.104}$$

where

$$W = e^{-j2\pi/N}$$

and $T = 1$ for simplicity. To calculate Eq.(2.104) in a straightforward manner requires $(N-1)^2$ complex multiplications and N complex additions. Cooley and Tukey [1965] introduced a method of decomposing the transform into smaller groups for calculation in such a way that the total number of complex multiplications is reduced to $N\log_2 N$. For $N = 1024 = 2^{10}$, $(N-1)^2/N\log_2 N \cong 100$; thus, there are approximately 100 times fewer multiplications. The details of this procedure can be found in Cooley and Tukey [1965], Brigham [1974] and Miller [1982]. A general introduction to how this algorithm works follows.

If $N = 2^m$, where m is a positive integer, then Eq.(2.104) can be written as

$$X(n) = \sum_{k=0,1}^{N-1} x(k)W^{nk} = \sum_{k=0,1}^{N/2-1} x(2k)W^{2nk} + W^n \sum_{k=0,1}^{N/2-1} x(2k+1)W^{2nk}$$

$$= X_1(n) + W^n X_2(n) \qquad n=0,1,\ldots,N/2-1 \tag{2.105}$$

where $X_1(n)$ and $X_2(n)$ are the $N/2$-point DFTs of the even and odd $x(k)$ points, respectively. Equation (2.105) only gives values of $X(n)$ for the half-range $N/2-1$. It can be shown that for $n > N/2-1$ (Miller [1982])

$$X(n+N/2) = X_1(n) - W^n X_2(n) \qquad n=0,1,2,\ldots,N/2-1. \tag{2.106}$$

To illustrate this consider the case $N = 8$. Equations (2.105) and (2.106) give:

$$X(0) = X_1(0) + W^0 X_2(0) \qquad X(4) = X_1(0) - W^0 X_2(0)$$

$$X(1) = X_1(1) + W^1 X_2(1) \qquad X(5) = X_1(1) - W^1 X_2(1)$$

$$X(2) = X_1(2) + W^2 X_2(2) \qquad X(6) = X_1(2) - W^2 X_2(2)$$

$$X(3) = X_1(3) + W^3 X_3(3) \qquad X(7) = X_1(3) - W^3 X_2(3).$$

The efficiency of these results can be improved if the two half sequences $X_1(n)$ and $X_2(n)$ are combined from two new $N/4$-point sequences $X_{11}(n)$ and $X_{12}(n)$, where

$$X_1(n) = X_{11}(n) + W^n X_{12}(n)$$

$$X_1(n+N/4) = X_{11}(n) - W^n X_{12}(n) \qquad n=0,1,\ldots,N/4-1$$

and

$$X_2(n) = X_{21}(n) + W^n X_{22}(n)$$

$$X_2(n+N/4) = X_{21}(n) - W^n X_{22}(n) \qquad n=0,1,\ldots,N/4-1.$$

The new series $X_{1j}(n)$ and $X_{2j}(n)$, $j=1,2$ are the same as $X_1(n)$ and $X_2(n)$ except that the series are only summed to $N/4-1$ instead of $N/2-1$. In the FFT algorithm this process is repeated until we are left with an array of 2-point transforms of the data itself. The 8-point transform using the FFT algorithm is summarized in Figure 2.15.

The practical implementation of the FFT algorithm has been attained by numerous manufacturers of self-contained microprocessor-controlled digital signal analyzers and plug-in boards to PC's. An application of the use of such a device is given in Section 8.6.2.

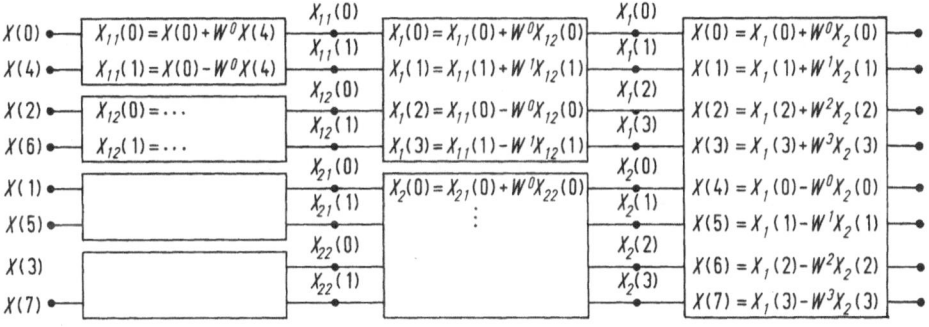

Figure 2.15   Representation of an 8-point FFT

2.6 Random Signals

We shall now introduce several definitions of different types of random signals. Consider Figure 2.16. Each time record is called a sample function and is denoted

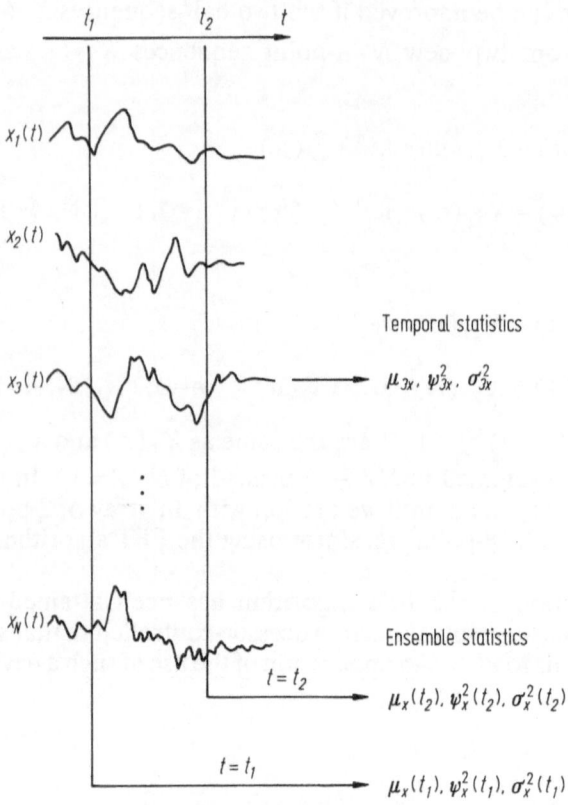

Figure 2.16 Ensemble and temporal statistics for N sample functions $x_k(t)$, $k = 1, 2, ..., N$

$x_k(t)$. The collection of sample functions is called the ensemble of the sample functions. For a given time $t = t_j$ the mean value over the ensemble is

$$\mu_x(t_j) = \lim_{K \to \infty} \frac{1}{K} \sum_{k=1}^{K} x_k(t_j). \tag{2.107a}$$

The corresponding ensemble average squared amplitude $\psi_x^2(t_j)$ and the ensemble variance $\sigma_x^2(t_j)$ are, respectively,

$$\psi_x^2(t_j) = \lim_{K \to \infty} \frac{1}{K} \sum_{k=1}^{K} x_k^2(t_j) \tag{2.107b}$$

and

$$\sigma_x^2(t_j) = \lim_{K \to \infty} \frac{1}{K} \sum_{k=1}^{K} [x_k(t_j) - \mu_x(t_j)]^2 = \psi_x^2(t_j) - \mu_x^2(t_j). \tag{2.107c}$$

On the other hand the temporal properties of the kth record are

$$\mu_{kx} = \lim_{T \to \infty} \frac{1}{T} \int_{-T/2}^{T/2} x_k(t) dt \qquad (2.108a)$$

$$\psi_{kx}^2 = \lim_{T \to \infty} \frac{1}{T} \int_{-T/2}^{T/2} x_k^2(t) dt \qquad (2.108b)$$

$$\sigma_{kx}^2 = \lim_{T \to \infty} \frac{1}{T} \int_{-T/2}^{T/2} [x_k(t) - \mu_{kx}]^2 dt = \psi_{kx}^2 - \mu_{kx}^2. \qquad (2.108c)$$

Using the definitions given by Eqs.(2.107) and (2.108) we can now define several types of random processes. If, for any choice of t_1 and t_2,

$$\mu_x(t_1) \neq \mu_x(t_2), \qquad \psi_x^2(t_1) \neq \psi_x^2(t_2), \qquad \sigma_x^2(t_1) \neq \sigma_x^2(t_2)$$

then the process is *non-stationary*. On the other hand, if for any choice of t_1 and t_2,

$$\mu_x(t_1) = \mu_x(t_2) = \mu_{kx}$$

$$\psi_x^2(t_1) = \psi_x^2(t_2) = \psi_{kx}^2$$

$$\sigma_x^2(t_1) = \sigma_x^2(t_2) = \sigma_{kx}^2$$

the process is *stationary*. If, further, for all k

$$\mu_{kx} = \mu_x, \qquad \psi_{kx}^2 = \psi_x^2, \qquad \sigma_{kx}^2 = \sigma_x^2$$

the process is *ergodic*. In many applications the ergodic assumption is reasonable. It is mentioned that strictly speaking these definitions should also include all higher order statistics. In practice, however, the above conditions are usually adequate.

We now introduce several "new" definitions for correlation and spectra. It is important to remember in these definitions that the random signal $f_r(t)$ cannot be mathematically described with an explicit function. It is employed only to indicate the operation that must be performed to arrive at the particular measure used to describe the signal.

Average Cross-correlation Function
The cross-correlation function for an ergodic random signal is obtained from the following definition:

$$\bar{R}_{21}(\tau) = \lim_{T \to \infty} \frac{1}{T} \int_{-T/2}^{T/2} f_{r1}(t) f_{r2}(t + \tau) dt. \qquad (2.109)$$

The autocorrelation function is obtained from Eq.(2.109) by setting $f_{r1}(t) = f_{r2}(t) = f_r(t)$. Notice that because of the averaging in Eq.(2.109) the units of the correlation function for random signals, when $f_{r1}(t)$ and $f_{r2}(t)$ have been converted by S_s to represent a physical quantity, are (physical units)^2, which are the same as those obtained for periodic signals [recall Eq.(2.35).]

Average Power Spectral Density

Employing the same idea incorporated in the definition of the correlation function, the spectrum of an ergodic signal is defined as

$$\overline{P}(\omega) = \lim_{T \to \infty} \frac{1}{T} \left| \int_{-T/2}^{T/2} f_r(t)e^{-j\omega t}dt \right|^2 . \qquad (2.110)$$

If again $f_r(t)$ represents a physical quantity, then the units of $\overline{P}(\omega)$ are (physical units)^2-s or (physical units)^2/Hz. Consequently $\overline{P}(\omega)$ is called the average power density of $f_r(t)$. The corresponding rms amplitude spectrum is obtained from the square root of $\overline{P}(\omega)$; that is, (physical units)$/\sqrt{Hz}$.

Relation Between Correlation and Power Spectra

The average power spectrum and the autocorrelation function are related by the Weiner-Khintchine theorem, which states that they are Fourier transforms of each other. Thus

$$\overline{P}_{11}(\omega) = \int_{-\infty}^{\infty} \overline{R}_{11}(\tau)e^{-j\omega\tau}d\tau \qquad (2.111a)$$

and

$$\overline{R}_{11}(\tau) = \frac{1}{2\pi} \int_{-\infty}^{\infty} \overline{P}_{11}(\omega)e^{j\omega\tau}d\omega . \qquad (2.111b)$$

These relations are extended to the cross properties in Section 3.4.

Random Error

Explicitly stated in the relations in this section is that the duration over which one takes data should approach infinity. In actuality, of course, this time has to be finite. However, of interest is the shortest time one can use to stay within a pre-scribed error for the magnitude of the rms amplitude measured in a given band-width. It has been shown that a rough estimate of the random error ϵ_n of the rms amplitude in a frequency band B is (Bendat and Piersol [1966])

$$\epsilon_n = \frac{1}{\sqrt{BT_n}} \qquad (2.112)$$

where ϵ_n is the ratio of the standard deviation divided by the mean and is sometimes called the coefficient of variation; T_n is the total averaging time. For a digital process, such as a DFT or FFT, Eq.(2.112) becomes

$$\epsilon_n = \sqrt{\frac{2}{N_T}} \tag{2.113}$$

where N_T is the number of time slices averaged in the frequency domain on a frequency band by frequency band basis.

Example 11: Autocorrelation Function of Broadband White Noise. One of the characteristics of white noise is that its average power spectral density is a constant over all frequencies; thus $\overline{P}(\omega) = N_o$. Substituting this value into Eq.(2.113b) yields

$$\overline{R}_{11}(\tau) = \frac{N_o}{2\pi} \int_{-\infty}^{\infty} e^{j\omega\tau} d\omega = N_o \delta(\tau) \tag{2.114}$$

where we have used Eq.(2.71b). This result will be used as one means of determining the characteristics of a linear system in Section 3.4.

References

1. Beck, M. S., and Plaskowski, A., *Cross-Correlation Flowmeters- Their Design and Application*, Adam Hilger, Bristol, 1987.

2. Bendat, J. S., and Piersol, A. G., *Measurement and Analysis of Random Data*, 2nd Ed., John Wiley and Sons, New York, 1986.

3. Bendat, J. S., and Piersol, A. G., *Engineering Applications of Correlation and Spectral Analysis*, John Wiley and Sons, New York, 1980.

4. Braun, S., Editor, *Mechanical Signature Analysis*, Academic Press, Inc., London, 1986.

5. Brigham, E. O., *The Fast Fourier Transform*, Prentice-Hall, Englewood Cliffs, NJ, 1974.

6. Cooley, J. W., and Tukey, J. W., "An Algorithm for Machine Calculation of Complex Fourier Series", Math. Computation, Vol. 19, pp. 297-301, April 1965.

7. Hsu, H. P., *Applied Fourier Analysis*, Harcourt Brace Jovanovich, San Diego, CA, 1984.

8. Jolley, L. B. W., *Summation of Series*, 2nd Ed., Dover Publications, Inc., New York, 1961.

9. Lee, Y. W., *Statistical Theory of Communication*, John Wiley and Sons, New York, 1967.

10. Magrab, E. B., and Blomquist, D. S., *The Measurement of Time-Varying Phenomena: Fundamentals and Applications*, John Wiley and Sons, New York, 1971.

11. Miller, M. J., "Discrete Signals and Frequency Spectra", in Sydenham, P. H., Ed., *Handbook of Measurement Science, Vol. 1, Theoretical Fundamentals*, John Wiley and Sons, New York, 1982.

12. Otnes, R. K., and Enochson, L., *Applied Time Series Analysis, Vol 1: Basic Techniques*, John Wiley and Sons, New York, 1978.

13. Papoulis, A., *The Fourier Integral and Its Applications*, McGraw-Hill Book Co., New York, 1962.

14. Papoulis, A., *Signal Analysis*, McGraw-Hill Book Co., New York, 1977.

Exercises

1. For the power and phase spectra of a periodic signal shown below determine: a) the original signal, b) the %THD, c) the bandwidth and d) the total average power.

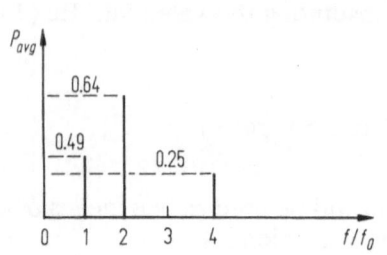 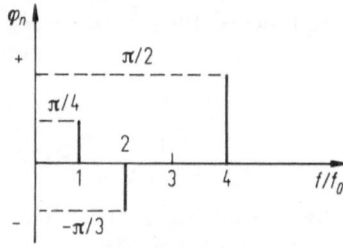

2. Obtain the power spectrum of the periodic signal $f(t) = A_o \, | \sin(2\pi f_1 t) |$. Plot the results.

3. Obtain an expression for the power spectrum of the periodic tone burst shown below. Sketch the results for $N/(N+M) = 1/9, 1/5$ and $31/63$. Compare the envelope of the results to the results shown in Figure 2.10.

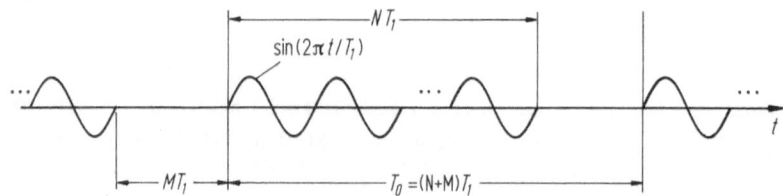

4. Find the cross-correlation function of a sine wave and square wave (see Figure 2.1a) of the same period and of amplitudes A_s and A_t, respectively, using (a) Eq.(2.30) and (b) Eq.(2.31).

5. For the pulse waveform shown below determine: (a) its amplitude density and phase spectra, (b) its energy spectral density and (c) an expression from which the maximum amplitude as a function of bandwidth can be obtained.

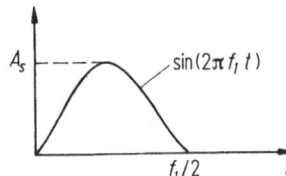

6. From the energy spectral density plot shown below determine the total energy in the signal. What are the units of A_o if the sensitivity S_s is $V/(m/s)$.

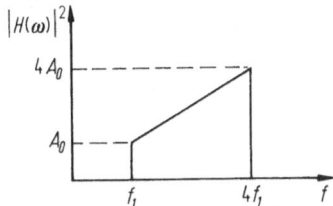

7. For the pair of rectangular pulses shown below obtain the amplitude density spectrum. Generalize the results to N such pulses and compare the results to those obtained for a single pulse. What is happening to the spectrum as N increases.

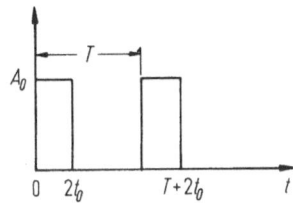

8. Obtain an expression for the energy density spectrum for the pulse shown below. Compare it with the results given in Example 8.

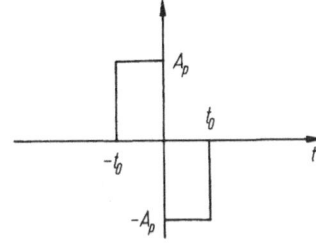

9. If the Fourier transform is

$$F(\omega) = \frac{j\omega}{(\alpha + j\omega)(\beta + j\omega)}$$

then using the results of Sections 2.4.2 and 2.4.4 obtain $f(t)$.

10. If the *sampling* frequency is 200 Hz, what frequencies below 700 Hz are aliased to 45 Hz.

11. Bandlimited white noise is defined as $\overline{P}(\omega) = N_o$, $|\omega| \leq \omega_c$. Determine its autocorrelation function. Compare the result to Eq.(2.114).

12. If one is measuring velocity in m/s what are the units of the corresponding amplitude spectra when the velocity signal is (a) periodic, (b) transient (aperiodic) and (c) stationary random.

3 Linear Systems and Filters

3.1 Introduction

In this chapter we shall introduce the definition of a linear system and develop in detail its effect on signals. These concepts will then be used to examine special linear systems that are expressly designed to modify its input signal in such a way that only signals of certain frequencies pass through it unaltered, while all other frequencies are greatly attenuated. Such systems are called filters. The actual design of filters, however, will not be discussed. The results of this chapter have far-reaching implications on the importance of knowing ahead of time certain properties of each individual electrical and mechanical component comprising an instrumentation system.

3.2 Definition of a Linear System

Consider a linear system shown in Figure 3.1. The input signal is $s_i(t)$ and its output signal $s_o(t)$. A linear system can be regarded as transforming the input signal in the following manner:

$$s_o(t) = \int_{-\infty}^{\infty} h(\tau) s_i(t-\tau) d\tau \tag{3.1}$$

Figure 3.1 Linear system $h(t)$ with input signal $s_i(t)$ and output signal $s_o(t)$

where $h(t)$ is called the *impulse response* of the system. Recalling Eq.(2.53) it is seen that Eq.(3.1) is the convolution integral. Therefore, if

$$s_i(t) \leftrightarrow S_i(\omega), \quad s_o(t) \leftrightarrow S_o(\omega), \quad h(t) \leftrightarrow H(\omega) \tag{3.2}$$

are Fourier transform pairs, then from Eq.(2.58)

$$S_o(\omega) = H(\omega)S_i(\omega) \tag{3.3}$$

where $H(\omega)$ is the *transfer function*; that is,

$$H(\omega) = \frac{S_o(\omega)}{S_i(\omega)}. \tag{3.4}$$

Notice that $h(t)$ has the units of $1/s$ whereas $H(\omega)$ is usually dimensionless. The exception is when the input signal represents a physical quantity and the output signal is a voltage. In this case $H(\omega)$ is the transfer function of S_s, the sensitivity introduced in Section 1.2.3.

Using Eqs.(2.44) and (3.3) we see that Eq.(3.1) can also be written as

$$s_o(t) = \frac{1}{2\pi} \int_{-\infty}^{\infty} H(\omega)S_i(\omega)e^{j\omega t} d\omega \tag{3.5}$$

which, we shall see, is a useful form.

To see why $h(t)$ is called the impulse response of the system we let $s_i(t) = \delta(t)$. Then from Eq.(2.24), Eq.(3.1) yields

$$s_0(t) = \int_{-\infty}^{\infty} h(\tau)\delta(t-\tau)d\tau = h(t).$$

Furthermore, from Eq.(2.71a), Eq.(3.4) becomes

$$S_o(\omega) = H(\omega).$$

Considering the form of Eq.(2.46), Eq.(3.4) can be written as

$$\frac{S_o(\omega)}{S_i(\omega)} = |H(\omega)|e^{-j\theta(\omega)} \tag{3.6}$$

where, if

$$H(\omega) = R(\omega) - jX(\omega) \tag{3.7}$$

then

$$|H(\omega)| = [R^2(\omega) + X^2(\omega)]^{1/2} \tag{3.8a}$$

and

$$\theta(\omega) = \tan^{-1}[X(\omega)/R(\omega)]. \tag{3.8b}$$

It is seen from Eq.(3.6) that at each frequency ω the input signal's amplitude and phase are modified by $|H(\omega)|$ and $\theta(\omega)$, respectively. The quantity $|H(\omega)|$ is the amplitude response of the system and $\theta(\omega)$ its phase response. Notice that in the definition of $\theta(\omega)$ in Eq.(3.6) a positive $\theta(\omega)$ given by Eq.(3.8b) indicates a phase *lag*.

Equation (3.6) suggests that $H(\omega)$ can be obtained by measuring the amplitude change and phase lag of a sine wave at each frequency ω as ω is incremented through the frequency range of interest. Thus, if

$$s_i(t) = Ae^{j\omega t}$$

then Eq.(3.1) gives

$$s_o(t) = A \int_{-\infty}^{\infty} h(\tau)e^{j\omega(t-\tau)}d\tau = Ae^{j\omega t}\int_{-\infty}^{\infty} h(\tau)e^{-j\omega\tau}d\tau$$

$$= A|H(\omega)|e^{j[\omega t - \theta(\omega)]}. \tag{3.9}$$

The preceding results indicate that there are two distinctly different signals available from which the characteristics of a linear system can be determined: an impulse and a sine wave. Note that only one of the methods has to be used on the system since $h(t)$ and $H(\omega)$ are the Fourier transforms of each other. Which method is used depends primarily the physical principles governing the measurement and the instrumentation available. A third type of signal using broadband noise as the input signal is discussed in Section 3.4.

If, in Eq.(3.9), the time delay of a sine wave of frequency $f_1 = 1/T_1$ is t_d, then the corresponding phase response in radians is given by

$$\theta(\omega_1) = \omega_1 t_d = 2\pi f_1 t_d = 2\pi(t_d/T_1). \tag{3.10}$$

Therefore a lag of $1°$ ($\pi/180$ radians) corresponds to the time delay $t_d = 1/(360 f_1)$ which, for example, at 1000 Hz is 2.8 μs. The lag (or lead) given in Eq.(3.10) has another interpretation when applied to the sampling of signals in multi-channel digital data acquisition systems. This is discussed in Section 5.6.

Ideal Linear Systems
Examination of Eq.(3.9) suggests the properties of an ideal linear system. Consider a system whose transfer function is given by

$$|H(\omega)| = k \tag{3.11a}$$

$$\theta(\omega) = \omega t_d. \tag{3.11b}$$

Substituting these values into Eq.(3.5) yields

$$s_o(t) = \frac{k}{2\pi}\int_{-\infty}^{\infty} S_i(\omega)e^{j\omega(t-t_d)}d\omega = ks_i(t-t_d). \tag{3.12}$$

Thus as seen from Eqs.(3.11) and (3.12) a linear system will not affect the shape of its input signal provided that the system has infinite bandwidth with constant amplitude response and a linear phase response that is proportional to frequency. The phase proportionality constant is the time delay of the system; that is, the amount of time that it takes the input signal to appear at the system's output. Consequently, any system having either amplitude distortion (k not constant with frequency) or phase distortion ($\phi(\omega) \neq \omega t_d$), or both, will alter the input signal.

In practice some alteration always occurs. The important consideration is how much will it affect the resulting signal and will this amount lead to incorrect results. As discussed in Section 1.2.3 the variation in k is usually specified as varying by no more than a certain percentage, or a certain number of decibels, between two frequencies. This frequency span is sometimes referred to as the system's linear frequency response region.

Equations (3.11) and (3.12) now lead to the definition of ideal filters (or linear systems). There are three kinds (Javid and Brenner [1963]):

Low Pass

$$H(\omega) = ke^{-j\omega t_d} \qquad |\omega| < \omega_c$$
$$= 0 \qquad |\omega| > \omega_c \qquad (3.13)$$

High Pass

$$H(\omega) = ke^{-j\omega t_d} \qquad |\omega| > \omega_c$$
$$= 0 \qquad |\omega| < \omega_c \qquad (3.14)$$

Bandpass

$$H(\omega) = ke^{-j\omega t_d} \qquad \omega_1 < |\omega| < \omega_2$$
$$= 0 \qquad |\omega| < \omega_1, \ |\omega| > \omega_2 \qquad (3.15)$$

where ω_c, ω_1 and ω_2 are called the *cutoff frequencies* of the filter. These three types of ideal filters are illustrated in Figure 3.2. (A fourth filter, called a notch filter, is not relevant for our purposes and is omitted.)

It is seen that the cutoff frequency defines the bandwidth of these filters, with the bandwidth of the lowpass filter being ω_c, the high pass filter infinity and the bandpass filter $\omega_2 - \omega_1$. Unfortunately these ideal filters are physically unattainable because we have specified the amplitude and phase responses independently. It can be shown that these quantities are not independent, but are related to each other by their Hilbert transform (Papoulis [1962]).

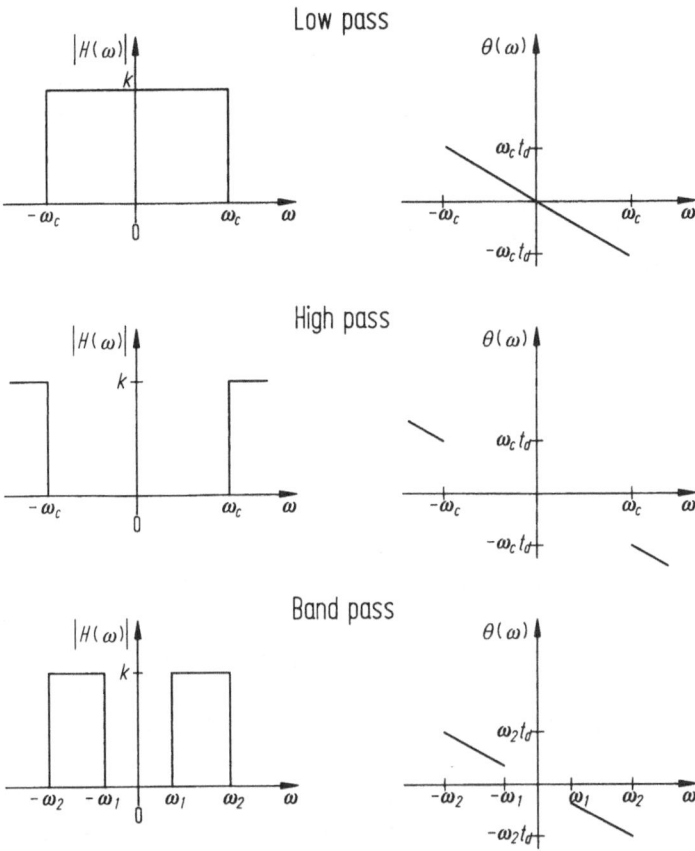

Figure 3.2   Ideal low pass, high pass and bandpass filters

Multiple Linear Systems

Returning to Eq.(3.3) it is seen that it is possible to determine the output spectrum of a signal that passes through two, or more, linear systems. Referring to Figure 3.3 we find that for two systems

$$S_o(\omega) = H_2(\omega)S_{o1}(\omega) = H_1(\omega)H_2(\omega)S_i(\omega). \qquad (3.16)$$

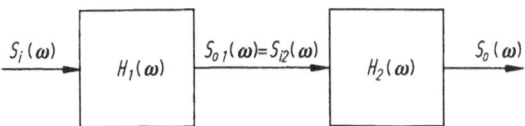

Figure 3.3   Two non-interacting linear systems connected together

Equation (3.16) is valid *only if the two systems do not influence each other, either electrically or mechanically*; that is, the two systems, when connected to each other, maintain the same frequency response characteristics they had prior to being joined together. The cases when the interconnections influence each other are considered in Sections 4.5, 8.6.1, 8.6.2, 9.2 and 9.5.

Periodic Signals and Linear Systems

We now determine the output of a linear system when the input signal is periodic. Using Eq.(2.75) the Fourier transform of the periodic input signal is given by

$$S_i(\omega) = 2\pi \sum_{n=-\infty}^{\infty} c_n \delta(\omega - n\omega_0) \tag{3.17}$$

where c_n is given by Eq.(2.2). Substituting Eq.(3.17) into Eq.(3.5) yields

$$s_o(t) = \sum_{n=-\infty}^{\infty} c_n \int_{-\infty}^{\infty} H(\omega)\delta(\omega - n\omega_0)e^{j\omega t}d\omega = \sum_{n=-\infty}^{\infty} c_n H(n\omega_0)e^{jn\omega_0 t}$$

$$= |c_0||H(0)| + 2\sum_{n=1}^{\infty} |c_n||H(n\omega_0)|\cos[n\omega_0 t - \phi_n - \theta(n\omega_0)] \tag{3.18}$$

where ϕ_n is given by Eq.(2.6b) and $\theta(n\omega_0)$ by Eq.(3.8b). From Eq.(3.18) we see that the components of the amplitude spectrum of the output periodic signal are now given by $2|c_n H(n\omega_0)|$ and the phase spectrum by $\phi_n + \theta(n\omega_0)$.

Example 1: Rise Time of an Ideal Low Pass Filter. Consider the ideal low pass filter given by Eq.(3.13) that is subjected to a step function change in its input. The transfer function $H(\omega)$ is given by Eq.(3.13) and the Fourier transform of the unit step function is given by Eq.(2.85). Substituting these equations in Eq.(3.5) yields (Javid and Brenner [1963])

$$s_o(t) = \frac{k}{2}\left[1 + \frac{2}{\pi}Si[\omega_c(t - t_d)]\right] \tag{3.19}$$

where $Si(y)$ is the sine integral

$$Si(y) = \int_0^y \frac{\sin(x)}{x}dx$$

which has the limiting values

$$\lim_{y \to 0} \to y, \qquad \lim_{y \to \pm\infty} \to \pm\frac{\pi}{2}.$$

Equation (3.19) is plotted in Figure 3.4. Using our definition of rise time it is seen from this figure that $\omega_c\tau_r = 0.88\pi$ and therefore,

$$f_c\tau_r = 0.44 \tag{3.20}$$

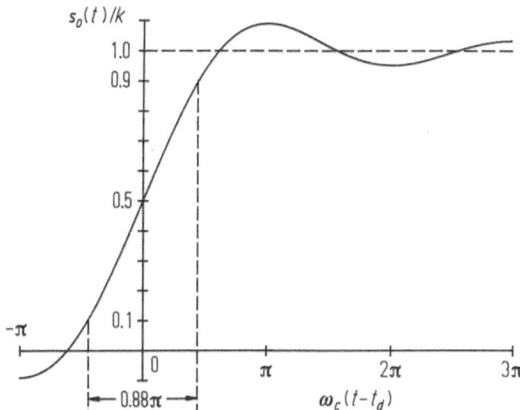

Figure 3.4   Response of an ideal low pass system to a unit step function

which is almost exactly what was obtained previously in Eqs.(2.21) and (2.68). Thus a linear system that has the frequency response given by Eq.(3.13) is a system that limits (filters) the input signal's frequency spectrum and, therefore, governs the rise time of the output signal, provided of course that the input signal's spectrum appreciably exceeds the system's cutoff frequency f_c.

Examination of Eq.(3.19) shows that there is an output signal prior to the application of the input signal at $t = 0$. This non-realizable anticipatory response is a direct consequence of assuming the form of the amplitude and phase responses independently in the definitions of the ideal filters.

3.3 Analysis of Several Linear Systems

3.3.1 Second-Order System

Consider a second-order system that is subjected to a unit step function. The transfer function of second-order system can be expressed as

$$H(\omega) = \frac{\omega_n^2}{\omega_n^2 - \omega^2 + 2jc\omega\omega_n} = |H(\omega)|\, e^{-j\theta(\omega)} \tag{3.21}$$

where

$$|H(\omega)| = \left[(1 - \overline{\omega}^2)^2 + (2c\overline{\omega})^2\right]^{-1/2}$$

$$\theta(\omega) = \tan^{-1} \frac{2c\overline{\omega}}{1 - \overline{\omega}^2}$$

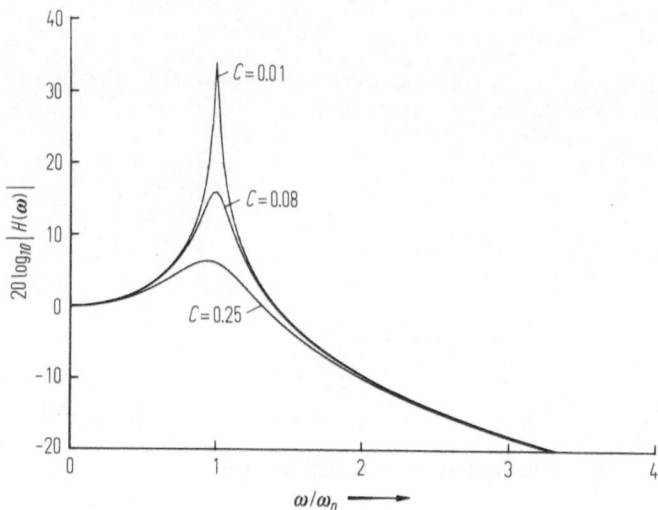

Figure 3.5 Normalized transfer function of a second-order system

and ω_n is the natural frequency of the system, c is the dimensionless damping coefficient and $\overline{\omega} = \omega/\omega_n$. The magnitude of Eq.(3.21) is plotted in Figure 3.5. The system's impulse response, which is the inverse Fourier transform of Eq.(3.21), can be obtained from Eq.(2.84) by noting that in Eq.(2.84)

$$\alpha = c\omega_n, \quad \omega_1^2 = \omega_n^2(1-c^2).$$

Consequently, with these definitions we obtain,

$$h(t) = \frac{\omega_n e^{-c\omega_n t}}{\sqrt{1-c^2}} \sin\left(\sqrt{1-c^2}\,\omega_n t\right) u(t).$$ (3.22)

Substituting Eq.(3.22) into Eq.(3.1) yields, since $s_i = u(t)$,

$$s_o(t) = \frac{\omega_n}{\sqrt{1-c^2}} \int_0^t e^{-c\omega_n \tau} \sin\left(\sqrt{1-c^2}\,\omega_n \tau\right) d\tau$$

$$= 1 - e^{-c\omega_n t}\left[\cos\left(\sqrt{1-c^2}\,\omega_n t\right) + \frac{c}{\sqrt{1-c^2}} \sin\left(\sqrt{1-c^2}\,\omega_n t\right) \right]$$ (3.23)

and we have used Eq.(2.57). The maximum value of $s_o(t)$ occurs when $\omega_n t = \pi/\sqrt{1-c^2}$, which corresponds to the maximum percentage overshoot s_{os} of

$$s_{os} = 100 e^{-c\pi/\sqrt{1-c^2}} \quad \% \qquad c < 1.$$

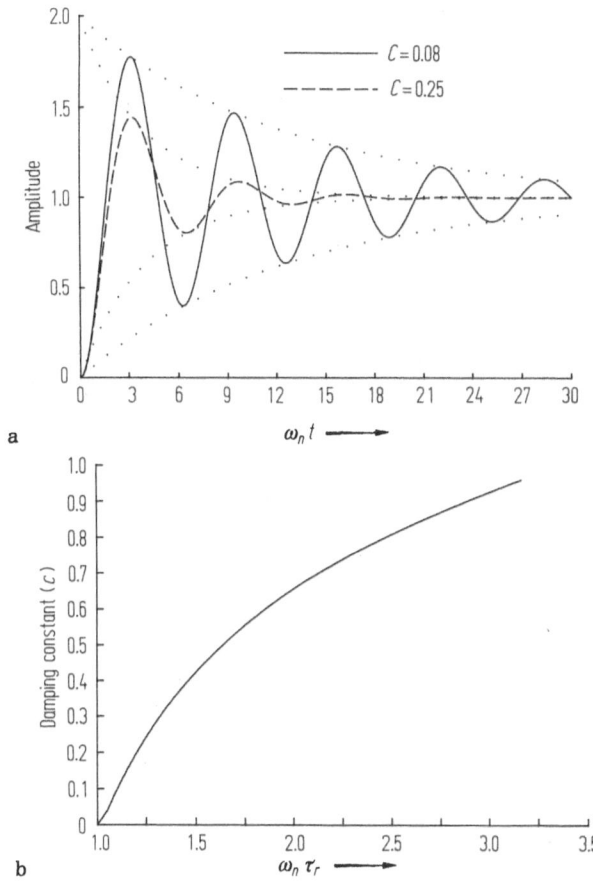

a

b

Figure 3.6 (a) Amplitude response of a second-order system to a unit step function for two values of damping c (b) rise time τ_r as a function of damping c and the natural frequency ω_n of a second-order system

Equation (3.23) is shown in Figure 3.6a for several values of damping coefficient c. In addition, Eq.(3.23) is also solved numerically to determined its rise time as a function of damping and the results are plotted in Figure 3.6b. As might be expected, for a given natural frequency the rise time increases as the damping increases. It could be concluded, therefore, that to decrease the rise time one simply has to decrease the damping. Unfortunately there is another characteristic of the output signal that is also of concern: the *settling time*. The settling time is the time it takes for the output signal of a system to reach a certain percentage of its final value after the system has been subjected to a step change in its input.

To determine this system's settling time we examine the *envelope* of the system's response given by Eq.(3.23). If x is the percentage of the system's final response and t_s is the time to reach this percentage (the settling time), then

$$\frac{s_o(t_s)}{s_o(\infty)} = \frac{x}{100} \qquad (3.24)$$

and we find that

$$\omega_n t_s \leq -\frac{1}{c}\ln[(1-x/100)\sqrt{1-c^2}]. \qquad (3.25)$$

Equation (3.25) is plotted in Figure 3.7 where it is clearly seen that for a given natural frequency any decrease in the damping is at the expense of increased settling time. The inescapable conclusion, therefore, is to increase the natural frequency of the system sufficiently so that both the rise time and settling time criteria can be satisfied simultaneously.

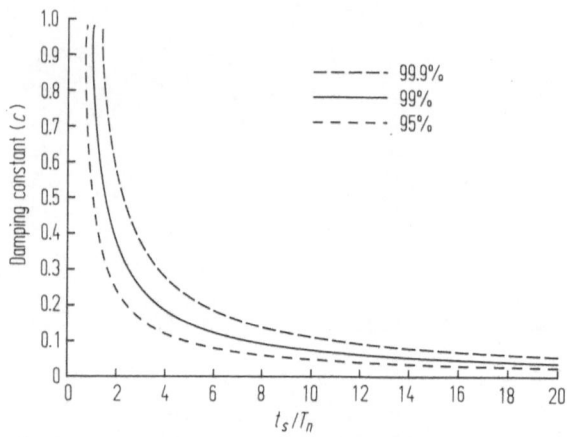

Figure 3.7 Settling time of a second-order system to within 95, 99, and 99.9% of its final value as a function of damping ($T_n = 2\pi/\omega_n$)

3.3.2 Low Pass RC Filter

Many electrical and electro-mechanical devices can be described by equivalent circuits comprised of resistors (R) and capacitors (C) connected in one of two ways: either as a low pass or a high pass filter. In this section the combination to be considered is that of the low pass RC filter shown in Figure 3.8a. The transfer function of this circuit is

$$H(\omega) = \frac{1}{1+j\omega\tau_0} = |H(\omega)|e^{-j\theta(\omega)} \qquad (3.26)$$

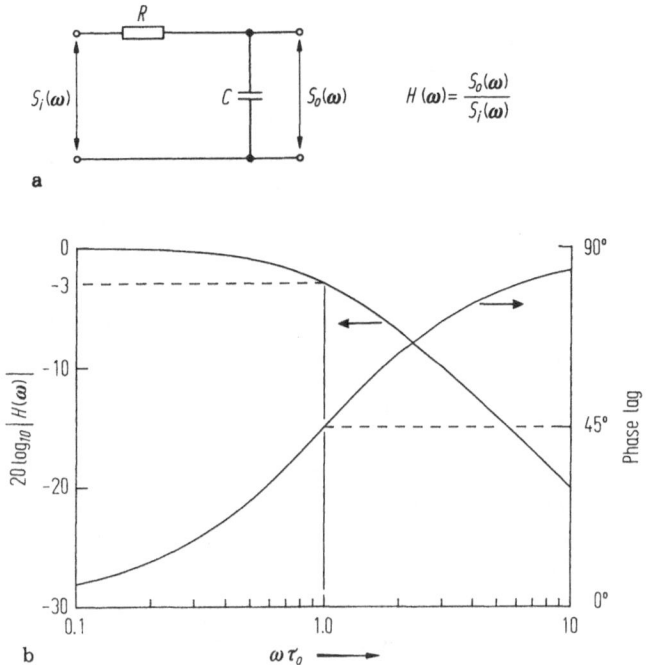

Figure 3.8 (a) Low pass RC filter   (b) transfer function of a low pass RC filter

where

$$|H(\omega)| = \frac{1}{\sqrt{1 + (\omega\tau_0)^2}} \qquad (3.27a)$$

$$\theta(\omega) = \tan^{-1}(\omega\tau_0) \qquad (3.27b)$$

and $\tau_0 = RC$ is called the *time constant* of the circuit. (Since R has the units of V/A and C the units of charge/V = A-s/V, RC has the units of seconds.) Equations (3.27) are shown in Figure 3.8b. The significance of the time constant will be apparent shortly.

Cutoff Frequency
We now introduce a definition for the cutoff frequency of a filter (or linear system) as the frequency at which the magnitude of its transfer function decreases by $1/\sqrt{2} = 0.707$ (-3 dB) from its maximum or mean value, whichever is applicable. In general, the cutoff frequency is not related to the amount of variability stipulated in the frequency response linearity specification discussed in Section 1.2.3. If the cutoff frequency is f_c, then Eq.(3.27a) gives

$$\frac{1}{\sqrt{2}} = \frac{1}{\sqrt{1 + (2\pi f_c \tau_0)^2}}$$

or

$$f_c = \frac{1}{2\pi\tau_0} = \frac{1}{2\pi RC}. \tag{3.28}$$

Therefore, the cutoff frequency of a low pass RC filter is proportional to the inverse of the circuit's time constant τ_0. For instruments and transducers the -3 dB cutoff frequencies define their bandwidth. Since this is a low pass filter its bandwidth, therefore, is f_c [recall Eq.(3.13)].

Limiting Cases: Integrator

We now examine two limiting cases of Eq.(3.26). When $\omega\tau_0 \ll 1$ we find from Eqs.(3.27) that Eq.(3.26) becomes

$$H(\omega) \cong e^{-j\omega\tau_0}$$

which we see from Eq.(3.13) is an ideal filter of unit magnitude. When $\omega\tau_0 \gg 1$ we find from Eqs.(3.27) that Eq.(3.26) becomes

$$H(\omega) \cong \frac{1}{j\omega\tau_0}. \tag{3.29}$$

To interpret this result let the input signal to the low pass filter in this region be a sine wave of frequency ω_0 such that $\omega_0\tau_0 \gg 1$. Substituting Eqs.(3.29) and (2.74) into Eq.(3.5) gives

$$s_o(t) = \int_{-\infty}^{\infty} \frac{\delta(\omega - \omega_0)}{j\omega\tau_0} e^{j\omega t} d\omega = \frac{e^{j\omega_0 t}}{j\omega_0\tau_0}.$$

Since the input signal was $e^{j\omega_0 t}$ we see that in this region the low pass filter acts as an integrator.

To illustrate this integrating function further consider the case where the input signal is the square wave shown in Figure 2.1a. The integration of a square wave results in a triangular wave shown in Figure 2.1b. The coefficients of the Fourier series expansion of the square wave are given in Example 1 of Chapter 2. Thus

$$|c_n| = \frac{2A}{n\pi} \qquad n=1,3,5,\ldots \tag{3.30a}$$

$$\phi_n = -(n-1)\pi/2 \qquad n=1,3,5,\ldots \tag{3.30b}$$

Since the input signal to the system is periodic, the output signal is given by Eq.(3.18), where in this frequency region,

$$|H(\omega)| = \frac{1}{\omega\tau_0} \tag{3.31a}$$

and

$$\theta(\omega) = \pi/2. \tag{3.31b}$$

Substituting Eqs.(3.30) and (3.31) into Eq.(3.18) yields, for $\omega_0 \gg \omega_c$

$$s_o(t) = \frac{4A}{\omega_0 \tau_0 \pi} \sum_{n=1,3,5}^{\infty} \frac{1}{n^2} \sin(n\omega_0 t + (n-1)\pi/2)$$

which, except for a constant factor, agrees with Eq.(2.13).

Example 2: Response of a Low Pass RC Filter to a Rectangular Pulse. We shall now examine the response of a low pass RC filter when a rectangular pulse of unit amplitude and of duration τ_d is applied to its input. In this case

$$s_i(t) = u(t) - u(t - \tau_d).$$

The impulse response of the low pass filter is obtained from the inverse transform of Eq.(3.26), which is given by Eq.(2.78). Therefore,

$$h(t) = \frac{1}{\tau_0} e^{-t/\tau_0} u(t).$$

Substituting this and the previous expression into Eq.(3.1) yields

$$s_o(t) = \frac{1}{\tau_0} \int_{-\infty}^{\infty} e^{-x/\tau_0} u(x)[u(t-x) - u(t - \tau_d - x)]dx$$

$$= (1 - e^{-t/\tau_0})u(t) - (1 - e^{-(t-\tau_d)/\tau_0})u(t - \tau_d) \qquad (3.32)$$

which is plotted in Figure 3.9 for several values of $\tau_d/\tau_0 = \omega_c \tau_d$. In examining Eq.(3.32) for $t < \tau_d$ it is seen that when $t = \tau_0$ the amplitude of the signal has reached

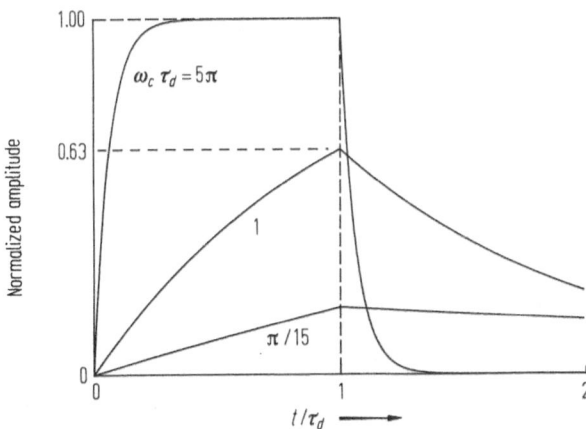

Figure 3.9 Normalized amplitude response of a low pass RC filter to a rectangular pulse: τ_d is the pulse duration and $\omega_c = 1/\tau_0 = 1/RC$

$(1 - 1/e) = 0.632$ of its final value. The rise time of this system is $\tau_r = 2.197\,\tau_0$ (see problem 10 at the end of the chapter) and, therefore, from Eq.(3.28)

$$f_c \tau_r = 0.35. \qquad (3.33)$$

Thus for a low pass RC filter the rise time-bandwidth product is 22% less than those obtained for the ideal bandlimited systems given by Eqs.(2.21), (2.68) and (3.20). However, Eq.(3.33) is the result most often used to estimate the rise time of actual systems based on their -3 dB cutoff frequency f_c.

Referring to Figure 3.9 it is seen that the time response of the low pass filter is as expected. When the pulse duration is long with respect to the filter's time constant ($\omega_c \tau_d = 5\pi$) the output waveform resembles the input waveform. When the situation is reversed the output waveform's shape and amplitude bear little resemblance to its input signal.

To determine the settling time of this filter we examine Eq.(3.32) for the case where $\tau_d \to \infty$. If we again let x be the percentage of the filter's final response and t_s be the time it takes to reach this percentage (the settling time), then from Eqs.(3.24) and (3.32) we find

$$\frac{t_s}{\tau_0} = -\ln(1 - x/100). \qquad (3.34)$$

Equation (3.34) is plotted in Figure 3.10. It is seen from this figure that for a low pass RC filter to settle to within 1% of its final value it takes 4.6 time constants ($t_s = 4.6\,\tau_o$); for 0.1%, 6.9 time constants; and for 0.01%, 9.2 time constants. It is also seen that for this filter to settle in the shortest possible time the time constant has to be as small as possible, or conversely, the cutoff frequency has to be as high as possible.

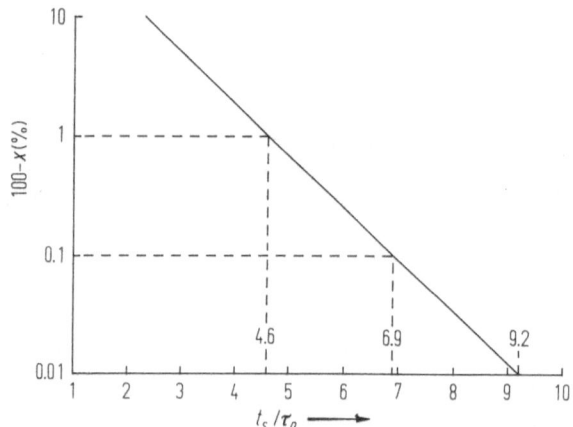

Figure 3.10 Settling time t_s of a low pass RC filter to be within $(100-x)$% of its final value ($\tau_o = RC$)

To further emphasize these points the product [recall Eq.(3.3)] of the spectrum of the rectangular pulse shown in Figure 2.8 and the transfer function of the filter shown in Figure 3.8b have been replotted in Figure 3.11. It is clearly seen in this new figure the degree of filtering that is done to each of the three pulses and its effects on the resulting shapes of the output signals.

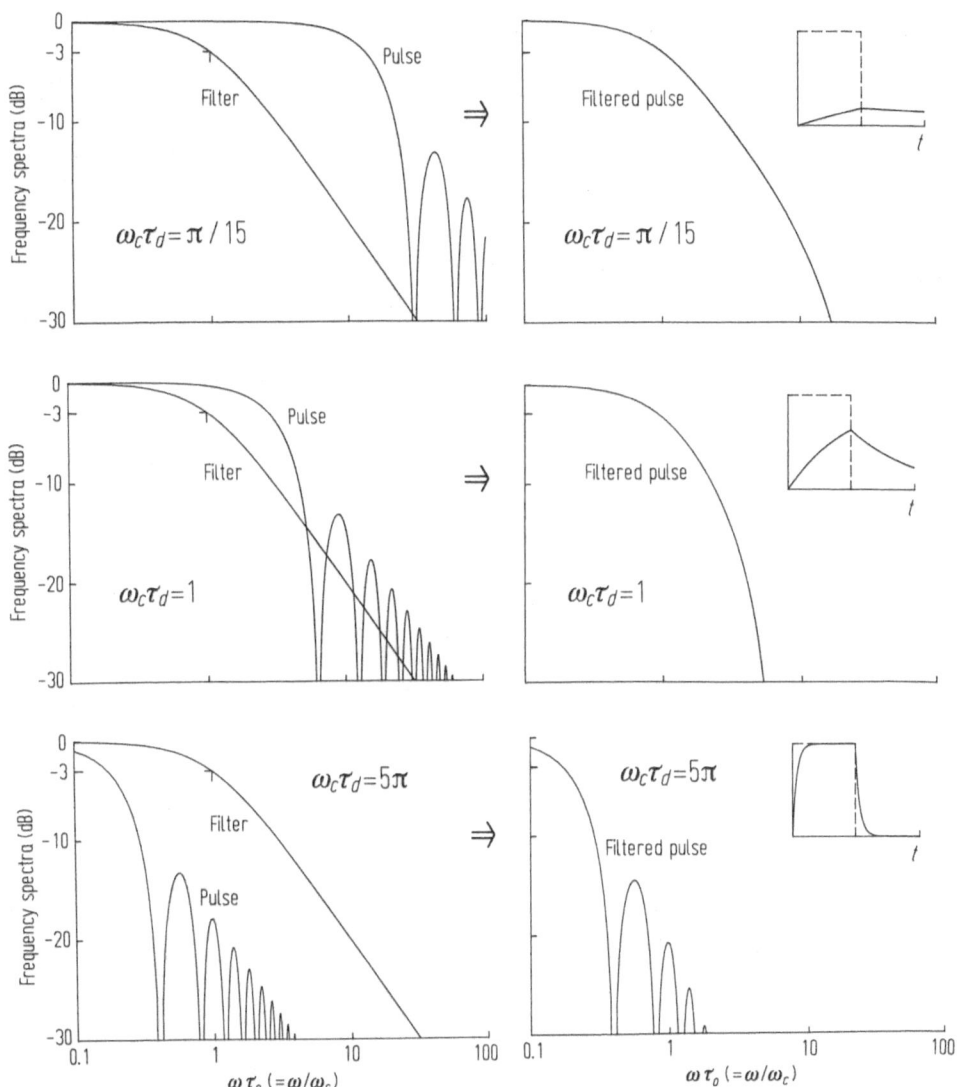

Figure 3.11 Normalized frequency spectrum of the output of a low pass RC filter when the input signal is a rectangular pulse: τ_d is the pulse duration and $\tau_o = RC = 1/\omega_c$

3.3.3 High Pass RC Filter

Consider now the high pass RC filter shown in Figure 3.12a. The transfer function of this circuit is

$$H(\omega) = \frac{j\omega\tau_0}{1 + j\omega\tau_0} = |H(\omega)| e^{-j\theta(\omega)} \tag{3.35}$$

where

$$|H(\omega)| = \frac{\omega\tau_0}{\sqrt{1 + (\omega\tau_0)^2}} \tag{3.36a}$$

$$\theta(\omega) = -\tan^{-1}(1/\omega\tau_0). \tag{3.36b}$$

It is seen from Eq.(3.36a) that the cutoff frequency for this filter is also given by Eq.(3.28). Equations (3.36) are shown Figure 3.12b.

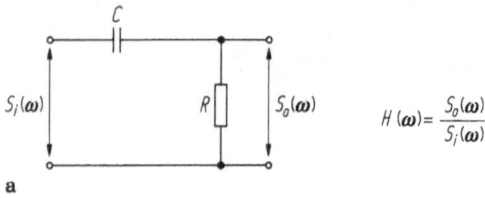

a

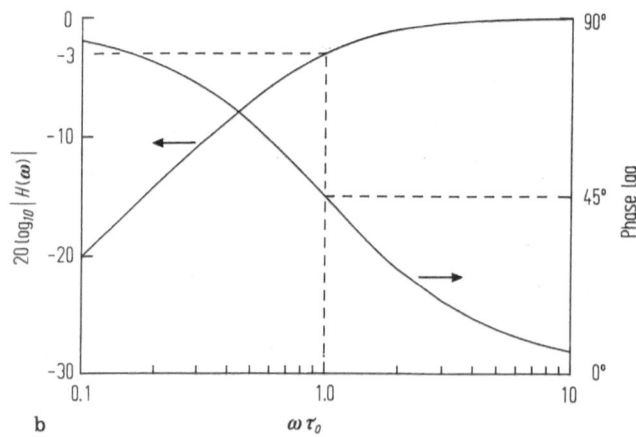

b

Figure 3.12 (a) High pass RC filter (b) transfer function of a high pass RC filter

Limiting Cases: Differentiator

We examine two limiting cases of Eq.(3.35). When $\omega\tau_0 \gg 1$ Eq.(3.35) becomes

$$H(\omega) \cong 1$$

which we see from Eq.(3.14) is an ideal high pass filter with zero phase lag. When $\omega\tau_0 \ll 1$ Eq.(3.35) becomes

$$H(\omega) \cong j\omega\tau_0.$$

To interpret this result let the input signal to high pass filter in this region be a sine wave of frequency ω_0 such that $\omega_0\tau_0 \ll 1$. Substituting Eq.(2.74) and the above equation into Eq.(3.5) gives

$$s_o(\omega) = \int_{-\infty}^{\infty} j\omega\tau_0\delta(\omega-\omega_0)e^{j\omega t}d\omega = j\omega_0\tau_0 e^{j\omega_0 t}.$$

Since the input signal was $e^{j\omega_0 t}$ we see that in this region the high pass filter acts as a differentiator. This conclusion is also immediately apparent from Eq.(2.80).

To illustrate this differentiating function further consider the case where the input signal is the square wave shown in Figure 2.1a. The differentiation of a square wave results in an impulse-like wave train shown in Figure 2.1c. The coefficients of the Fourier series expansion of the square wave are given in Eqs.(3.30). Since the input signal to the system is periodic, the output signal is given by Eq.(3.18), where in this frequency region,

$$|H(\omega)| = \omega\tau_0 \tag{3.37a}$$

and

$$\theta(\omega) = -\pi/2. \tag{3.37b}$$

Substituting Eqs.(3.30) and (3.37) into Eq.(3.18) yields, for $N_o\omega_0 \ll \omega_c$

$$s_o(t) = \frac{4A\omega_0\tau_0}{\pi} \sum_{n=1,3,5}^{N_o} \sin(n\omega_0 t + (n+1)\pi/2)$$

which, except for a constant scale factor and the finite bandwidth $N_o\omega_0$, agrees with Eq.(2.14). The finite bandwidth had to be introduced because it is unrealistic to assume that all the harmonics can be below the cutoff frequency. However, if a sufficient number of them are, then the above result is a reasonable approximation to Eq.(2.14).

Example 3: Response of a High Pass RC Filter to a Rectangular Pulse. We shall now examine the response of a high pass RC filter when a rectangular pulse of duration τ_d is applied to its input. In this case

$$s_i(t) = u(t) - u(t-\tau_d).$$

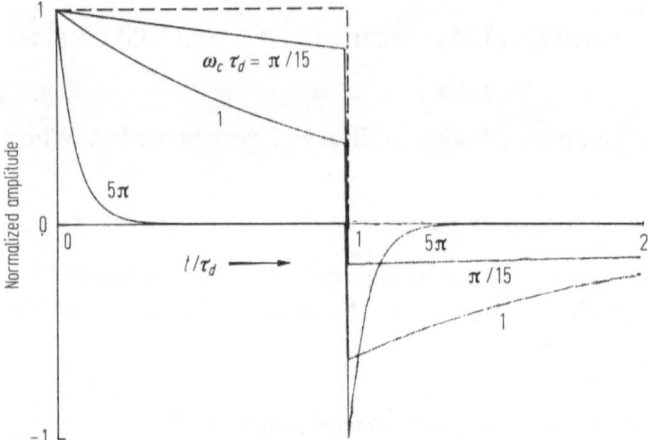

Figure 3.13   Normalized amplitude response of a high pass RC filter to a rectangular pulse: τ_d is the pulse duration and $\omega_c = 1/\tau_o = 1/RC$

The impulse response of the high pass filter is obtained from the inverse transform of Eq.(3.35), which is given by Eq.(2.81). Therefore,

$$h(t) = \left[\delta(t) - \frac{1}{\tau_0} u(t) \right] e^{-t/\tau_0} .$$

Substituting the above two expressions into Eq.(3.1) yields

$$s_o(t) = e^{-t/\tau_0} u(t) - e^{-(t-\tau_d)/\tau_0} u(t - \tau_d) \qquad (3.38)$$

which is plotted in Figure 3.13 for several values of $\tau_d/\tau_0 = \omega_c \tau_d$.

Examination of Figure 3.13 shows an entirely different type of response than that of the second-order system and the low pass RC filter. In both of these systems the bandwidth was limited and, therefore, so was the output signal's rise time. In this filter the bandwidth is infinite and consequently the rise time is zero. However, in order for the signal to maintain the amplitude reached at $t = 0^+$ the circuit's time constant must be very long. In other words the cutoff frequency must be made as low as possible. To further emphasize this point the product [recall Eq.(3.3)] of the spectrum of the rectangular pulse shown in Figure 2.8 and the transfer function of the filter shown in Figure 3.12b have been replotted in Figure 3.14. It is clearly seen in this new figure the degree of filtering (differentiating) that is done to each of the three pulses. In addition it was seen in Figures 2.8 and 2.9 that almost 95% of the signal's energy was contained in the frequency band from dc (0 Hz) to $2/\tau_d$.

Therefore, in order for a system not to modify a signal of relatively long duration it must have a low frequency response that approaches dc. If a substantial portion of the low frequency energy is filtered by a high pass RC filter the output signal will be differentiated. As seen in Figure 3.14 this is clearly the case when $\tau_d / \tau_0 = 5\pi$.

For the high pass RC filter we do not determine its settling time. Instead we determine the maximum duration of the pulse for which the initial amplitude at $t = 0^+$ stays to within a stated percentage of that value. If we let y be the percentage

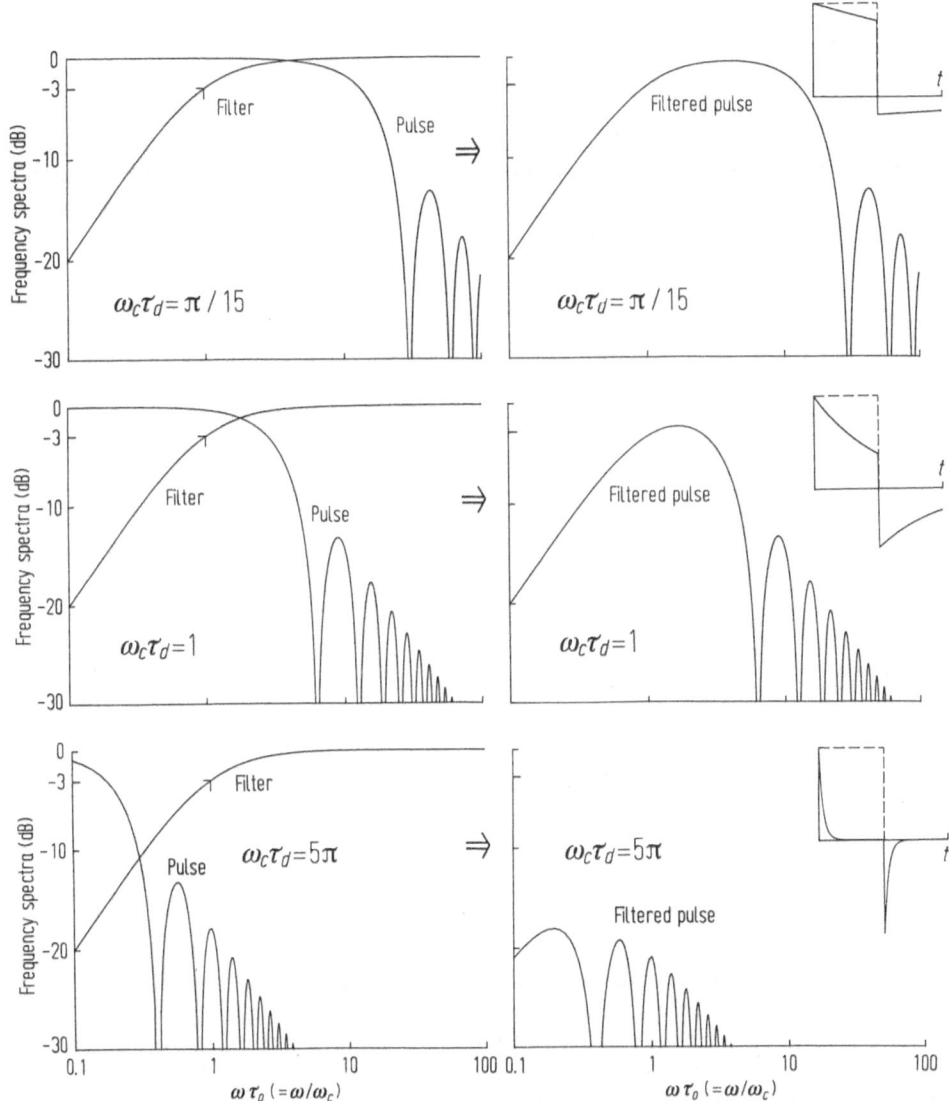

Figure 3.14 Normalized frequency spectrum of the output of a high pass RC filter when the input signal is a rectangular pulse: τ_d is the pulse duration and $\tau_o = RC = 1/\omega_c$

of the filter's initial response and t_D the length of time this response remains within this percentage, then

$$\frac{s_o(t_D)}{s_o(0)} = \frac{y}{100}.$$

Equation (3.38), therefore, gives

$$\frac{t_D}{\tau_o} = -\ln(y/100). \tag{3.39}$$

Equation (3.39) is plotted in Figure 3.15, where it is seen that to very closely maintain the initial amplitude for an appreciable amount of time requires a very large RC time constant.

It will be seen in subsequent chapters that all capacitive and capacitive-like sensors and transducers have an equivalent circuit equal to that of a high pass filter and, therefore, must be used with care when employed to measure long duration phenomena.

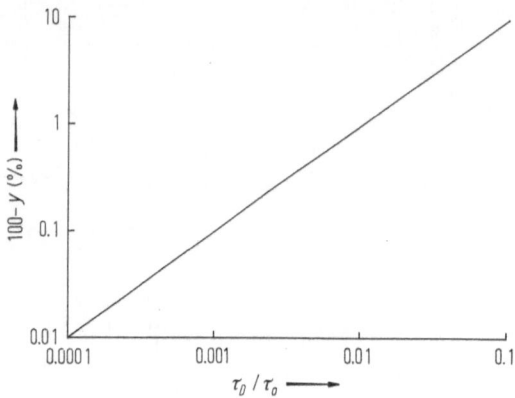

Figure 3.15 Settling time (droop) t_D of a high pass RC filter to be within $(100 - y)\%$ of its initial value

3.4 Correlation and Power Spectra

In many applications one has to deal with situations in which the input signals to a linear system are random. Therefore it is desirable to obtain expressions for the average output power spectrum as a function of the average input power spectrum when the input spectrum is applied to a linear system. From Section 2.6 it was seen that when the signals are random we have to determine the average correlation functions in order to obtain the signal's average spectral properties and *vice versa*.

Rewriting Eq.(2.109) in the present notation yields

$$\bar{R}_{oo}(\tau) = \lim_{T \to \infty} \frac{1}{T} \int_{-T/2}^{T/2} s_o(t)s_o(t+\tau)dt \qquad (3.40)$$

where, from Eq.(3.1),

$$s_o(t) = \int_{-\infty}^{\infty} h(\lambda)s_i(t-\lambda)d\lambda. \qquad (3.41)$$

Substituting Eq.(3.41) into Eq.(3.40) yields

$$\bar{R}_{oo}(\tau) = \lim_{T \to \infty} \frac{1}{T} \int_{-T/2}^{T/2} \int_{-\infty}^{\infty} h(\lambda)s_i(t-\lambda)d\lambda \int_{-\infty}^{\infty} h(\sigma)s_i(t+\tau-\sigma)d\sigma dt$$

$$= \int_{-\infty}^{\infty} h(\lambda) \int_{-\infty}^{\infty} h(\sigma) \left[\lim_{T \to \infty} \frac{1}{T} \int_{-T/2}^{T/2} s_i(t-\lambda)s_i(t+\tau-\sigma)dt \right] d\sigma d\lambda$$

$$= \int_{-\infty}^{\infty} h(\lambda) \int_{-\infty}^{\infty} h(\sigma)\bar{R}_{ii}(\tau+\lambda-\sigma)d\sigma d\lambda. \qquad (3.42)$$

We now employ the Weiner-Khintchine theorem, Eq.(2.111a), to obtain

$$\bar{P}_{oo}(\omega) = \int_{-\infty}^{\infty} \bar{R}_{oo}(\tau)e^{-j\omega\tau}d\tau$$

$$= \int_{-\infty}^{\infty} h(\lambda) \int_{-\infty}^{\infty} h(\sigma) \int_{-\infty}^{\infty} \bar{R}_{ii}(\tau+\lambda-\sigma)e^{-j\omega\tau}d\tau d\sigma d\lambda.$$

If we now let $\mu = \tau + \lambda - \sigma$, then

$$\bar{P}_{oo}(\omega) = \int_{-\infty}^{\infty} h(\lambda)e^{j\omega\lambda}d\lambda \int_{-\infty}^{\infty} h(\sigma)e^{-j\omega\sigma}d\sigma \int_{-\infty}^{\infty} \bar{R}_{ii}(\mu)e^{-j\omega\mu}d\mu$$

$$= H(-\omega)H(\omega)\bar{P}_{ii}(\omega)$$

or, since $h(t)$ is real, $H(-\omega) = H^*(\omega)$ [recall Eq.(2.48)], and therefore,

$$\bar{P}_{oo}(\omega) = |H(\omega)|^2 \bar{P}_{ii}(\omega). \qquad (3.43)$$

It is noted that although Eq.(3.43) gives us the magnitude of the transfer function, it does not give us a means to get its phase response. To obtain the phase information we have to measure the cross-correlation of $s_i(t)$ and $s_o(t)$. Thus

$$\bar{R}_{oi}(\tau) = \lim_{T \to \infty} \frac{1}{T} \int_{-T/2}^{T/2} s_i(t)s_o(t+\tau)dt \qquad (3.44)$$

where, as before, $s_o(t)$ is given by Eq.(3.41). Substituting Eq.(3.41) into Eq.(3.44) and using the same procedure that was used to obtain Eq.(3.43) yields

$$\bar{P}_{oi}(\omega) = H(\omega)\bar{P}_{ii}(\omega). \qquad (3.45)$$

Using Eqs.(2.111b) and (2.58) the inverse of Eq.(3.45) is

$$\bar{R}_{oi}(\tau) = \int_{-\infty}^{\infty} h(x)\bar{R}_{ii}(\tau-x)dx. \qquad (3.46)$$

We see that Eqs.(3.45) and (3.46) provide a third type if input signal by which one can obtain either the impulse response or the transfer function of a linear system: measure either the cross-correlation or the cross-power spectrum of the output and input signals when the input signal is random. This is demonstrated in the example that follows.

Example 4: Using Cross-correlation and White Noise to Obtain the Impulse Response and Transfer Function. Consider the case where the input signal is white noise. From Eq.(1.114)

$$\bar{R}_{ii}(\tau) = N_o\delta(\tau).$$

Thus Eq.(3.46) gives

$$\bar{R}_{oi}(\tau) = N_o \int_{-\infty}^{\infty} h(x)\delta(\tau-x)dx = N_o h(\tau)$$

or

$$h(\tau) = \bar{R}_{oi}(\tau)/N_o.$$

Using Eq.(3.2) the transfer function is

$$H(\omega) = \int_{\infty}^{\infty} h(\tau)e^{-j\omega\tau}d\tau.$$

3.5 Practical Filters and Their Attributes

In many applications filters are used to eliminate, or minimize, the effects of "unwanted" signals. The purposes for filtering signals varies widely. They are, for example, used to perform frequency analysis, to improve the amplitude ratio of the desired signal to the undesired signal (improve the signal-to-noise ratio), to obtain bandlimited signals prior to digital sampling and to shape a signal's frequency spectrum. It is recalled that an ideal filter is simply defined by its cutoff frequencies. A practical filter on the other hand has to be described by several attributes, all of which give an indication of how close the actual filter is to an ideal one. This is not meant to imply, however, that one should always seek a filter that is as close to an ideal filter as possible. This is undesirable, for example, when one wants to use a low pass RC filter as an integrator in a control circuit.

To describe a practical filter in a meaningful manner it is often best to have its complete transfer function available. Since this may not always be the case it is important to define a set of unambiguous descriptors that adequately characterize a filter. In Section 3.3.2 we introduced the definition of the cutoff frequency of a practical filter, which is also called the half power points of the filter's transfer function. However, when performing frequency analysis on broadband signals just knowing the cutoff frequency is insufficient. Consequently we introduce the *equivalent noise bandwidth* of a filter, which is defined as

$$B_e = \frac{1}{2\pi H_{max}^2} \int_0^\infty |H(\omega)|^2 \, d\omega \qquad (3.47)$$

where H_{max} is the maximum value of the filter's transfer function. It is seen that B_e is the bandwidth of an ideal filter that has H_{max} as the maximum amplitude of its transfer function and that passes the same average power from a white noise source as does the actual filter $H(\omega)$. The closer $H(\omega)$ is to an ideal filter, the closer the measured power through the actual filter is to the true power in that frequency band. These ideas are shown in Figure 3.16.

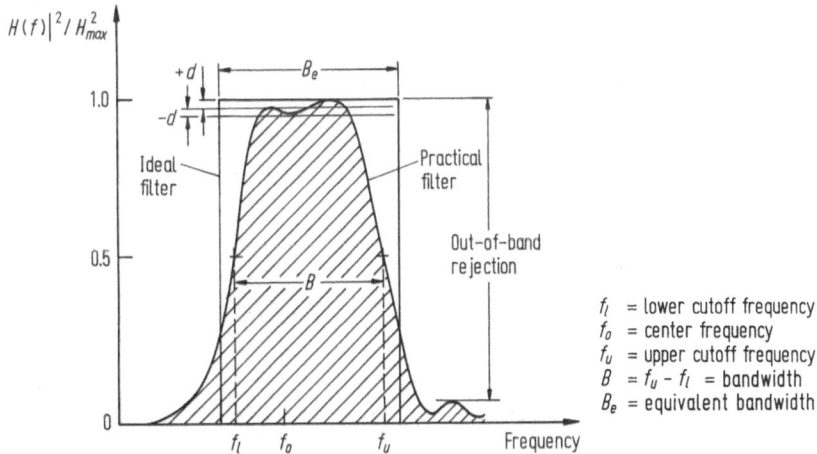

Figure 3.16 Typical characteristics of a practical bandpass filter and its relationship to an equivalent ideal bandpass filter

Another descriptor is the *selectivity* of the filter; that is, the rate at which the filter attenuates a signal beyond its cutoff frequencies. The selectivity is usually given in either dB/octave or dB/decade; that is, the amount of attenuation per doubling (or halving) of frequency or the amount of attenuation for each tenfold increase (or decrease) in frequency, respectively. The conversion from one to the other is (dB/decade) = 3.32(dB/octave). In some filter designs the filter selectivity can be misleading, for it implies that the farther one goes from the cutoff frequency the greater the attenuation of the signal. Therefore, to clarify those situations in which this is not the case the *maximum out-of-band rejection* is also stipulated as shown in Figure 3.16. A fourth descriptor is the filter *ripple* d, which is a means of defining the linearity of the frequency response of a filter within a portion of the filter's bandwidth (recall Figure 1.4). Ripple is an important characteristic for anti-aliasing filters.

The *center frequency* f_o of a bandpass filter is defined as its geometric mean frequency; that is,

$$f_o = \sqrt{f_u f_l} \tag{3.48}$$

where f_u is the upper cutoff frequency and f_l the lower cutoff frequency. This implies that for some constant $k > 1$,

$$f_u = k f_o \tag{3.49a}$$

$$f_l = f_o / k . \tag{3.49b}$$

The bandwidth of the filter is given by

$$B = f_u - f_l = \beta f_o \tag{3.50}$$

where

$$\beta = k - \frac{1}{k} \tag{3.51}$$

and, therefore,

$$k = \frac{\beta}{2} + \sqrt{\left(\frac{\beta}{2}\right)^2 + 1} . \tag{3.52}$$

Since k is a constant so is β. Consequently the filter is designated a *constant percentage* ($100\beta\%$) bandpass filter.

In many situations $\beta \ll 1$ and Eq.(3.52) simplifies so that $k \cong 1 + \beta/2$ and, consequently, $1/k \cong 1 - \beta/2$. Therefore, Eqs.(3.49) become

$$f_u = (1 + \beta/2) f_o , \quad f_l = (1 - \beta/2) f_o$$

and, hence

$$f_o = (f_u + f_l)/2 \qquad \beta \ll 1$$

which is the *arithmetic mean* of the filter's cutoff frequencies. To predict the center frequency using the arithmetic mean to within 1%, $\beta < 0.29$ (29%).

A special class of constant percentage bandwidth filters is the octave filter. For these filters

$$f_u = 2^n f_l$$

and, therefore,

$$k = 2^{n/2}.$$

When $n = 1$ the filter is called an octave filter; when $n = 1/3$ it is called a one-third octave filter. The octave filters are used in various aspects of acoustics.

Example 5: Equivalent Noise Bandwidth of a Low Pass RC Filter.   Substituting Eq.(3.27a) into Eq.(3.47) yields

$$B_e = \frac{1}{2\pi} \int_0^\infty \frac{1}{1+(\omega\tau_0)^2} d\omega = \frac{1}{4\tau_0} = \frac{\pi}{2} f_c$$

and we have used Eq.(3.28). Thus, in terms of the total power passing through it, the low pass RC filter is equivalent to an ideal filter with a cutoff frequency of $1.57 f_c$.

References

1. Javid, J., and Brenner, E., *Analysis, Transmission, and Filtering of Signals*, McGraw-Hill Book Co., New York, 1963.
2. Magrab, E. B., and Blomquist, D. S., *The Measurement of Time-Varying Phenomena: Fundamentals and Applications*, John Wiley and Sons, New York, 1971.
3. Papoulis, A., *The Fourier Integral and Its Applications*, McGraw-Hill Book Co., New York, 1962.

Exercises

1. A signal $s_i(t) = 2\sin(\omega_1 t) + 0.05\sin(3\omega_1 t)$ is passed through a second-order system whose natural frequency is $\omega_n = 3\omega_1$ and whose damping is $c = 0.025$. What is the %THD of the output signal.
2. Determine the impulse response of an ideal (a) low pass filter and (b) bandpass filter. What is implausible about these results.
3. If the maximum value of a second-order system is very closely equal to $1/(2c\omega/\omega_n)$ when $c \ll 1$, then determine the bandwidth and cutoff frequencies of this system in terms of c and ω_n.

4. What are the bandwidths and cutoff frequencies of an octave and a third-octave filter whose center frequencies are 2000 Hz.

5. A rectangular pulse of duration t_o is passed through a second-order system whose rise time is τ_r. Sketch the output pulse shapes for $c = 0.08$ and $c = 0.9$ when (a) $t_o \ll \tau_r$, (b) $t_o \cong \tau_r$ and (c) $t_o \gg \tau_r$.

6. Describe three different ways in which the transfer function of a linear system can be obtained.

7. A square wave of period T_o is passed through a second-order system whose natural frequency is N / T_o, where N is an integer. Obtain an expression for the output signal. Sketch the results for $c = 0.08$ and $c = 0.9$ and (a) $N = 1$ and (b) $N = 21$.

8. Given a filter with the transfer function shown below. The input signal to this filter is

$$f(t) = 10\cos(\omega_o t) + 5\cos(2\omega_o t) + 10\cos(4\omega_o t).$$

If the filter is centered at ω_o, $2\omega_o$ and $4\omega_o$, respectively, what would be the rms value of the filter's output signal at each of these frequencies.

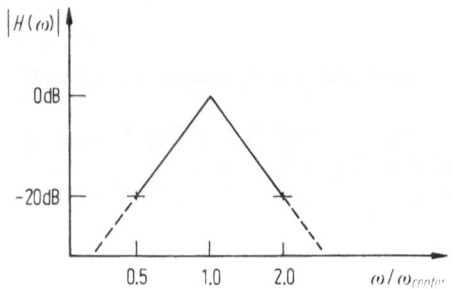

9. Given a high pass RC filter whose time constant is 0.1 s. What is the lowest frequency of the input signal for which (a) the phase angle of the output signal relative to the input signal is 45 ° and (b) the output amplitude is at least 90% of its input amplitude. For cases (a) and (b) what are the time delays between the output and input signals.

10. Use Eq.(3.32) to derive Eq.(3.33), the rise time of a low pass RC filter.

4 Amplifiers

4.1 Introduction

Amplifiers are usually the first electronic component encountered at the input to a signal conditioning or signal acquisition system. Their primary purpose is to provide an interface between one component of one system with that of another while (a) providing accurate voltage or current amplification, (b) not affecting the amplitude and frequency response of the input signal because of the interconnection and (c) without adding extraneous signals, such as noise, temperature drift, and nonlinearity, to the input signal. In practice instrumentation amplifiers are special combinations and configurations of a class of amplifiers called operational amplifiers. The properties of operational amplifiers will, therefore, be introduced along with their potential major sources of errors.

4.2 Some Definitions

Consider the amplifier shown in Figure 4.1 in which a source with resistance R_s is connected to its input. The input voltage across this source consists of a signal having an rms value V_i and additive noise whose rms value is N_i. Noise refers to any spurious currents or voltages extraneous to the signal of interest. The maximum (peak) value of V_i is V_{max}, which is that value of V_i that causes a stated value of %THD [recall Eq.(2.11)] at the amplifier's output. If the output voltage of the amplifier is V_{out}, then the voltage gain of the amplifier is

$$A_v = \frac{V_{out}}{V_i} \tag{4.1}$$

where V_{out} is the noise-free output voltage. The current gain of the amplifier is given as

$$A_c = \frac{i_{out}}{i_i} = \frac{V_{out}/R_o}{V_i/R_s} = A_v \frac{R_s}{R_o}. \tag{4.2}$$

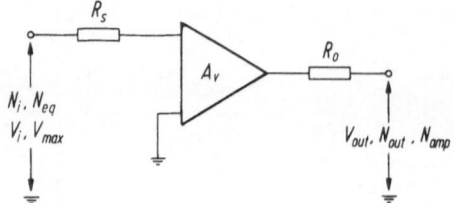

Figure 4.1 Nomenclature for an amplifier with voltage gain A_v, output resistance R_o and source resistance R_s

Consequently we can define the power gain of an amplifier as

$$A_p = A_v A_c = A_v^2 \frac{R_s}{R_o}. \tag{4.3}$$

The rms value of the inherent noise of the amplifier is N_{amp} and is the noise appearing at its output when the input is shorted. One could also refer N_{amp} to its input; that is,

$$N_{eq} = N_{amp}/A_v \tag{4.4}$$

and then consider the amplifier noiseless. Some amplifier specifications are stated in this manner. In general N_i and N_{amp} are functions of frequency.

Signal-to-Noise Ratio
The signal-to-noise ratio S/N is defined as the ratio of the signal power to the noise power. Therefore, when considering the gain of the amplifier in this definition one must use the power gain A_p of the amplifier. Since the square of the rms value is proportional to the power in the signal [recall Eqs.(2.10), (2.109) and (2.111b)], the input signal-to-noise ratio is

$$(S/N)_i = \frac{V_i^2}{N_i^2} \tag{4.5}$$

and the output signal-to-noise ratio is

$$(S/N)_{out} = \frac{V_{out}^2}{N_{out}^2} = \frac{A_p V_i^2}{A_p N_i^2 + N_{amp}^2} \tag{4.6}$$

where V_{out} and V_i are the noise-free output and input voltages, respectively.

Noise Figure
The noise figure is a useful amplifier figure-of-merit that is defined as

$$F = \frac{(S/N)_i}{(S/N)_{out}} = 1 + \frac{N_{amp}^2}{A_p^2 N_i^2} \geq 1 . \tag{4.7}$$

If the noise figure equals 2 ($F = 2$) then one can conclude that the amplifier and source combination is adding an amount of noise equal to that in the input signal. Although not explicitly stated in Eqs.(4.5) to (4.7), the signal-to-noise ratios are, as mentioned previously, functions of frequency.

If we have two amplifiers cascaded as shown in Figure 4.2 the total noise figure F_T is

$$F_T = \frac{V_i^2/N_i^2}{V_{out}^2[A_{p2}(A_{p1}N_i^2 + N_{amp1}^2) + N_{amp2}^2]^{-1}} = F_1 + \frac{F_2 - 1}{A_{p1}}. \qquad (4.8)$$

Since the power gain is usually large, F_T very closely equals F_1. Thus in using cascaded amplifiers the first amplifier should have the lowest noise figure and the highest power gain.

Dynamic Range
The dynamic range of an amplifier is defined as

$$\text{dynamic Range} = 20\log_{10}(V_{max}/N_{eq}). \qquad (4.9)$$

Notice that the definition assumes that the lowest input signal level has a signal-to-noise ratio of one. This is often of little practical use, so that N_{eq} should be replaced by N'_{eq}, where $N'_{eq} > N_{eq}$ is a value for which the signal-to-noise ratio is sufficiently greater than one so that small amplitude signals can be meaningfully used.

Linearity
Linearity is the maximum deviation of the output voltage from a straight line that passes through zero and its maximum value, which is usually $A_v V_{max}$ [recall Eq.(1.12)].

Stability
Stability is a measure of the ability of the amplifier's output voltage to remain, in a specified temperature range, within predefined error limits for a specified period of time.

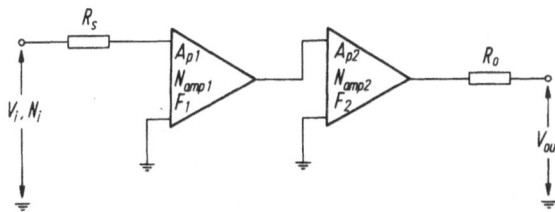

Figure 4.2 Nomenclature for determining the noise figure of two cascaded amplifiers

4.3 Noise in Amplifiers

There are several sources of noise in amplifiers: Johnson noise, flicker or "1/f" (read: one over f) noise, and "shot" noise. Johnson noise is due to the thermal agitation of electrons in resistive circuit elements which causes random currents to flow in the resistors, thus creating random voltages. The rms amplitude of the Johnson noise can be calculated from the relation

$$e_N = 7.41 \times 10^{-6} \sqrt{RTB} \qquad \mu V_{rms} \qquad (4.10)$$

where R is the value of the resistance in ohms, T is the temperature of the resistor in °K (see Section 8.7.1 for temperature scale conversions) and B is the bandwidth. If $R = R_k$ kohm, $T = 20\,°C$ and the bandwidth is unspecified, then Eq.(4.10) simplifies to

$$\hat{e}_N = \frac{e_N}{\sqrt{B}} \cong 4\sqrt{R_k} \qquad nV_{rms}/\sqrt{Hz}.$$

At room temperature, therefore, a 1 kohm resistor generates 4 $nV_{rms}/\sqrt{Hz}$ of noise, which in a 10 kHz bandwidth is 0.4 μV_{rms}. Thus in most situations the thermal noise is a small quantity. Although low-noise amplifiers are readily available it is good practice to have source resistances as low as possible and to limit the bandwidth of the signals to the frequency range of interest.

Flicker noise is usually the dominant noise source at frequencies below 100 Hz. In this frequency range the power spectral density is proportional to $1/f$ and the rms amplitude of the noise in the frequency band $f_2 - f_1$ is

$$e'_N = k_o \sqrt{\int_{f_1}^{f_2} \frac{df}{f}} = k_o \sqrt{\ln_e(f_2/f_1)} \qquad V_{rms} \qquad (4.11)$$

where k_o is the value of e'_N at $f = 1$ Hz. Thus, for example, if $e'_N = 0.1\,\mu V_{rms}$ at 1 Hz, then the total rms noise from 0.1 to 10 Hz is 0.21 μV_{rms} (= $0.1\sqrt{\ln_e 100}$). Above 100 Hz other sources of noise, such as Johnson noise and shot noise, become dominant. Usually the dc offset drift [see Example 1 of Section 4.4.2] is the low frequency limit of the "1/f" noise.

Shot noise arises whenever current passes through a transistor junction. It typically has a white noise-like spectrum and is a dominant contributor to amplifier noise at high frequencies.

Before leaving the subject of noise let us consider a broadband noise source whose amplitude distribution is gaussian as shown in Figure 4.3. For this type of noise a relationship can be developed that expresses the percentage of time the signal's amplitude will likely exceed a certain value with respect to its rms value (standard deviation). These results are presented in Table 4.1. Notice that the choice of the percentage of time in effect determines the lower end of the dynamic

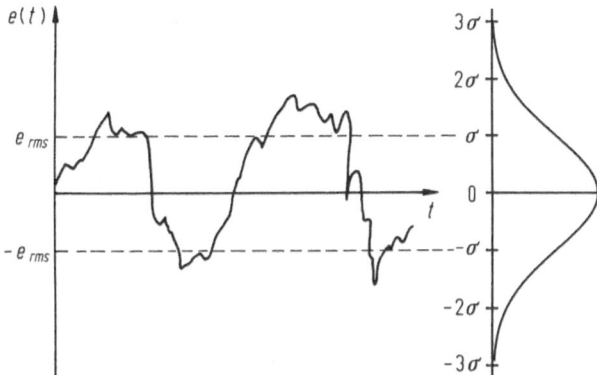

Figure 4.3 Interpretation of a gaussian amplitude probability distribution of a noise signal $e(t)$

Table 4.1 Percentage of time e_{rms} is exceeded by peak-to-peak noise amplitude ke_{rms}

% *	k
31.7	2.00
10.0	3.29
5.0	3.92
4.5	4.00
2.0	4.63
1.0	5.15
0.3	6.00
0.1	6.58
0.01	7.78

* Halve these values if signal is only positive or only negative.

range of an amplifier. Recall that the noise generated in a 1 kohm resistor at 20 °C and in a 10 kHz bandwidth was $0.4\,\mu V_{rms}$. In order for the noise interference to occur less than 0.01% of the time the input signal level would have to be greater than $3.1\,\mu V_{rms}$ ($=7.78\times0.4\,\mu V_{rms}$).

4.4 Operational Amplifiers

4.4.1 Ideal Operational Amplifiers

The equivalent circuit of an ideal operational amplifier is shown in Figure 4.4. An ideal operational amplifier is an amplifier that has infinite open loop gain

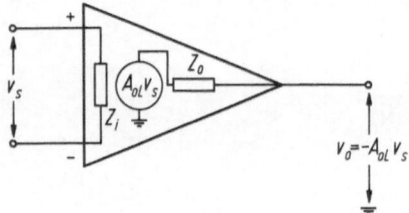

Figure 4.4   Equivalent circuit of an ideal operational amplifier

$(A_{OL} = \infty)$, infinite input impedance $(Z_i = \infty)$, and zero output impedance $(Z_o = 0)$. In the convention implied in Figure 4.4 the negative (-) input terminal is called the inverting input; that is, when a signal is connected to this terminal the output will be the negative of the input. The positive (+) input terminal is the non-inverting connection.

To obtain a good approximation to the characteristics of an ideal operational amplifier, high gain amplifiers are used with negative feedback. Consider the case of the voltage follower shown in Figure 4.5. Summing voltages gives

$$v_i + v_s = v_o. \tag{4.12}$$

But

$$\frac{v_o}{v_s} = -A_{OL}. \tag{4.13}$$

Therefore Eq.(4.12) becomes

$$v_i = v_o\left(1 + \frac{1}{A_{OL}}\right) \tag{4.14}$$

in which $v_i = v_o$ as A_{OL} approaches infinity. Thus, the output "follows" the input. It can also be shown for this configuration, and with the notation of Figure 4.4,

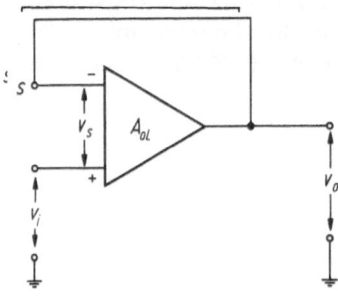

Figure 4.5   Voltage follower

that

$$Z'_i = A_{OL} Z_i$$

where Z'_i is the closed loop input impedance, and that

$$Z'_o = Z_o / A_{OL}$$

where Z'_o is the closed loop output impedance. Since A_{OL} is usually greater than 10^4 and frequently as high as 10^6, a Z'_i on the order of 10^{12} ohms and a $Z'_o < 0.1$ ohms are attainable. Consequently amplifiers with very high gain can be made to closely approximate the attributes of ideal operational amplifiers.

Now consider the inverting configuration shown in Figure 4.6. Summing currents at node "1" yields

$$i_i = i_s + i_f. \tag{4.15}$$

Under the assumption of an ideal operational amplifier $i_s = 0$ (since Z_i approaches infinity) and Eq.(4.15) becomes

$$i_i = i_f. \tag{4.16}$$

But

$$i_i = \frac{v_i - v_s}{R_1} \tag{4.17}$$

and

$$i_f = \frac{v_s - v_o}{R_2}. \tag{4.18}$$

Substituting Eqs.(4.17) and (4.18) into Eq.(4.16) and using Eq.(4.13) gives

$$\frac{v_o}{v_i} = \frac{-R_2 / R_1}{1 + \beta / A_{OL}} \tag{4.19}$$

where

$$\beta = 1 + R_2 / R_1. \tag{4.20}$$

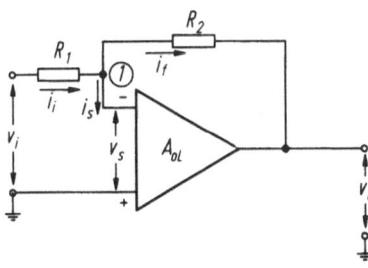

Figure 4.6 Inverting amplifier

When $\beta / A_{OL} \ll 1$, Eq.(4.19) reduces to

$$\frac{v_o}{v_i} = -R_2 / R_1 \tag{4.21}$$

which is called the closed loop gain and is determined solely by the ratio of R_1 and R_2 and not by the open loop gain A_{OL}. In the general case one can replace R_1 and R_2 with any combination of electronic elements; in the frequency domain this ratio R_2 / R_1 is replaced by the ratio Z_2 / Z_1, where Z_1 and Z_2 are the elements' impedances (see Section 4.5.2).

By selecting the appropriate input and output elements in place of R_1 and R_2 in Figure 4.6, these amplifiers can be used to perform a wide variety of mathematical operations on their input signals: e.g., summation, subtraction, integration, differentiation, and logarithms. The details of these configurations can be found in Diefenderfer [1978], Brophy [1983], Horrocks [1983], Nelson [1986] and Sanderson [1987]. However, two configurations are presented in Figure 4.7 for future reference: a summing amplifier and an integrator. (See Sections 5.4.2, 5.4.3 and 9.5.) Also, a special purpose amplifier called a charge amplifier is discussed in Section 8.4.

A less commonly used configuration is the non-inverting negative feedback configuration shown in Figure 4.8. For this case

$$i_1 = i_s + i_f = i_f \tag{4.22}$$

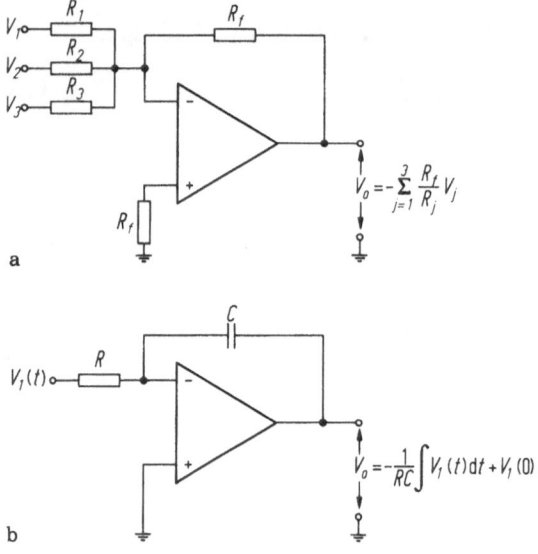

a

b

Figure 4.7 (a) Summing amplifier   (b) integrator

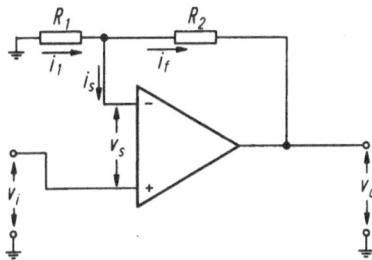

Figure 4.8 Non-inverting amplifier

and

$$i_1 = \frac{v_i + v_s}{R_1} \tag{4.23}$$

$$i_f = \frac{v_o - v_s - v_i}{R_2}. \tag{4.24}$$

Substituting Eqs.(4.23) and (4.24) into Eq.(4.22) and using Eq.(4.13) gives

$$\frac{v_o}{v_i} = \frac{\beta}{1 + \beta / A_{OL}}. \tag{4.25}$$

Since β and A_{OL} are positive quantities v_o has the same sign as v_i; hence the term non-inverting. When $\beta / A_{OL} \ll 1$ Eq.(4.25) becomes

$$\frac{v_o}{v_i} = \beta = 1 + \frac{R_2}{R_1} \tag{4.26}$$

which is the gain for the closed loop non-inverting configuration.

In addition to the properties cited, feedback can be shown to decrease amplifier distortion, extend its frequency response and decrease drift and noise.

Gain-Bandwidth Product
Operational amplifiers respond to a dc (0 Hz) input voltage. However the range of linearity of its frequency response is a function of both the open loop gain A_{OL} and the closed loop gain. A typical (closed loop) gain *versus* frequency relationship for an operational amplifier is shown in Figure 4.9. The gain at dc for this amplifier ranges from 1 (0 dB) to 10^5 (100 dB). However, since the maximum total energy over a given bandwidth that can appear at its output is a constant, this amplifier is governed by the relation

$$GB = f_c G \qquad 1 \leq G \leq A_{OL}$$

where GB is the gain-bandwidth product and f_c is the cutoff frequency at each value of the gain G. Since the amplifier responds to dc, f_c is also the amplifier's

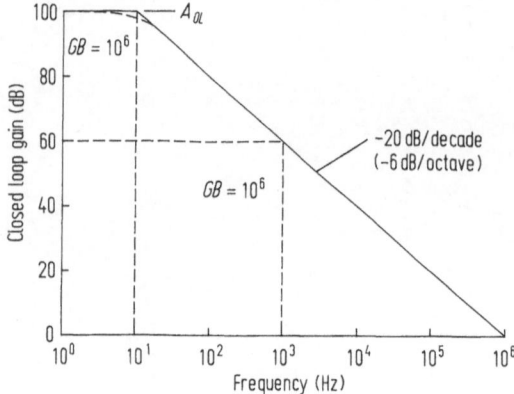

Figure 4.9 Typical gain *versus* bandwidth of an operational amplifier

bandwidth. For the operational amplifier shown in Figure 4.9 the gain-bandwidth product equals 10^6.

4.4.2 Non-Ideal Operational Amplifiers: Sources of Errors

Common Mode Rejection
In many applications, in particular those using strain gages and thermocouples, a non-inverting operational amplifier is used to amplify the electrical difference between one side and the other side of the device (resistor in this case). Thus each side of the device is connected to one of the inputs of the amplifier as shown in Figure 4.10a. Such an amplifier is called a differential amplifier. In an ideal differential amplifier, if the same signal were to appear at the positive and negative inputs of the amplifier the output voltage would be zero. In practice this is not the case. The ability of the amplifier to reject (subtract) signals common to both inputs is called common mode rejection, usually expressed in dB.

The amplifier has essentially two different gains: the open loop gain A_{OL} and the intrinsic, small common mode gain A_{CM}. We therefore have an output voltage V_o from the differential amplifier that is given by

$$V_o = A_{OL} V_{DM} + A_{CM} V_{CM} \qquad (4.27)$$

where

$$V_{DM} = V_1 - V_2, \qquad V_{CM} = (V_1 + V_2)/2. \qquad (4.28)$$

We now define the common mode rejection ratio (CMRR) as

$$CMRR = \frac{A_{OL}}{A_{CM}}. \qquad (4.29)$$

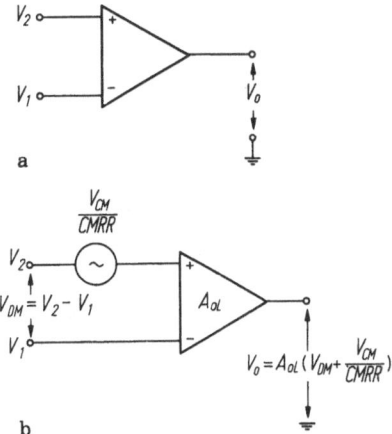

a

b

Figure 4.10 (a) Non-inverting amplifier used as a differential amplifier (b) differential amplifier with equivalent common mode voltage at its input

Therefore, Eq.(4.27) can be written as

$$V_o = A_{OL}\left(V_{DM} + \frac{V_{CM}}{CMRR}\right).$$ (4.30)

Thus, when $V_1 = V_2$, $V_{DM} = 0$ and Eq.(4.30) becomes

$$V_o = A_{OL}\frac{V_1}{CMRR} = A_{OL}\frac{V_2}{CMRR}.$$ (4.31)

Equation (4.30) implies that the common mode voltage divided by the CMRR can be considered an input to the positive terminal of the amplifier as shown in Figure 4.10b. The CMRR usually depends on temperature, source impedance of each input, cable capacitance, stray capacitance and frequency (Lang [1987]). An application using a differential amplifier is given in Section 8.2.11.

For inverting operational amplifier configurations V_{CM} is virtually zero and, therefore, common mode voltages are not a source of error.

Input Offset Voltage
Input offset voltage is the dc voltage that must be applied between the input terminals to force the quiescent dc output to zero. This voltage is due to slight differences the amplifier's transistor components have after manufacture. This source of error can be minimized at a given temperature (usually ambient), but not over a range of temperatures. This error is usually referred to the input of the amplifier. Any offset voltage not compensated for will be amplified by the gain of the amplifier and appear at its output at this amplified magnitude.

Slew Rate

Slew rate S_r is the maximum rate of change of the output signal that can be produced while still maintaining linear operation. For a single frequency sine wave it is given by

$$S_r = V_{op-p} \pi f_s \quad V/s \qquad (4.32)$$

where V_{op-p} is the peak-to-peak voltage of the output sine wave of frequency f_s.

Noise

Noise in amplifiers has been discussed in Section 4.3. It is recalled that for a given bandwidth its peak voltage level can be approximated from the appropriate factor given in Table 4.1 and the square root of the bandwidth. Noise is typically presented as noise referred to input (RTI).

Total Error

The major amplifier errors are summarized in Table 4.2. The total error referred to input is

$$E_T = \sqrt{\sum_1^6 E_n^2} \qquad (4.33)$$

and the total error as a percentage of the full scale output e_{oFS} is

$$e_{oFS} = \frac{E_T A_{CL}}{V_{FS}} 100 \ \% \qquad (4.34)$$

where A_{CL} is the closed loop gain of the configuration and V_{FS} is the full scale output voltage.

Table 4.2 Summary of amplifier errors

Error, E_n	Error Source	Typical Units
E_1	Gain Error	%
E_2	Gain Instability	$\mu V/°C$
E_3	Input Offset Voltage, RTI[*]	mV
E_4	Input Offset Voltage Drift, RTI	$\mu V/°C$
E_5	CMRR @ Stated Input Voltage	dB
E_6	Noise, RTI	$V_{rms}/\sqrt{Hz}$

[*]Referred to input

Example 1: Determination of Total Amplifier Error. A differential amplifier has an output voltage that varies from 0 to 10 V ($V_{FS} = 10$ V) and is configured to have a closed loop gain of 40 dB (A_{CL} = 100). Thus the full scale input voltage is 100 mV (= 10V/100). The amplifier is to be used over a temperature range of 25 to 60 °C and over a frequency range of dc to 10,000 Hz. The amplifier's specifications at 1 kHz and 25 °C are as follows:

Gain error: ±0.25 %
Gain instability: ±0.0025 %/°C
Input offset voltage, RTI: ±100 µV
Input offset voltage drift, RTI: ±2 µV/°C
CMRR @ 10 V_{dc}: 110 dB
Noise, RTI: 60 $nV_{rms}/\sqrt{Hz}$

The gain error is

$$E_1 = (\pm.0025)(100\ mV) = \pm250\ \mu V.$$

Since the temperature range is 35 °C the gain instability is

$$E_2 = (\pm.000025\ /°C)(100\ mV)(35\ °C) = \pm87.5\ \mu V.$$

The input offset voltage is as stated. Thus

$$E_3 = \pm100\ \mu V.$$

The input offset voltage drift for the 35 °C temperature range is

$$E_4 = (\pm2\ \mu V/°C)(35\ °C) = \pm70\ \mu V.$$

The common mode error is

$$E_S = \frac{10\ V}{CMRR} = \frac{10\ V}{10^{110/20}} = 31.6\ \mu V.$$

Lastly, the peak-to-peak noise is selected to be a factor of 6 times the rms value in the 10 kHz bandwidth (recall Table 4.1). Thus

$$E_6 = (6)(60\ nV_{rms}/\sqrt{Hz})\sqrt{10000\ Hz} = 36\ \mu V_{rms}.$$

Using Eq.(4.33) the total error referred to input is

$$E_T = \pm296\ \mu V$$

so that the output error with respect to full scale is, from Eq.(4.34),

$$e_{oFS} = \frac{(296 x\ 10^{-6}\ V)(100)}{10\ V}100 = \pm0.296\%.$$

4.5 Impedance Considerations

4.5.1 Introduction

The interconnection of one electrical device to another can, if done improperly, create a new electrical circuit whose transfer function alters the signal in an undesirable manner. This coupling of devices is directly related to their respective impedances: the output impedance of one and the input impedance of the other. In this section we shall examine the consequences of impedance matching and mismatching and determine the conditions under which the desired results can be obtained.

4.5.2 Impedance Mismatching

Consider the general case shown in Figure 4.11 where we have a signal source with output voltage e_g and with a complex impedance Z_g that is connected to the input of an amplifier whose complex input impedance is Z_i. This source representation is an appropriate model for both a transducer and for the output stage of another amplifier.

It is a simple matter to show that

$$\frac{e_i}{e_g} = \left(1 + \frac{Z_g}{Z_i}\right)^{-1}.\tag{4.35}$$

The objective is to determine under what conditions $e_i = e_g$, and to consider any deviation from this equality an error caused by the relative value of Z_i to Z_g. If we let

$$\frac{Z_g}{Z_i} = \frac{a + jb}{c + jd}\tag{4.36}$$

then it is straightforward to show that

$$\left|(1 + Z_g/Z_i)^{-1}\right|^2 = \frac{c^2 + d^2}{(a+c)^2 + (b+d)^2}.\tag{4.37}$$

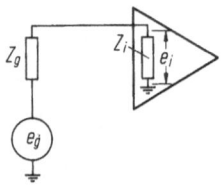

Figure 4.11 Signal source of impedance Z_g connected to an amplifier of input impedance Z_i

We now define the error Δ caused by the effects of Z_i and Z_g as

$$\Delta = 20 \log_{10} \left| \frac{e_i}{e_g} \right| = 20 \log_{10} | (1 + Z_g / Z_i)^{-1} |$$

$$= 10 \log_{10} \left[\frac{c^2 + d^2}{(a+c)^2 + (b+d)^2} \right] \quad dB. \tag{4.38}$$

Several special cases of Eq.(4.36) are now examined.

Case 1
Let $Z_g = R_g$ and $Z_i = R_i$. This corresponds to the case where the source belongs to a family of resistive type transducers such as strain gages, thermocouples, hot wire anemometers, thermistors and potentiometers. Then Eq.(4.36) gives

$$a = R_g, \qquad b = 0, \qquad c = R_i, \qquad d = 0$$

and, therefore, Eq.(4.38) becomes

$$\Delta = 10 \log_{10} \left[\frac{R_i^2}{(R_i + R_g)^2} \right] = -20 \log_{10} \left(1 + \frac{R_g}{R_i} \right) \quad dB. \tag{4.39}$$

Equation (4.39) is plotted in Figure 4.12. It is seen that for $R_i / R_g > 100$ $\Delta < -0.1$ dB, or the error is less than a 1% (recall Table 1.2). Consequently, to virtually eliminate the effects of R_g one should select the value of R_i such that R_i / R_g is considerably greater than 100 (1000 or more).

Case 2
For this case we again let $Z_g = R_g$, but place a capacitor C_i in parallel with R_i. This situation more closely represents the input stage of an amplifier at high fre-

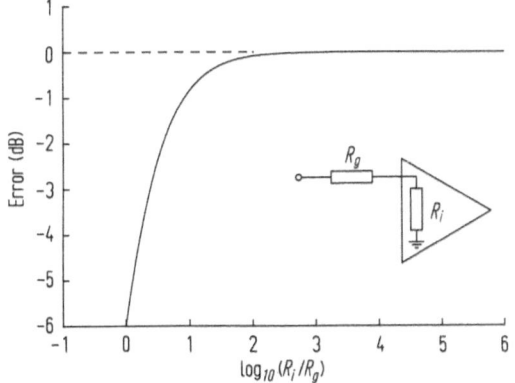

Figure 4.12 Impedance mismatching error when $Z_g = R_g$ and $Z_i = R_i$

quency. Then

$$Z_i = \frac{R_i}{1 + j\omega R_i C_i}.$$ (4.40)

Thus, Eq.(4.36) becomes

$$\frac{Z_g}{Z_i} = \frac{R_g}{R_i}(1 + j\omega R_i C_i)$$

and, therefore,

$$a = R_g/R_i, \qquad b = \omega R_g C_i, \qquad c = 1, \qquad d = 0.$$

Equation (4.38) now becomes

$$\Delta = -10\log_{10}\left[\left(1 + \frac{R_g}{R_i}\right)^2 + (\omega R_g C_i)^2\right] \quad dB$$ (4.41)

which is plotted in Figure 4.13. It is seen from this figure that to incur the smallest error over the widest frequency range once again $R_i/R_g > 100$. However the introduction of the shunt capacitor has formed a low pass filter that has a cutoff frequency of [recall Eq.(3.28)]

$$f_c = \frac{1}{2\pi R_g C_i}.$$

Thus, for a fixed R_g, C_i will have to be as small as possible in order to make f_c large. When $C_i = 0$, Eq.(4.41) reduces to Eq.(4.39).

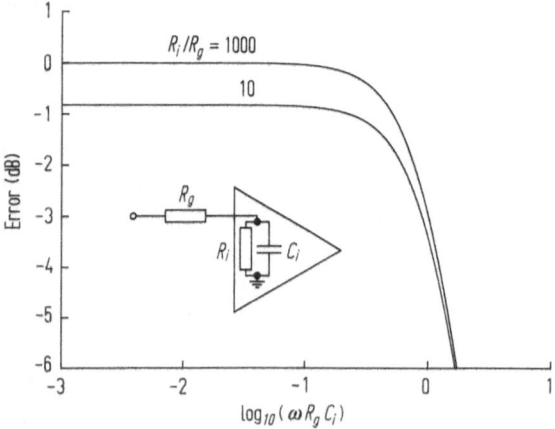

Figure 4.13 Impedance mismatching error when $Z_g = R_g$ and $Z_i = R_i/(1 + \omega R_i C_i)$

Case 3

In this case we let Z_i again be represented by Eq.(4.40) and we let Z_g be capacitive; therefore, $Z_g = 1/j\omega C_g$. This configuration corresponds to the case where the source belongs to a family of capacitive type transducers such as capacitance gages, condenser and electret microphones, and piezoelectric accelerometers and force gages. Then

$$\frac{Z_g}{Z_i} = \frac{1 + j\omega R_i C_i}{j\omega R_i C_g}$$

and, therefore,

$$a = 1, \qquad b = \omega R_i C_i, \qquad c = 0, \qquad d = \omega R_i C_g.$$

Equation (4.38) becomes

$$\Delta = -10\log_{10}\left[\frac{1}{(\omega R_i C_g)^2} + \left(1 + \frac{C_i}{C_g}\right)^2\right] \quad \text{dB}. \tag{4.42}$$

Equation (4.42) is plotted in Figure 4.14 where it is seen that in order to reduce the error $C_g/C_i > 100$. In addition, to obtain the lowest cutoff frequency for a given C_g, R_i has to be made as large as possible. This is clearly seen if we set $C_i = 0$ in Eq.(4.42), which then becomes equal to the logarithm of the transfer function of a high pass filter given by Eq.(3.36a) and discussed in detail in Section 3.3.3.

Case 4

In this case we continue to let Z_i be given by Eq.(4.40), but now let Z_g represent a resistor and inductor in series. Thus $Z_g = R_g + j\omega L_g$. This case corresponds to the family of inductive type transducers such as electrodynamic velocity sensors, eddy current gages, and linear variable differential transformers (LVDTs). Then

$$\frac{Z_g}{Z_i} = \frac{R_g}{R_i}\left[1 - \left(\frac{\omega L_g}{R_g}\right)^2\beta\right] + j\frac{R_g}{R_i}\left(\frac{\omega L_g}{R_g}\right)(1 + \beta) \tag{4.43}$$

where

$$\beta = \frac{R_g}{L_g}R_i C_i \tag{4.44}$$

and, therefore,

$$a = \frac{R_g}{R_i}\left[1 - \left(\frac{\omega L_g}{R_g}\right)^2\beta\right], \qquad b = \frac{R_g}{R_i}\left(\frac{\omega L_g}{R_g}\right)(1 + \beta)$$

$$c = 1, \qquad d = 0.$$

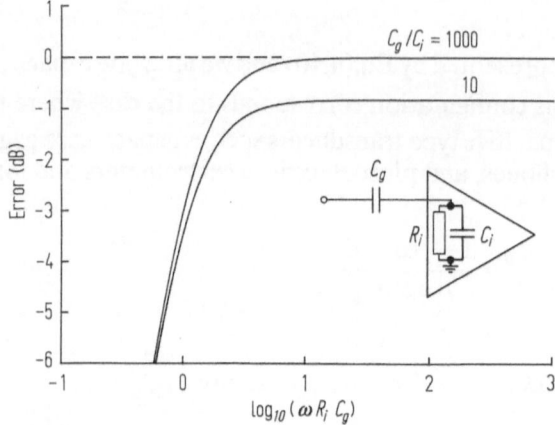

Figure 4.14 Impedance mismatching error when $Z_g = 1/\omega C_g$ and $Z_i = R_i/(1 + j\omega R_i C_i)$

Equation (4.38) becomes

$$\Delta = -10\log_{10}[\left\{ 1 + \frac{R_g}{R_i}\left[1 - \left(\frac{\omega L_g}{R_g}\right)^2 \beta \right]\right\}^2 +$$

$$\left(\frac{R_g}{R_i}\right)^2\left(\frac{\omega L_g}{R_g}\right)^2 (1+\beta)^2] \quad dB. \tag{4.45}$$

When $C_i = 0$, $\beta = 0$ and Eq.(4.45) reduces to

$$\Delta = -10\log_{10}\left[\left(1 + \frac{R_g}{R_i}\right)^2 + \left(\frac{R_g}{R_i}\right)^2\left(\frac{\omega L_g}{R_g}\right)^2 \right] \quad dB. \tag{4.46}$$

If, further, $R_g = 0$ then Eq.(4.46) reduces to

$$\Delta = -10\log_{10}\left[1 + \left(\frac{\omega L_g}{R_i}\right)^2 \right] \quad dB. \tag{4.47}$$

Equation (4.45) is shown in Figure 4.15 where it is seen once again that a necessary condition for minimum error is for R_i/R_g to be much greater than 100. It is also seen that the introduction of L_g creates a second order system [recall Section 3.3.1] that under certain conditions exhibits moderately damped resonances. Furthermore, the inductive element forms a low pass filter with the input stage of the amplifier irrespective of the value of C_i.

Case 5
For a final case we consider a more complex arrangement of impedances as shown in Figure 4.16. This will permit us to evaluate the effects of both cable and/or

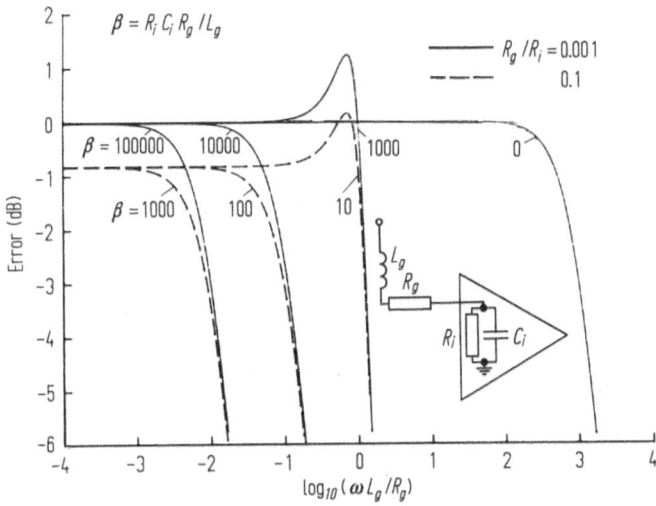

Figure 4.15 Impedance mismatching error when $Z_g = R_g + j\omega L_g$ and $Z_i = R_i/(1 + j\omega R_i C_i)$

stray capacitance $(Z_c = 1/j\omega C_c)$ and the effects of a blocking capacitor $(Z_b = 1/j\omega C_b)$, which is used to block any dc component in the input signal. It can be shown that (Kirwan and Grodzinsky [1980])

$$\frac{e_i}{e_g} = \left(1 + \frac{Z_N}{Z_D}\right)^{-1} \tag{4.48}$$

where

$$\frac{Z_N}{Z_D} = \frac{Z_b}{Z_i} + \frac{Z_g}{Z_c}\left(1 + \frac{Z_b}{Z_i} + \frac{Z_c}{Z_i}\right). \tag{4.49}$$

When $Z_b \rightarrow 0(C_b \rightarrow \infty)$ and $Z_c \rightarrow \infty(C_c \rightarrow 0)$ Eqs.(4.48) and (4.49) reduce to Eq.(4.35). Equation (4.49) is now examined for the following somewhat general

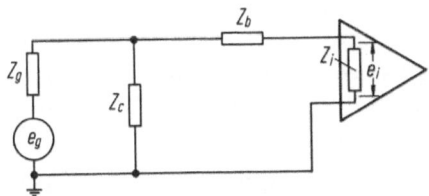

Figure 4.16 Signal source of impedance Z_g connected to an amplifier of input impedance Z_i and including the effects of cable and stray capacitance Z_c and a blocking capacitance Z_b

case:

$$Z_g = R_g + \frac{1}{j\omega C_g}, \qquad Z_c = \frac{1}{j\omega C_c}$$

$$Z_b = \frac{1}{j\omega C_b}, \qquad Z_i = \frac{R_i}{1 + j\omega R_i C_i}. \tag{4.50}$$

If we let

$$\frac{Z_N}{Z_D} = \frac{a + jb}{c + jd} \tag{4.51}$$

then Eqs.(4.49) to (4.51) give

$$a = 1 + \frac{C_c}{C_b} + \frac{C_g}{C_b} - (\omega R_i C_g)^2 \left(\frac{R_g}{R_i}\right)\left(\frac{C_i}{C_g} + \frac{C_c}{C_g} + \frac{C_c C_i}{C_b C_g}\right)$$

$$b = (\omega R_i C_g)\left[\frac{C_i}{C_b}\left(1 + \frac{C_c}{C_g}\right) + \frac{R_g}{R_i}\left(1 + \frac{C_c}{C_b}\right) + \frac{C_i}{C_g} + \frac{C_c}{C_g}\right]$$

$$c = 0, \qquad d = \omega R_i C_g. \tag{4.52}$$

Two special cases of Eqs.(4.50) and (4.52) will be examined: (a) $Z_g = R_g$ and (b) $Z_g = 1/j\omega C_g$. For each case the effects of the cable and/or stray capacitance C_c and the blocking capacitance C_b will be determined.

(a) $Z_g = R_g$ In this case $C_g \to \infty$ and Eq.(4.52) simplifies to

$$a = \frac{C_c}{C_b} - (\omega R_i C_c)^2 \left(\frac{R_g}{R_i}\right)\left(1 + \frac{C_i}{C_c} + \frac{C_i}{C_b}\right)$$

$$b = (\omega R_i C_c)\left[\frac{C_i}{C_b} + \frac{R_g}{R_i}\left(1 + \frac{C_c}{C_b}\right)\right]$$

$$c = 0, \qquad d = \omega R_i C_c. \tag{4.53}$$

Further specialization of Eq.(4.53) can be obtained as follows: to ignore the effects of Z_b let $C_b \to \infty$; to remove the cable capacitance let $C_c \to 0$. If both of these effects are omitted from the analysis Eqs.(4.53) and (4.38) will result in Eq.(4.41).

Substituting Eq.(4.53) into Eq.(4.38) and numerically evaluating the result yields the curves shown in Figure 4.17 for the case $R_g/R_i = 0.001$. Only this ratio is investigated because the previous cases all showed that this makes the effects of R_g/R_i negligible; any errors will therefore be due to the interaction of the other components. Examination of Figure 4.17 shows that the ratio C_c/C_b governs the location of the lower cutoff frequency, with the smaller value of this ratio producing

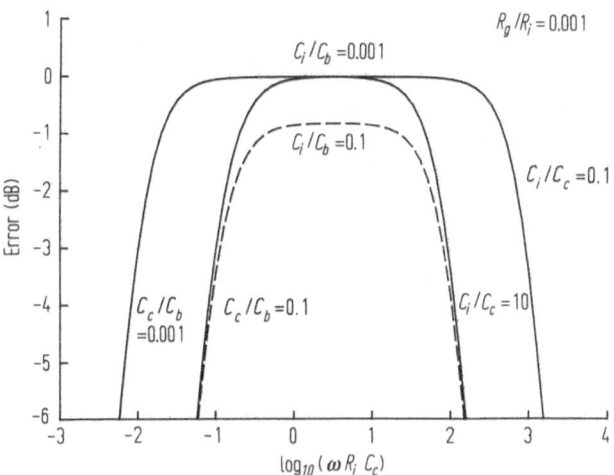

Figure 4.17 Impedance mismatching error for the configuration shown in Figure 4.16 when $Z_g = R_g$ and the other quantities are as defined in Eq.(4.50)

the lower cutoff frequency. The upper cutoff frequency is determined by the ratio of C_i/C_c, with the smaller value of this ratio producing a higher cutoff frequency. It is again seen that the smaller the value of C_i the higher the cutoff frequency. The error in the region between these two cutoff frequencies is governed by the ratio C_i/C_b, with the smaller value of this ratio producing the smaller error. In particular, for $C_i/C_b \leq 0.001$ the error is negligible.

(b) $Z_g = 1/j\omega C_g$ In this case $R_g = 0$ and Eq.(4.52) simplifies to

$$a = 1 + \frac{C_c}{C_b} + \frac{C_g}{C_b}, \qquad b = (\omega R_i C_g)\left[\frac{C_i}{C_b}\left(1 + \frac{C_c}{C_g}\right) + \frac{C_i}{C_g} + \frac{C_c}{C_g}\right]$$

$$c = 0, \qquad d = \omega R_i C_g. \tag{4.54}$$

Further specialization is obtained in the same manner as for case (a). If the effects of C_c and C_b are ignored these results reduce to Eq.(4.42).

Substituting Eq.(4.54) into Eq.(4.38) and numerically evaluating the result yields the curves shown in Figure 4.18 for the case $C_i/C_g = 0.001$. Only this ratio is investigated because the results of Case 3 showed that this makes the effects of C_i/C_g negligible; any errors will therefore due to the interaction of the other components. Examination of Figure 4.18 shows that the ratios C_g/C_b and C_c/C_b govern the location of the lower cutoff frequency, with the smaller values of each ratio producing the lower cutoff frequency. The error above the cutoff frequency

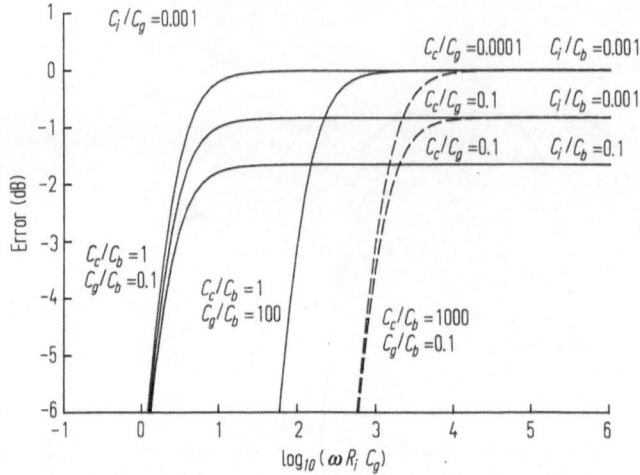

Figure 4.18 Impedance mismatching error for the configuration shown in Figure 4.16 when $Z_g = 1/j\omega C_g$ and the other quantities are as defined in Eq.(4.50)

is governed by the ratios C_i/C_b and C_c/C_g, with the smaller values of these ratios producing the smaller error. In particular, for $C_c/C_g \leq 0.0001$ and $C_i/C_b \leq 0.001$ the error is negligible.

4.5.3 Impedance Matching: Maximum Power Transfer

We now reverse the interpretation of the loads shown in Figure 4.11 and let $Z_g = Z_o$ be the output impedance of an amplifier and $Z_i = Z_L$ be the impedance of a load; e.g., heating element, motor, loudspeaker, electrodynamic and piezoelectric vibration exciters, etc. The new configuration is shown in Figure 4.19. The objective in this situation is to determine the conditions under which the maximum power can be transferred to Z_L.

In this new notation Eq.(4.35) becomes

$$E_L = \frac{E_o Z_L}{Z_o + Z_L}$$

and the current to Z_L is

$$I_L = \frac{E_o}{Z_o + Z_L}.$$

The power is the time average of the square of the current into the resistive part of Z_L. Thus, if

$$Z_L = R_L + jX_L, \qquad Z_o = R_o + jX_o$$

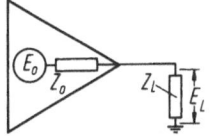

Figure 4.19 Amplifier arrangement to determine impedance matching requirement for the maximum power transfer into a load of impedance Z_L

then

$$P_{avg} = |I_L|^2 R_L = \frac{E_{orms}^2 R_L}{|Z_o + Z_L|^2} = \frac{E_{orms}^2 R_L}{(R_o + R_L)^2 + (X_o + X_L)^2} .$$

The maximum value of P_{avg} is obtained from the conditions at which

$$\left(\frac{\partial P_{avg}}{\partial X_L}\right)_{R_L} = 0, \qquad \left(\frac{\partial P_{avg}}{\partial R_L}\right)_{X_L} = 0 .$$

These conditions give that

$$X_o = -X_L, \qquad R_o = R_L$$

or

$$Z_o = Z_L^* .$$

Therefore, in order to obtain the maximum power transfer the output impedance of the driving system must equal the complex conjugate of the load impedance.

It is pointed out that this maximum power transfer is between the source and the load. In transmission line theory the match is between the line and the load, for which the maximum power transfer is attained when $Z_o = Z_L$ (Javid and Brenner [1963]). The impedance Z_o is the characteristic impedance of the line defined as

$$Z_o = \sqrt{\frac{R + j\omega L}{G + j\omega C}}$$

where R, G, L and C are the resistance, leakage conductance, inductance and capacitance per unit length of the transmission line, respectively. The conductance G depends on the material and geometry *between* conductors only, while R depends on the conductors only.

References

1. Brophy, J. J., *Basic Electronics for Scientists*, 4th Ed., McGraw-Hill Book Co., New York, 1983.
2. Buckingham, M. J., *Noise in Electronic Devices and Systems*, Ellis Horwood Ltd., Chichester, England, 1983.

3. Diefenderfer, A. J., *Principles of Electronic Instrumentation*, 2nd Ed., W. B. Saunders Co., Philadelphia, 1978.

4. Garrett, P. H., *Analog I/O Design: Acquisition, Conversion, Recovery*, Reston Publishing Co., Reston, Virginia, 1981.

5. Gayakwad, R. A., *Op-Amps and Linear Integrated Circuits*, 2nd Ed., Prentice-Hall, Englewood Cliffs, New Jersey, 1988.

6. Higgins, R. J., *Electronics with Digital and Analog Integrated Circuits*, Prentice-Hall, Englewood Cliffs, New Jersey, 1983.

7. Horrocks, D. H., *Feedback Circuits and Op Amps*, Van Nostrand Reinhold (UK), 1983.

8. Javid, J., and Brenner, E., *Analysis, Transmission and Filtering of Signals*, McGraw-Hill Book Co., New York, 1963.

9. Kirwan, G. J., and Grodzinsky, S. E., *Basic Circuit Analysis*, Houghton Mifflin Co., Boston, 1980.

10. Lang, T. T., *Electronics of Measuring Systems*, John Wiley & Sons, New York, 1987.

11. Nelson, J. C. C., *BASIC Operational Amplifiers*, Butterworths, London, 1986.

12. Sanderson, M. L., "Signal Processing", in *Jone's Instrument Technology, Vol 4*, 4th Ed., B. E. Noltingk, Ed., Butterworths, London, 1987.

13. Travers, D., *Precision Signal Handling and Converter-Microprocessor Interface Techniques*, Instrument Society of America, Research Triangle Park, North Carolina, 1984.

14. Vergers, C. A., *Handbook of Electrical Noise Measurement and Technology*, 2nd Ed., TAB Books, Inc., Blue Ridge Summit, Pennsylvania, 1987.

15. Wilmshurst, T. H., *Signal Recovery from Noise in Electronic Instrumentation*, Adam Hilger Ltd., Bristol, England, 1985.

Exercises

1. An amplifier has a voltage gain of 60 dB and a noise referred to input of 2 μV_{rms}. An input signal to the amplifier has an amplitude of 3 mV_{rms} and a signal-to-noise ratio of 30 dB. If the noise figure is 2, what is the power gain of the amplifier.

2. The noise figures of two cascaded amplifiers are equal. It is desired that the total noise figure does not increase by more than 10% of the individual values. What is the minimum power gain of the first amplifier in terms of the amplifiers' noise figures.

3. A signal has a signal-to-noise ratio of 20 dB. What is the rms value of the signal when the signal without the noise has an amplitude of $2V_{rms}$.

4. An amplifier with 60 dB of gain is to have a dynamic range of 70 dB. The maximum allowable output voltage is 5 V_{rms}. What is the value of the amplifier's maximum rms noise level referred to input if its peak noise level is not to interfere with the signal more than 10% of the time.

5. An amplifier has an input resistance of 10 MΩ shunted by a 10 pF capacitance. A resistive source is connected to the amplifier. What is the largest value of

the source's resistance if the resulting configuration is to have an amplitude error of less than -0.1 dB up to 1000 Hz.

6. Repeat problem 5 for a capacitive source.

7. A differential amplifier has a CMRR of 90 dB. The common mode voltage is 5 V_{dc}. What is the minimum value of the input signal if the input signal is to be 54 dB greater than the common mode voltage.

5 Analog-to-Digital Conversion

5.1 Introduction

Since naturally occurring physical phenomena such as temperature, pressure, displacement, and so on, are analog, and since most practical methods of data collection, manipulation, and analysis are digital, a conversion from the analog quantities to digital quantities must take place. This conversion is called digitization and has virtually unlimited application. The device that converts the analog signal into a digital representation is called an analog-to-digital (A/D) converter. In this chapter we shall discuss the fundamental aspects of the conversion process that are common to most applications. Although there are numerous methods and techniques for converting an analog signal to a digital number only three will be discussed. The details of the implementation of the various methods can be found in many of the references given at the end of the chapter.

5.2 A/D Conversion Process

Consider an integer (decimal) number M such that M can be expressed as the sum

$$M = a_N 2^{N-1} + a_{N-1} 2^{N-2} + \dots + a_2 2^1 + a_1 2^0 \le 2^N - 1 \qquad (5.1)$$

where the weights a_n, $n = 1, 2, \dots, N$ have the values of either 0 or 1. The a_n are called bits and N is the number of bits. We can express the weights themselves as a number, called a binary number, in the form $a_N a_{N-1} a_{N-2} \dots a_2 a_1$. If we normalize (scale) M such that $M = 2^N V / V_{FS}$, where $0 \le V \le V_{FS}$ and V_{FS} is the full scale (maximum) value of V, then Eq.(5.1) becomes

$$\frac{V}{V_{FS}} = a_N 2^{-1} + a_{N-1} 2^{-2} + \dots + a_2 2^{1-N} + a_1 2^{-N} \le 1 - 2^{-N}. \qquad (5.2)$$

It is seen that a_N multiplies the most significant number, which is always 1/2, and a_1 the least significant one. Therefore a_N is called the most significant bit (MSB) and α_1 the least significant bit (LSB), where

$$LSB = 2^{-N} \qquad (5.3)$$

is the resolution of the A/D conversion process. Equation (5.2) is illustrated with a 3-bit ($N = 3$) binary representation in Figure 5.1. It should be noted in both Figure 5.1 and Eq.(5.2) that the maximum value of V ($= V_{FS}$) is never reached, but that it is always $2^{-N} V_{FS}$ less than V_{FS}. The bit weights in various forms are tabulated in Table 5.1 for binary numbers up to 20 bits. It should be noted from Table 5.1 that it requires 10 binary bits to get 3 decimal digits, 14 bits to get 4 decimal digits and 17 bits to attain 5 decimal digits. Although only a straight binary code is shown in Figure 5.1 many other codes are used as outputs from digital conversion devices. A description of several of the most common codes can be found in Hnatek [1976], Clayton [1982], and Sheingold [1986].

Before proceeding it is important to clarify the meaning of V_{FS}. We see in Eq.(5.2) that the binary number as we have defined it is always positive. When V is always positive (or always negative) the quantity is called *unipolar*. On the other hand when the quantity varies between positive and negative values it is called *bipolar*. For a given N, therefore, Eq.(5.2) represents either the digitization of a unipolar quantity in N bits or the digitization of a bipolar quantity in $N-1$ bits, where one bit is required for the sign. Thus in the bipolar case this is equivalent to considering a unipolar quantity whose full scale magnitude is given by $V_B = 2V_{FS}$, where V_B is the peak-to-peak value of the bipolar signal. Thus in

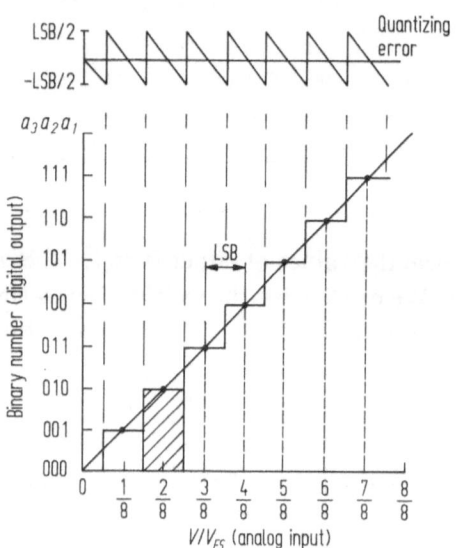

Figure 5.1   Ideal 3-bit A/D conversion of a unipolar signal and the associated quantization error

Table 5.1 Value of a binary bit

Bit, n	2^n	2^{-n}*	dB
1 (MSB)	2	0.5	-6.0
2	4	0.25	-12.0
3	8	0.125	-18.1
4	16	0.0625	-24.1
5	32	0.03125	-30.1
6	64	0.015625	-36.1
7	128	0.0078125	-42.1
8	256	0.0039063	-48.2
9	512	0.0019531	-54.2
10	1024	0.00097656	-60.2
11	2048	0.00048828	-66.2
12	4096	0.00024414	-72.2
13	8192	0.00012207	-78.3
14	16,384	0.000061035	-84.3
15	32,768	0.0000305176	-90.3
16	65,536	0.0000152588	-96.3
17	131,072	0.00000762939	-102.3
18	262,144	0.00000381469	-108.4
19	524,288	0.00000190735	-114.4
20	1,048,576	0.000000953674	-120.4

*To obtain the percentage of full scale multiply by 100

referring to an A/D conversion process one must determine whether or not the number of bits cited includes the sign. In the particular case where the bipolar quantity is a sine wave of frequency f_1, its representation in this notation is

$$V = \frac{V_B}{2} \sin(2\pi f_1 t) \tag{5.4}$$

and its rms value is [recall Eq.(2.38)]

$$V_{rms} = \frac{V_B}{2\sqrt{2}}. \tag{5.5}$$

Returning to Figure 5.1 it is seen that every value of V/V_{FS} that lies between, say,

$$\frac{2}{8} - \frac{1}{16} \le \frac{V}{V_{FS}} < \frac{2}{8} + \frac{1}{16}$$

is assigned the same binary value. However, the error caused by this assignation varies linearly from LSB/2 to -LSB/2, with the error being zero when V/V_{FS} is exactly 2/8. This type of conversion error is called the quantization error and its magnitude is a direct function of the number of bits. This is shown at the top of Figure 5.1. The average value of this error is zero. However, if it is assumed that

116

it is equally probable for V/V_{FS} to take on any value within each LSB, then its rms value (standard deviation) can be shown to be (Otnes and Enochson [1978])

$$\sigma = \frac{LSB}{\sqrt{12}}. \qquad (5.6)$$

Thus, on the average, the error is not the LSB but $LSB/\sqrt{12}$. Consequently, this is the effective resolution of the system.

Using Eq.(5.6) we can now determine the dynamic range (signal-to-error ratio) of an A/D converter for a sine wave, which is given by

$$Dynamic\ Range = 20\log_{10}\left(\frac{Maximum\ rms\ value}{Quantitization\ noise}\right) \quad dB. \quad (5.7)$$

Using Eqs.(5.5) and (5.6), Eq.(5.7) yields

$$Dynamic\ Range = 20\log_{10}\left(\frac{V_{FS}/\sqrt{2})}{V_{FS}2^{-N}/\sqrt{12}}\right)$$

$$= 6.02N + 1.76 \quad dB \qquad (5.8)$$

where N is the number of bits not including the sign. Equation (5.8) assumes that the input sine wave is noise-free.

If the signal does contain noise the dynamic range of the A/D conversion process may become degraded. Recall Table 4.1 which related the rms value of gaussian noise to its peak value. In order for the error due to noise to be less than the LSB a stated percentage of the time $LSB < ke_{rms}$, where k is the factor obtained from the right hand column of Table 4.1. Thus if the noise is to interfere less than 0.01% of the time, we find that for, say, a 12-bit (including sign) bipolar converter with a full scale voltage of 10 V that $e_{rms} < 0.63$ mV $(= 20 \times 2^{-12}/7.78)$.

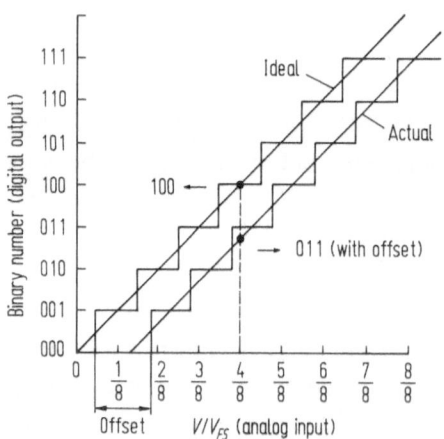

Figure 5.2 Effect of dc offset on a 3-bit A/D conversion process

There are several sources of errors that an A/D converter can introduce. Two common ones are shown in Figures 5.2 and 5.3. In Figure 5.2 are shown the effects of a dc shift (offset) on the signal while in Figure 5.3 the effects of gain (or linearity) error are shown. One can conclude that both the offset and linearity of an A/D converter should each cause an error that is less than LSB/2 of the converter itself.

There is one useful exception to the linearity requirement and that is to employ a logarithmic compression amplifier to those unipolar signals that have wide dynamic range but are capable of tolerating constant fractional error; that is, constant resolution throughout the dynamic range at the expense of high resolution at any point within the range. This is shown in Figure 5.4.

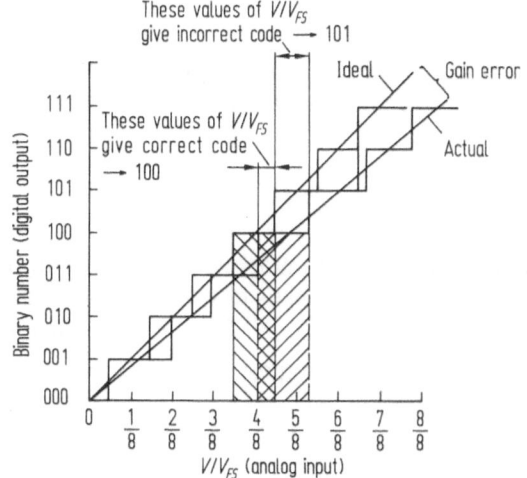

Figure 5.3 Effect of gain (linearity) error on a 3-bit A/D conversion process

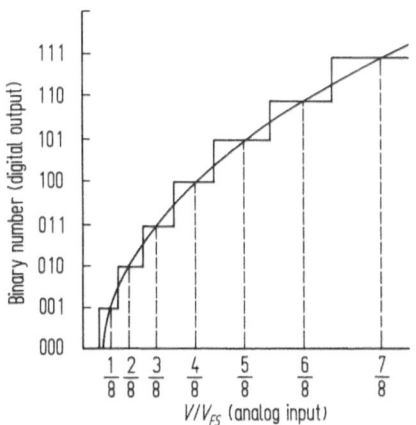

Figure 5.4 Example of a 3-bit nonlinear A/D conversion of a unipolar signal

As a final remark it is pointed out that an A/D converter is a fixed amplitude responding device. Thus, to utilize its entire dynamic range one must insure that the maximum value of the input signal lies within the range $V_{FS}/2$ to V_{FS}. If not, the MSB is not being used and the A/D converter is essentially an $N-1$ bit converter, or less.

Several applications requiring the A/D conversion process can be found in Sections 8.2.11, 8.5.4, 8.5.5 and 9.5.

5.3 A/D Conversion Speed

Let the input signal to an A/D converter be the sine wave given by Eq.(5.4). Referring to Figure 5.5 it is seen that the maximum rate of change of the amplitude of this signal is

$$\frac{\Delta V}{\Delta t} = \pi f_1 V_B.$$ (5.9)

If the conversion process is to take place correctly, the sampled signal's amplitude must change by less than the LSB during the time it takes to convert the voltage level into a digital number. Therefore, if we let $\Delta V / V_{FS} = $ LSB, $\Delta t = t_{conv}$, where t_{conv} is the conversion time, then Eq.(5.9) yields the highest sine wave frequency that can satisfy these conditions:

$$f_1 = \frac{LSB}{\pi t_{conv}} = \frac{2^{-N}}{\pi t_{conv}}$$ (5.10)

where N is the number of bits including the sign bit. If the conversion time is $1\,\mu s$ and $N = 12$, then Eq.(5.10) gives $f_1 = 77.7\,Hz$! For these conditions any attempt to digitize a sine wave of frequency greater than $77.7\,Hz$ will give erroneous results.

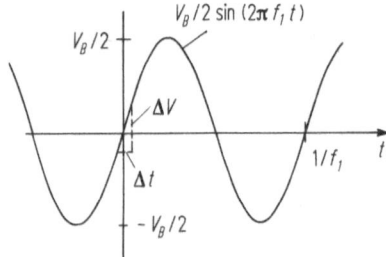

$\Delta V/\Delta t$ = maximum rate of change of sine wave of peak amplitude $V_B/2$

Figure 5.5 Determination of the maximum rate of change of a sine wave of frequency f_1

Consequently, something further must be done to increase the maximum signal frequency.

Sample-and-Hold

Ideally one would like to be able to hold the amplitude of the analog signal constant from the instant the A/D conversion starts until the conversion has completed. To closely approximate this ideal condition a *sample-and-hold (S/H) amplifier* is introduced before the A/D converter. A S/H amplifier is essentially a signal memory device that can extremely rapidly store a charge on a capacitor and hold it for a relatively long period of time. It has two modes of operation: sample and hold. In the sample mode of operation its output signal is equal to its input signal and it is tracking, or following, the input signal. On receipt of a hold command its output signal is constant at the value of the input signal at the instant the command was received. There are, however, several errors that are introduced by this process. One error is called the aperture time t_{ap}, which is the time between the application of the "holding" of the signal amplitude and the time at which the output signal is no longer affected by changes in the input signal. Another error is the sample-to-hold delay error t_{apd}, which is the time it takes for the S/H amplifier to actually respond to the hold command. A third error is a delay t_{apt} that is required to allow sufficient time for the transient at the output of the S/H amplifier to settle prior to the start of the conversion process. This transient is caused by the holding (switching) action of the S/H amplifier and is sometimes referred to as the sample-to-hold transient recovery time. The total S/H time is, therefore, $t_a = t_{ap} + t_{apd} + t_{apt}$. The fourth error refers to what is meant by holding the signal constant. In this application the signal can be considered held constant if, after conversion, the S/H amplifier's output signal amplitude has changed (drooped) by less than the LSB. This is not unlike the response in Figure 3.13.

The S/H process is shown in Figure 5.6. The settling time t_{SH} is typically the time it takes, upon release from the hold condition, for the S/H amplifier to be tracking to within 0.01% of the true value. This settling time is different than the transient recovery time t_{apt}, which is the recovery time prior to the start of the conversion. The minimum sampling interval is the sum of the conversion time, the settling or recovery time and the total S/H amplifier delays, t_a. The significance of these times are more fully discussed in Section 5.6.

Thus, with a S/H amplifier preceding the A/D converter, Eq.(5.10) can be written as

$$f_{SH} = \frac{2^{-N}}{\pi t_{ap}} \tag{5.11}$$

where f_{SH} is the maximum frequency with a S/H amplifier and N is the number of bits including the sign bit. Equation (5.11) is plotted in Figure 5.7. Implied in Eq.(5.11) is the restriction that the sampling interval is greater than the total processing time; that is, $1/f_{SH} > t_a + t_{conv} + t_{SH}$. These points will be examined further when two system configurations are discussed in Section 5.6.

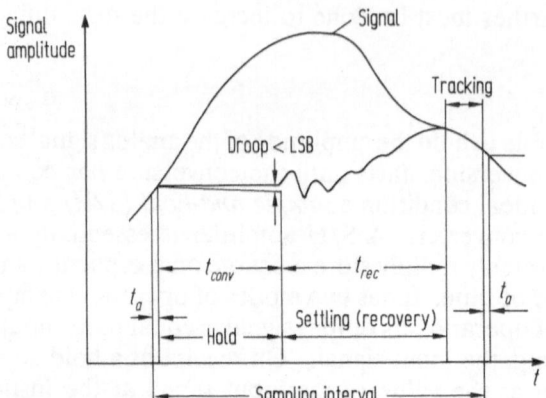

Figure 5.6 Sample-and-hold process: t_a is the total S/H time, t_{conv} is the A/D conversion time and t_{rec} the recovery (settling) time to tracking

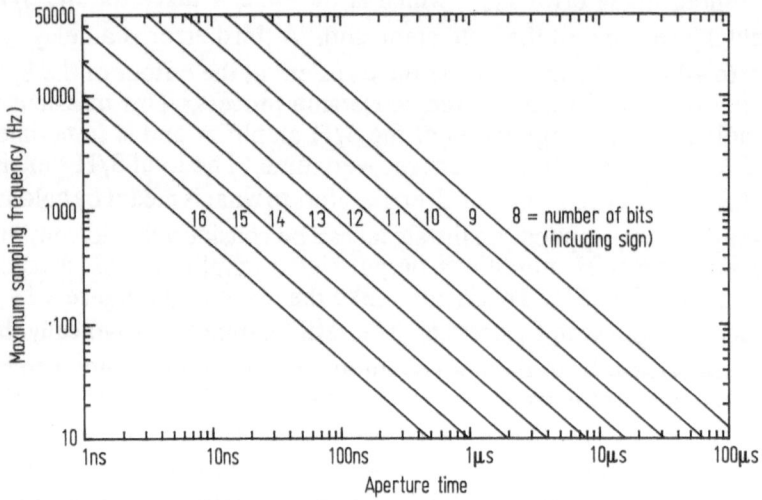

Figure 5.7 Estimated maximum aperture time for an N-bit A/D converter as a function of the maximum sampling frequency for the change in the amplitude of a sine wave to be less than the LSB

It should be mentioned that there are severe requirements placed on the S/H amplifier in order that it perform in a manner consistent with the desired LSB of the A/D converter that follows it. Namely, the combination of its amplitude and frequency linearity errors and its offset errors must be less than the LSB. Furthermore, the noise of the S/H amplifier must be considerably less than the LSB.

Example 1: Determination of Maximum A/D Conversion Time From S/H Amplifier Properties. Consider a S/H amplifier that has an aperture time of 10 ns and a recovery (settling) time to within 0.01% of 800 ns. To be consistent with this recovery time percentage it is found in Table 5.1 that the A/D converter need not have more than 13 bits, not including sign. Consequently from Eq.(5.11) the maximum frequency for a unipolar signal is $f_u = 2^{-13}/(\pi x 10^{-8}) = 3885$ Hz, whereas for a bipolar signal it is $f_b = 2^{-14}/(\pi x 10^{-8}) = 1943$ Hz, since another bit is needed to determine sign. Therefore the maximum A/D conversion time for a unipolar signal is $t_{conv} \leq 1/3885 - 810 \times 10^{-9} = 256.6$ μs while that for a bipolar signal is $t_{conv} \leq 1/1943 - 810 \times 10^{-9} = 513.9$ μs.

5.4 A/D Conversion Methods

5.4.1 Successive Approximation A/D Converter

Referring to Figure 5.8a the successive approximation A/D converter is a fast comparison method in which the input voltage V_i is compared with the output voltage V_{out} from the D/A converter (digital-to-analog converter: see Section 5.8). The sign of the error signal $V_i - V_{out}$ sets the successive bits of input to the D/A converter. The bits are set and tested in turn starting with the MSB, which is half scale. As illustrated in Figure 5.8b the process proceeds as follows: When the start command is received the MSB bit of the D/A is set. If V_i is greater than the half scale voltage the bit is kept, if not it is set to zero. On the next clock cycle the next MSB is set and a new (either larger than or less than half scale depending on the results of the previous comparison) V_{out} is compared to V_i. If the difference is negative the bit is turned off and the next MSB is set. The process is repeated until all N-bits have been examined.

This type of conversion process is fast (1 MHz for 12-bits), capable of high resolution (16-bits) and has a fixed conversion time independent of the amplitude of the signal. This latter attribute is advantageous when multiplexing is used (see Section 5.6). A disadvantage of this type of converter is that there is no inherent noise immunity and, consequently, some prefiltering must be done for those converters with high resolution.

5.4.2 Dual Slope A/D Converter

The dual slope A/D method of conversion is shown in Figure 5.9a. When the input voltage to the integrator is a constant, its output voltage V_o increases at a rate proportional to the magnitude of the input voltage. At the start of the cycle V_i is connected to the integrator's input. At the end of a fixed time interval T_1

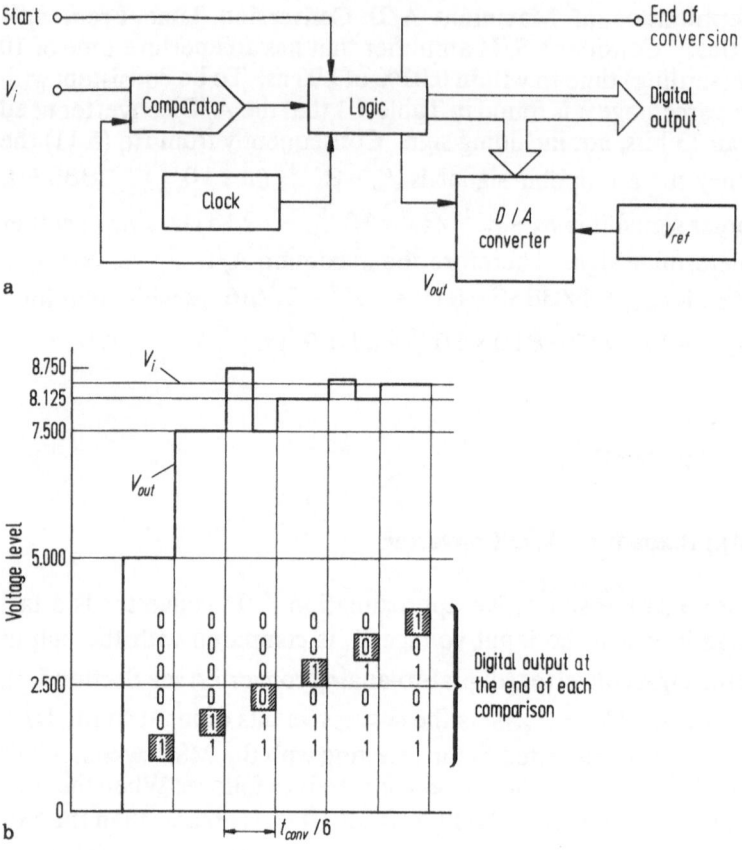

Figure 5.8 (a) Successive approximation A/D converter (b) A/D output for a 6-bit conversion

the input to the integrator in switched to the reference voltage V_{ref}, which has the opposite polarity of V_i. At the same time the switching logic starts the counter, which continues counting until the integrator's output voltage reaches zero. Thus the dual slope method measures the voltage by converting it to a time interval T_2. Let $N_c = T_1 f_c$ be the total number of counts used to set T_1 at a clock frequency f_c and $n = T_2 f_c$ be the total number of counts occurring in the second half of the cycle. Referring to Figure 5.9b it is seen that

$$V_i = \frac{n}{N_c} V_{ref}.$$

Thus the accuracy of the method depends on the quality of V_{ref} and its resolution on the value of f_c.

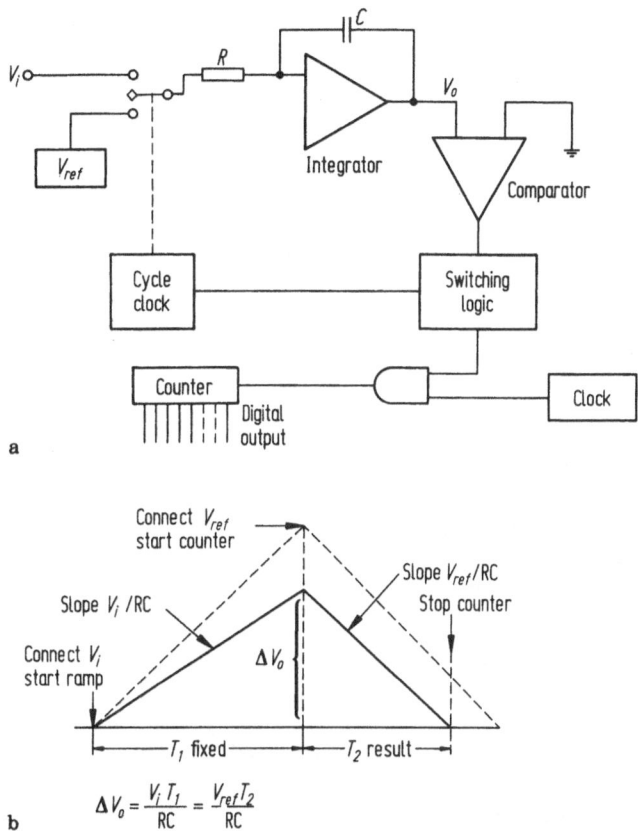

a

b

$$\Delta V_o = \frac{V_i T_1}{RC} = \frac{V_{ref} T_2}{RC}$$

Figure 5.9 (a) Dual slope A/D converter (b) integrator output during one measuring cycle

The dual slope method has very high noise immunity because of its integrating properties. In addition its linearity is good because the output digital number is a function only of the clock frequency f_c and the counter. For these same reasons the dual slope converter can easily attain 5 and 6 decimal digits (17- and 20-bit resolution). See Section 6.2 for a discussion of digital counters.

5.4.3 Pulse-Width Conversion

A pulse-width conversion system is shown in Figure 5.10. The dc input signal is inverted and summed with a fixed amplitude square wave of constant frequency (usually 1 kHz) and with a constant voltage determined by either of two comparators. The summed voltages are integrated and fed into the inputs of two comparators. One comparator tests for a positive voltage level and the other for a negative voltage level. The output of the positive voltage comparator goes to the

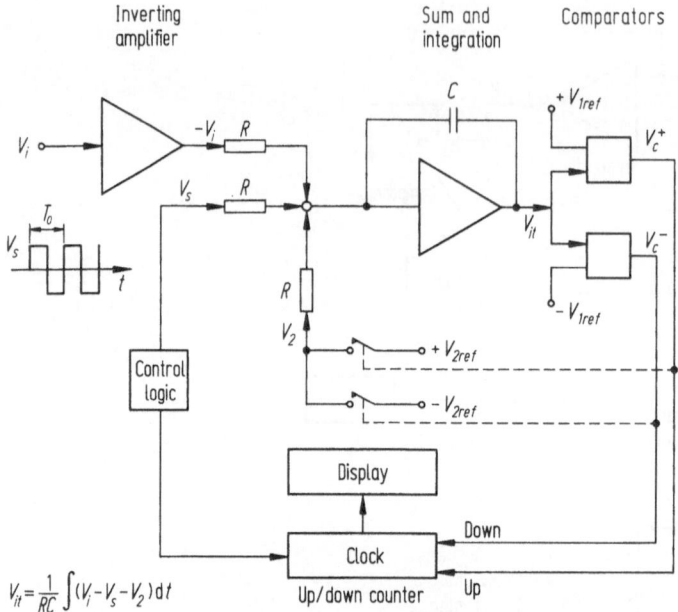

Figure 5.10 Pulse-width modulation A/D converter

"count up" input of an up/down counter and closes a switch so that $+V_{2ref}$ goes to the summing junction of the integrator. The output of the negative voltage comparator goes to the "count down" input of the up/down counter and closes a switch so that $-V_{2ref}$ goes to the summing junction of the integrator. An up/down counter is one in which a voltage to its "count up" input causes the counter to increment at the clock frequency rate until this input is removed. A voltage to its "count down" input decrements the counter in a similar fashion. The output of the counter is a digital value of the net count.

If the input voltage V_i is zero and the comparator reference voltage V_{1ref} were set equal to the voltage of the square wave V_s, then the $\pm V_{2ref}$ switches would always remain open and the output voltage of the integrator V_{it} would be that shown in Figure 5.11a (recall Figure 2.1b). When V_{1ref} is set so that $V_{1ref} < V_s$ the switch for V_{2ref}, in response to the voltage comparator's output, closes for the period of time in which $V_{it} > V_{1ref}$. This places a voltage V_{2ref} at the summing junction, resulting in a V_{it} shown in Figure 5.11b. This procedure is repeated in the negative sense for the next half period of the square wave. In this case the duration of the "up" and "down" comparator voltages (their pulse widths) is equal and, therefore, the net output of the up/down counter is zero.

When the input voltage is a positive voltage V_i, V_{it} becomes that shown in Figure 5.11c. In this case the pulse width of the "up count" voltage is greater than that of

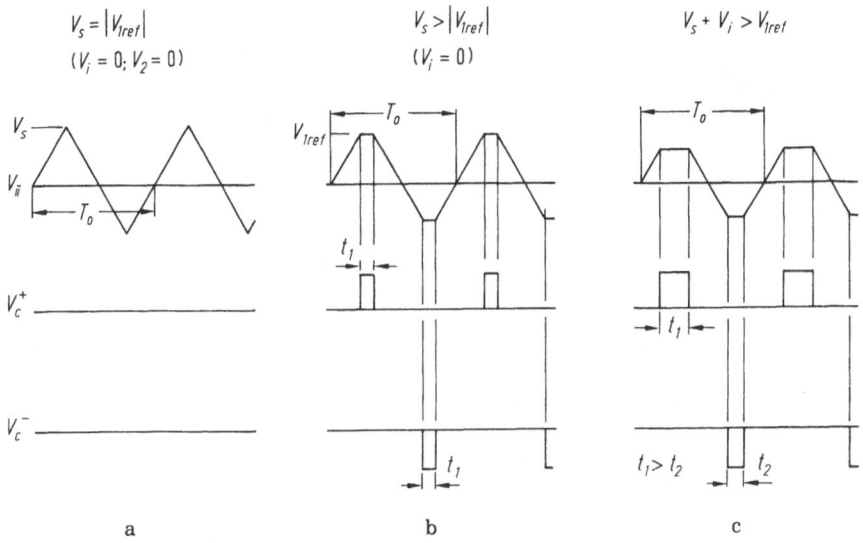

Figure 5.11 Up/down counter input pulses as a function of V_i and V_{1ref} (a) $V_s = |V_{1ref}|$ (b) $V_s > |V_{1ref}|$ (c) $V_s + V_i > V_{1ref}$

the "down counter" and there is a net positive count from the up/down counter, which is proportional to V_i.

A pulse width conversion process responds continually to changes in the input signal. Periodic noise is averaged out and abrupt changes in signal amplitude are immediately available to activate autoranging circuits.

5.5 Aliasing and Filter Selectivity

We shall now develop a relationship between filter selectivity, A/D converter resolution and the minimization of aliasing errors. Consider a white noise spectrum [recall Example 11 of Chapter 2] that is passed through an idealized low pass filter shown in Figure 5.12. The purpose of the filter is to limit the effects of the spectrum beyond the filter's cutoff frequency f_c. A filter used specifically for this purpose with A/D converters is called an *anti-aliasing filter*. The attenuation of this idealized filter is

$$L_o = 20 \log_{10} \left[\frac{1}{\sqrt{2}} \left(\frac{f}{f_c} \right)^{-y/6.02} \right] \quad \text{dB} \qquad f \geq f_c \qquad (5.12)$$

where y is the filter selectivity in dB/octàve.

To minimize the aliasing error we recall that if the sampling frequency is $f_s = 2f_c$, then the frequencies that will be aliased are those for which $f > f_c$. Recalling Figure 2.12 we now redraw Figure 5.12 in Figure 5.13a for the case where $f_s = 2f_c$. We see the large shaded portion that will be aliased (folded) into the region $f < f_c$. If, on the other hand, we raise the sampling frequency such that

$$f_s = \alpha f_c \qquad \alpha > 2 \tag{5.13}$$

then we can reduce the portion of the spectrum that will be aliased. Thus the aliased spectrum of Figure 5.13a becomes that shown on Figure 5.13b. It is seen in this latter figure that the maximum amplitude of the aliased portion, denoted L_o in Figure 5.13b, can be made as small as desired by selecting the proper combination of sampling frequency αf_c and filter selectivity y.

We now establish the relationships among α, y and L_o. Referring to Figure 5.13b we define the folding frequency f_f as the geometric mean frequency of f_c and f_o, where f_o will be determined subsequently. Thus

$$f_f = \sqrt{f_o f_c} = f_c \sqrt{\frac{f_o}{f_c}}. \tag{5.14}$$

Since $f_s = 2f_f$ Eq.(5.14) becomes

$$\frac{f_s}{f_c} = 2\sqrt{\frac{f_o}{f_c}}. \tag{5.15}$$

From Eq.(5.12) and Figure 5.13b we find that

$$\sqrt{\frac{f_o}{f_c}} = 10^{-3.01 L_o/(20y)}$$

and, therefore, Eq.(5.15) yields the final result

$$\frac{f_s}{f_c} = 2^{(1-0.5 L_o/y)}. \tag{5.16}$$

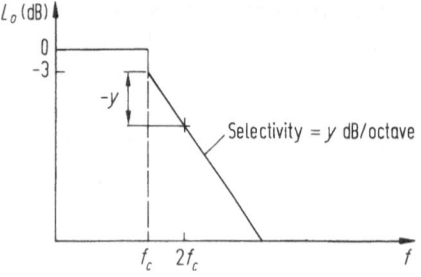

Figure 5.12 Transfer function of an idealized low pass anti-aliasing filter

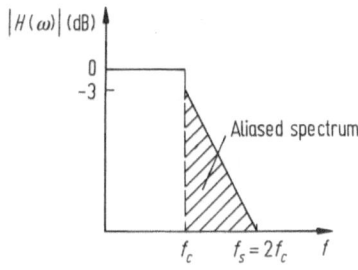

a

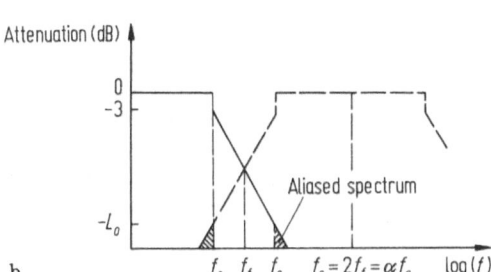

b

Figure 5.13 (a) Aliased spectrum of the filter shown in Figure 5.12 when $f_s = 2f_c$ (b) aliased spectrum of the filter shown in Figure 5.12 when $f_s = \alpha f_c$ and $\alpha > 2$

Referring to Table 5.1 we see that the relationship between the number of bits and the number of dB down from full scale is

$$L_0 = -6.02n \quad \text{dB} \tag{5.17a}$$

where n is the number of bits without the sign bit. In terms of the percentage full scale ϵ,

$$L_0 = 20 \log_{10} \frac{\epsilon}{100} \quad \text{dB}. \tag{5.17b}$$

Equations (5.16) and (5.17b) are used to obtain the results presented in Figure 5.14, where it is seen that to use the slowest sampling rate for a given aliasing error for a white noise spectrum, the filter selectivity should be as high as possible. It should also be noted that for this spectrum the error is maximum at f_c^+ and decreases as frequency increases.

As a final remark concerning the significance of the anti-aliasing filter we return to the DFT (Section 2.5.3) where it was found that its frequency resolution is $\Delta f = 1/NT = f_s/N$. From Eq.(5.13) $f_s = \alpha f_c$ ($\alpha > 2$); thus $\Delta f = \alpha f_c/N$. In many practical implementations of the DFT the aliasing filter selectivities are 120 dB/octave, N is 1024 and $\alpha=2.56$; therefore, $\Delta f = f_c/400$. Notice that in this case 1024 samples of the time waveform $f(t)$ provide a DFT with frequency resolution of only 1/400 th of the signal's bandwidth. This apparent loss of resolution happened as follows: The 1024 samples were used by the DFT to obtain 512 complex pairs of numbers that contain the signal's amplitude and phase information in the frequency domain. Additionally, to produce these 512 complex pairs the sampling frequency had to be increased 28%, from 2 to 2.56 ($f_f = 1.28 f_c$), in order to obtain reasonably alias-free results (recall Figure 5.14). This is equivalent to retaining only 78.125% (2/2.56×100) of the data samples; hence, 0.78125×512=400.

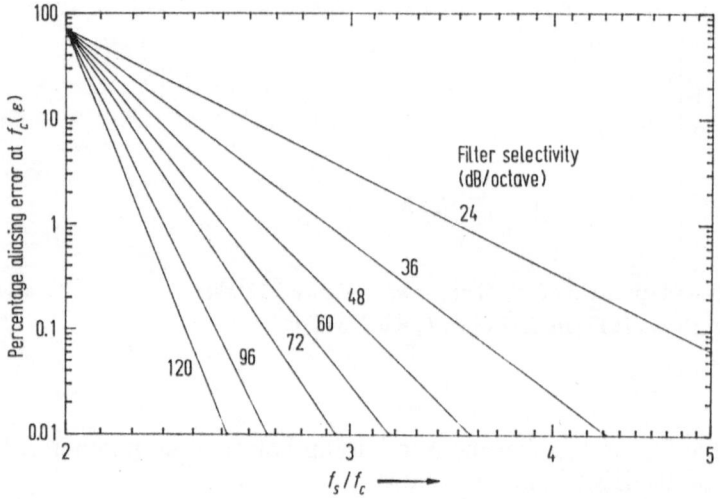

Figure 5.14 Maximum aliasing error of an idealized practical low pass spectrum of Figure 5.12 as a function of filter selectivity and the ratio of the sampling frequency f_s to the filter's cutoff frequency f_c

5.6 Multi-channel A/D Converter Configurations

We now examine the overall characteristics of two types of multi-channel digital acquisition systems. Both of these systems consist of transducers and signal conditioning amplifiers, anti-aliasing filters, S/H amplifiers, an analog multiplexer, an A/D converter, a control unit, and a storage device, usually a computer. The control unit and the computer are, in some configurations, the same. An analog multiplexer is typically an array of high-speed solid state switches that permit, upon application of an appropriate digital code, the selection of any one of its input channels to appear at its output.

The fundamental difference between the two systems is the manner in which the signals at each channel are sampled: either sequentially or simultaneously. For only one channel both systems are equivalent.

Sequential Sampling of Channels

The equipment block diagram of this system is shown in Figure 5.15. The maximum speed at which this system can operate is determined with reference to Figure 5.16. It is noted that in this configuration the sampling rate for each channel is determined by the switching speed of the multiplexer and the number of channels. If t_a is the S/H amplifier's total delay time (recall Section 5.3), t_{conv} the A/D conversion time and t_{rec} the recovery, or settling time, (to within a stated percentage) to return to the tracking condition, then the fastest each channel can be sequentially sampled is

$$t_s = t_a + t_{conv} + t_{rec} \quad s \tag{5.18}$$

where t_s is the minimum sampling interval per channel and t_{rec} is the greater of the multiplexer switching and settling times (t_{mux}) and the S/H amplifier recovery time to tracking condition (t_{SH}). Thus the maximum throughput rate per channel for this system is

$$f_{TP} = \frac{1}{N_c t_s} = \frac{1}{N_c(t_a + t_{conv} + t_{rec})} \quad \text{samples/s/channel} \tag{5.19}$$

where N_c is the number of channels. Thus, in this configuration the sampling rate is governed by the multiplexer's switching rate. As seen in Eq.(5.19) as the number of channels increases there is a decrease in the maximum sampling rate of each

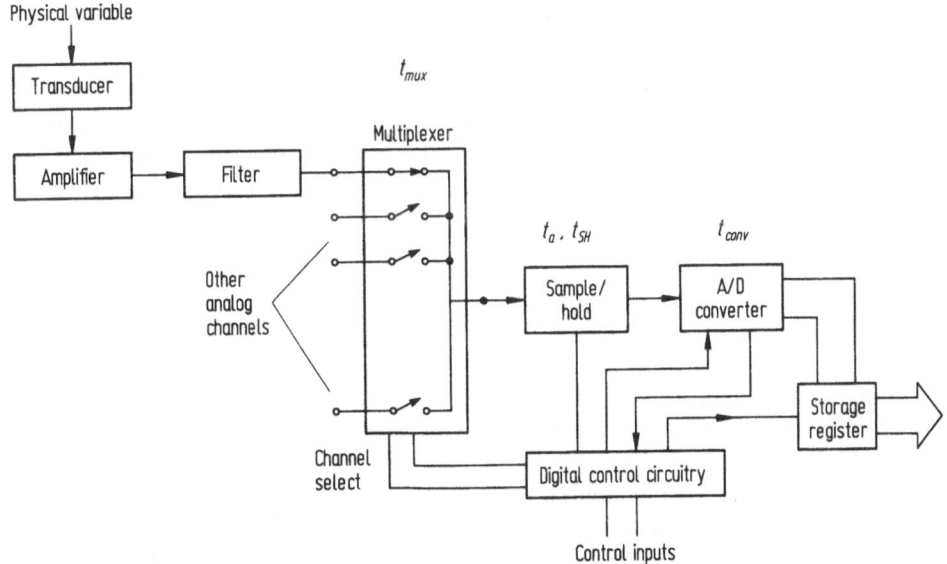

Figure 5.15 Typical components of a sequentially sampled multi-channel A/D conversion system

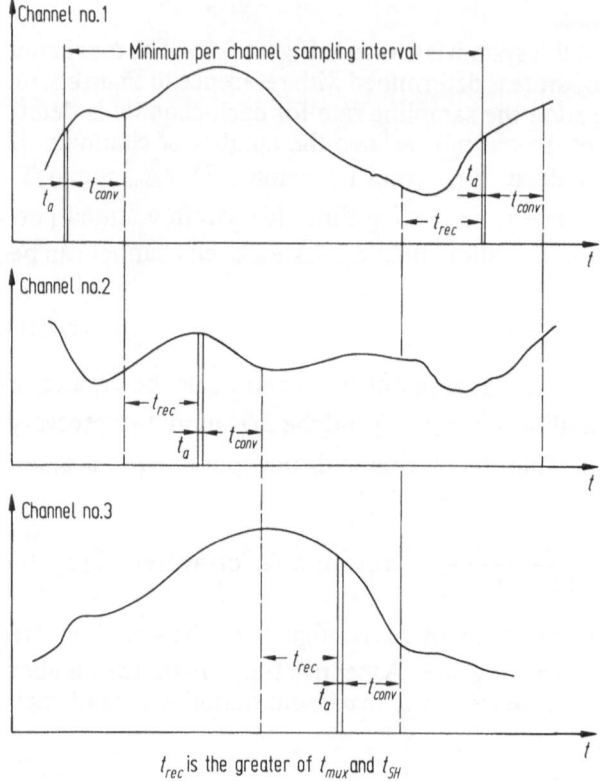

Channel no.1

Minimum per channel sampling interval

t_a t_{conv} t_a t_{conv} t_{rec}

t

Channel no.2

t_{rec} t_a t_{conv}

t

Channel no.3

t_{rec} t_a t_{conv}

t

t_{rec} is the greater of t_{mux} and t_{SH}

Figure 5.16 Timing relationships of a 3-channel sequentially scanned A/D conversion system

channel. Depending on the bandwidth of a channel this decrease may not be permissible because the resulting decreased sampling rate may not satisfy the minimum rate required for that channel. Lastly it should be noted that, in general, this configuration is unable to adequately preserve time synchronization (phase) among the channels' signals.

Simultaneous Sampling of Channels
The equipment block diagram for this system is given in Figure 5.17. This configuration is different from the sequentially sampled one in that the per channel sampling rate is, up to a point, independent of the multiplexer's switching rate since the multiplexer's switching rate is fixed (usually the fastest speed at which the system can operate). The maximum speed of this system is determined with reference to Figure 5.18, which shows that the maximum rate at which the N_c channels can be simultaneously sampled is

$$f'_{TP} = \frac{1}{N_c(t_a + t_{conv} + t_{mux}) + t_x} \qquad \text{samples/s/channel} \qquad (5.20)$$

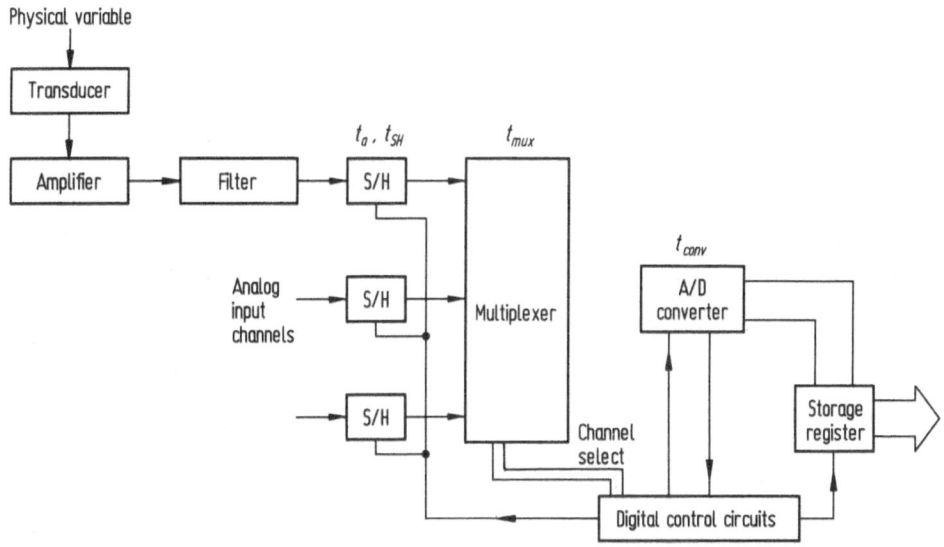

Figure 5.17   Typical components of a simultaneously sampled multi-channel A/D conversion system

where t_x is equal to the greater of t_{SH} and t_{mux}. As with sequential sampling the per channel throughput rate is essentially inversely proportional to the number of channels. Notice, however, that the capabilities of the S/H amplifier have to be such that the last channel sampled (N_c) must hold the voltage level constant to within less than the LSB of the A/D converter for $N_c t_{conv} + (N_c - 1)t_{mux}$ s.

The major differences between the two methods is that simultaneous sampling preserves phase information among the channels, which in many applications is important, and that, except at the maximum sampling rate, the sampling rate is independent of the multiplexer switching rate. As seen from Eq.(3.10), to have a phase delay of less than, say, 0.5° at 2000 Hz requires the phase lag between the first and the N_c th channels to be less than 0.7 µs. In current A/D systems the combination of $t_{conv} + t_{rec}$ for *each* channel exceeds this value. Therefore, except for very low bandwidth signals and a small number of channels N_c, sequentially sampled A/D systems can not be expected to preserve the phase relation among the channels. On the other hand the preservation of phase for the simultaneously sampled channels is just a function of the differences in the S/H delays t_{apd} of each S/H amplifier. These differences are typically on the order of 10-20 ns. For a 2000 Hz signal this corresponds to a phase delay of less than 0.014°. Sequentially sampled channels also require a decrease in the sampling interval for each increase in the number of channels in order for the per channel sampling interval to remain constant. In the case of the simultaneously sampled system the sampling rate is independent of the multiplexer switching rate, although the multiplexer does govern the maximum per channel sampling rate. In either case the maximum

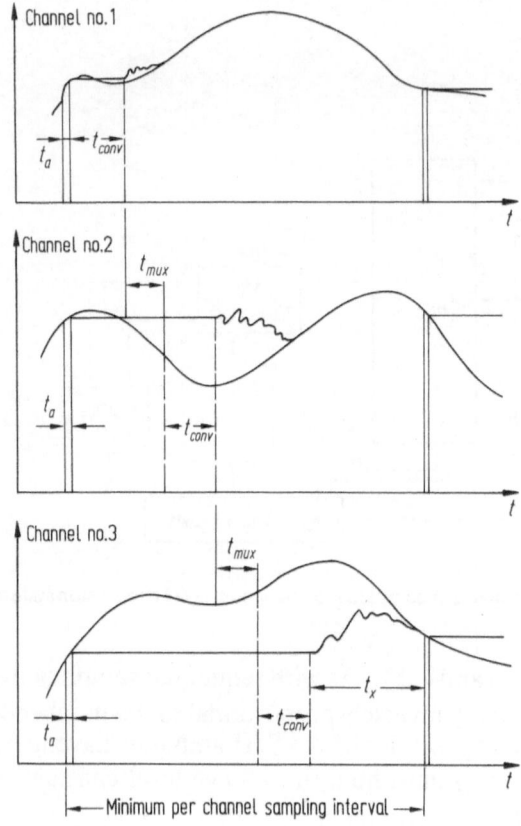

t_x is greater of t_{mux} and t_{SH}

Figure 5.18 Timing relationships of a 3-channel simultaneously sampled A/D conversion system

throughput rate cannot be exceeded. Finally there is an additional limitation that has nothing to do with either of these two methods *per se*, but is implied in their usage. There are many instances in which the device to which the data finally reside dictate both the speed and the amount of data that can be converted. This is especially true for a large N_c and/or large bandwidth signals.

Applications of sequential sampling are given in Sections 8.8.6 and 9.4 and for simultaneous sampling in Section 9.5.

5.7 Calibration of a Multi-channel Digital Data Acquisition System

Consider the simultaneously sampled N-channel A/D data acquisition system shown in Figure 5.17. The storage register is replaced by a computer system that has, as part of its analysis programs, an FFT implementation of the DFT (recall Sections 2.5.3 and 2.5.4). The Fourier transform of the output of the S/H amplifier of the kth channel is $V_{ok}(f)$, $k = 1,2,...,K$ and is related to the transfer functions $H_{ATk}(f)$ of the transducers and transducers' amplifiers, $H_{Fk}(f)$ of the anti-aliasing filters, and $H_{Sk}(f)$ of the S/H amplifiers by [recall Eq.(3.16)]

$$V_{ok}(f) = q_{ok} S_k 10^{A_k/20} H_{ATk}(f) H_{Fk}(f) H_{Sk}(f)$$

$$k = 1,2,...,K \qquad (5.21)$$

where q_{ok} is the physical quantity of interest as measured by the transducer, S_k is the transducer's sensitivity and A_k is the gain (in dB) of the amplifier. It is implied in this formulation that any deviation of q_{ok} from the true value is accounted for in $H_{ATk}(f)$. Also, the sensitivity S_k is a constant at only one frequency, which is arbitrarily chosen, but not independent of the definition of $H_{ATk}(f)$.

The numerical value that gets stored in the computer at each sampled time interval is $M_k(m\Delta t)$. A collection of N of these $\{M_k(m\Delta t)\}$ are transformed into the frequency domain using the FFT algorithm. After transformation we have stored in the computer

$$FFT\{M_k(m\Delta t)\} \rightarrow X_{km} = R_k(m\Delta f) + \chi_k(m\Delta f) \quad k = 1,2,...,K$$

$$m = 0,1,2,...,N/2-1$$

and $\Delta f = f_s/N$ (recall the end of Section 5.5).

If the full scale voltage of the A/D converter is V_{FS} and the number of bits is N_B, not including sign, then at each $m\Delta f$,

$$V_{okm} = \frac{V_{FS} X_{km}}{2^{N_B}\Delta f} \qquad m = 0,1,2,...,N/2-1 \qquad (5.22)$$

where the X_{km} have been divided by Δf to obtain the correct units, amplitude density (recall Section 2.4.1). Therefore, from Eqs.(5.21) and (5.22),

$$q_{ok}(m\Delta f) = \frac{B_k X_{km}}{G_k(m\Delta f)} \qquad (5.23)$$

where

$$B_k = \frac{V_{FS}}{2^{N_B}\Delta f S_k 10^{A_k/20}} \qquad (5.24)$$

is independent of frequency and

$$G_k(f) = H_{ATk}(f) H_{Fk}(f) H_{Sk}(f) \qquad (5.25)$$

is a function of frequency. Thus, in order to relate the transformed values X_{km} to the voltages representing the physical quantities of interest q_{ok}, the transformed value of each channel at each frequency has to be scaled by B_k and adjusted for the complex frequency response of each component in the instrumentation chain by $G_k(m\Delta f)$.

The determination of B_k and $G_k(f)$ are dependent on the experimental setup. However, a procedure that can be used in some cases is described. The first part of the procedure determines

$$G_{1k}(f) = H_{Fk}(f) H_{Sk}(f). \qquad (5.26)$$

Place band-limited white noise into the inputs of each channel $k = 1, 2, \ldots, K-1$. By-pass the Kth anti-aliasing filter and place the noise directly into the Kth channel of the multiplexer. After taking the Fourier transform of N_T time slices and averaging the results of each channel [recall Eq.(2.112)] one has obtained and stored in the computer the quantity $G'_{1k}(f)$, $k = 1, 2, \ldots, K-1$. To obtain $G_{1k}(f)$ each channel up to the $K-1$ channel is divided by $G'_{1K}(f)$ to obtain $G_{1k}(f)$. Thus

$$G_{1k}(f) = \frac{G'_{1k}(f)}{G'_{1K}(f)} \qquad k = 1, 2, \ldots, K-1.$$

The Kth channel remains unknown.

To obtain the transfer function $H_{ATk}(f)$ we assume for simplicity that all the transducers measure the same physical quantity. We now subject $K-1$ transducers simultaneously to the same physical stimulus over the frequency range of interest. The Fourier transform of each of the transducer's output is taken, averaged and stored in the computer. These quantities are denoted X'_{km}, $k = 1, 2, \ldots, K-1$. Equations (5.23), (5.25) and (5.26) yield

$$H_{ATk}(m\Delta f) = \frac{B_k X'_{km}}{q_{ok}(m\Delta f) G_{1k}(m\Delta f)}. \qquad (5.27)$$

If we now assign transducer 1 as the reference transducer, then Eq.(5.27) gives

$$H_{ATk}(m\Delta f) = H_{AT1}(m\Delta f) \frac{B_k X'_{km} G_{11}(m\Delta f)}{B_1 X'_{1m} G_{1k}(m\Delta f)}$$

$$k = 2, 3, \ldots, K-1 \qquad m = 0, 1, \ldots, N/2-1 \qquad (5.28)$$

where the ratio $q_{o1}(m\Delta f)/q_{ok}(m\Delta f) = 1$, $k = 1, 2, \ldots, K-1$ since all the transducers were subjected to the same physical quantity. The B_k contain the transducer sensitivities S_k. In order for Eq.(5.28) to be useful the transducer connected to channel 1 has to be calibrated independently. Also, this procedure assumes that the A/D converter has been calibrated by other means. The application of this procedure is given in Section 9.5.

5.8 Digital-to-Analog Converters

A digital-to-analog (D/A) converter transforms binary patterns of 1's and 0's into discrete analog voltages or currents. The transfer function of a D/A converter consists of a set of discrete points for which each output voltage is a fraction of a reference quantity. The fraction is determined by how the input binary number was coded. Depending on its configuration the output analog output voltage can be either unipolar or bipolar. The resolution and accuracy characteristics of a D/A converter are very similar to what was discussed for A/D converters, and, in fact, Figures 5.1 to 5.3 apply as well to D/A converters.

As shown in Figure 5.19 the D/A converter involves a network of precision resistors, a set of switches, some form of voltage scaling to adapt the switch outputs to the appropriate logic levels, and a (usually) temperature compensated voltage reference. Each switch closure adds a binary-weighted current increment to an output summing device. The output signals from the D/A converter are then smoothed by a low pass filter that has an appropriately chosen cutoff frequency.

An application of an D/A converter is given in Section 8.6.2.

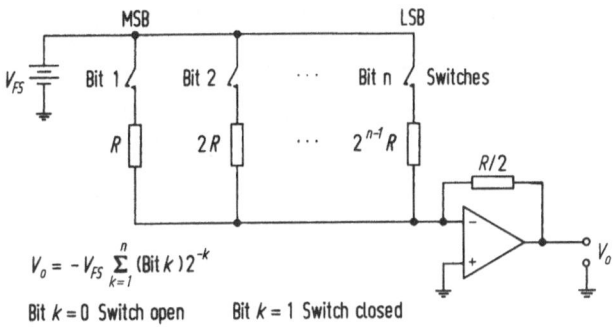

$$V_o = -V_{FS} \sum_{k=1}^{n} (\text{Bit } k) 2^{-k}$$

Bit $k = 0$ Switch open   Bit $k = 1$ Switch closed

Figure 5.19   D/A converter concept

References

1. Clayton, G. B., *Data Converters*, John Wiley and Sons, New York, 1982.
2. Garrett, P. H., *Analog I/O Design: Acquisition, Conversion, Recovery*, Reston Publishing Co., Reston, Virginia, 1981.
3. Higgins, R. J., *Electronics with Digital and Analog Integrated Circuits*, Prentice-Hall, Englewood Cliffs, New Jersey, 1983.
4. Hnatek, R., *A User's Handbook of D/A and A/D Converters*, John Wiley and Sons, New York, 1976.
5. Kinstler, A., "Are All High Resolution DMMs Created Equal?", Test & Measurement World, June, 1987.

6. Lang, T. T., *Electronics of Measuring Systems*, John Wiley and Sons, New York, 1987.

7. Otnes, R. K., and Enochson, L., *Applied Time Series Analysis, Vol 1*, John Wiley and Sons, New York, 1978.

8. Sheingold, D. H., Ed., *Analog-Digital Conversion Handbook*, 3rd Ed., Prentice-Hall, Englewood Cliffs, New Jersey, 1986.

9. Taylor, J., *Computer-Based Data Acquisition Systems*, Instrument Society of America, Research Triangle Park, North Carolina, 1986.

10. Zuch, E. L., "Signal Data Conversion", in *Handbook of Measurement Science, Vol 1*, P. H. Sydenham, Ed., John Wiley and Sons, New York, 1982.

Exercises

1. A bipolar signal is being sampled at 48,570 Hz with a 16-bit, including sign, A/D converter. What is the maximum aperture time of the S/H amplifier.

2. A seven channel simultaneous sample-and-hold A/D conversion system has a conversion time to M bits (including sign) of 2 µs, a multiplexer switching and settling time to within 0.01% of 1.1 µs and a S/H settling (recovery) time to within 0.01% of 0.1 µs. The full scale voltage of the A/D converter is ± 1 0 V. To be consistent with the above system's capabilities what is (a) the maximum value of M, (b) the system's maximum throughput and (c) the longest aperture time of the S/H amplifier.

3. An idealized anti-aliasing filter has a selectivity of 80 dB/octave. For a maximum aliasing error of 0.05%, what is the percentage increase in the sampling rate compared to that of a filter with infinite selectivity.

6 Some General Purpose Instrumentation

6.1 Voltage Detectors

6.1.1 Types of Detection

Voltage detectors are electronic circuits that convert a time-varying input voltage into a slowly varying dc output voltage, which represents some aspect of the input signal. There are three common types of detectors: true rms, rectified average and peak. Each of these shall now be discussed from a mathematical point of view.

Peak Detection
Peak detection of a signal $s(t)$ gives the maximum absolute value of $s(t)$ over a time interval T. Thus

$$s(t)_{peak} = \max[|s(t)|] \qquad 0<t<T \qquad (6.1)$$

where T is effectively the rise time of the circuit. Since the peak value of $s(t)$ can be fleeting with respect to the interval T, some peak detectors have circuitry that holds the peak value for an extended period of time ($\gg T$) so that a proper conversion to a readable quantity can be made.

The peak value of several common waveforms will now be determined.

1. Sine Wave. In this case

$$s(t) = A \sin \omega t \qquad (6.2)$$

and, hence,

$$s(t)_{peak} = A. \qquad (6.3)$$

2. Sine Wave plus Third Harmonic. In this case

$$s(t) = A[\sin \omega t + b \sin 3\omega t] \qquad (6.4)$$

where $|b| \leq 1$. It is seen from Eq.(6.4) that the peak value depends on the magnitude and phase of b. If $b < 0$ then the third harmonic is 180° out of phase with the fundamental. Solving Eqs.(6.1) and (6.4) numerically yields for $b = 0.5$, $s(t)_{peak} = 1.08 A$ whereas for $b = -0.5$, $s(t)_{peak} = 1.50 A$.

3. Square Wave. Recalling Figure 2.1a

$$s(t)_{peak} = A.\qquad(6.5)$$

It should be realized from Eq.(2.12) that for the peak detector to correctly measure this amplitude the circuitry must have a sufficiently broad bandwidth.

4. Gaussian Noise. For this signal the peak value is a continually randomly changing quantity within each interval T. Hence a peak measurement of noise does not yield useful results.

The above examples imply that to correctly interpret a peak value measurement one must know the properties of the waveform *a priori*.

Rectified Average Detection
A rectified average detector measures the following quantity:

$$\{|s(t)|\}_T = \frac{1}{T}\int_0^T |s(t)|\,dt.\qquad(6.6)$$

The rectified average value of several common waveforms will now be determined.

1. Sine Wave. Substituting Eq.(6.2) into Eq.(6.6) gives

$$\{|s(t)|\}_T = \frac{2A}{T}\int_0^{T/2}\sin(2\pi t/T)\,dt = \frac{2}{\pi}A = 0.637\,A.\qquad(6.7)$$

2. Sine Wave plus Third Harmonic. Substituting Eq.(6.4) into Eq.(6.6) yields

$$\{|s(t)|\}_T = \frac{A}{T}\int_0^T |\sin\omega t + b\sin 3\omega t|\,dt.\qquad(6.8)$$

It is seen that the rectified average value is dependent on the magnitude and phase of b. If $b = 0.5$ then a numerical evaluation of Eq.(6.8) gives $\{|s(t)|\}_T = 0.743\,A$ and for $b = -0.5$, $\{|s(t)|\}_T = 1.315\,A$.

3. Square Wave. From Figure 2.1a

$$\{|s(t)|\}_T = \frac{2}{T}\int_0^{T/2} A\,dt = A.\qquad(6.9)$$

Thus the rectified average value of a square wave is the same as that obtained by a peak detector.

4. Gaussian Noise. If the noise has a zero mean and a standard deviation σ_o, then it can be shown that (Oliver and Cage [1971])

$$\{|s(t)|\}_T = 0.798\sigma_o.\qquad(6.10)$$

True RMS Detection

The rms value is defined in Eq.(2.10) as

$$\{s(t)\}_{rms} = \sqrt{\frac{1}{T} \int_0^T s^2(t)dt} \,. \tag{6.11}$$

From the results of Section 2.3.1 it should be realized that the rms value has the strongest theoretical basis and consequently in many situations is the most meaningful of the three types of detection. In discussing rms detectors a quantity called the crest factor is defined to describe one measure of its capability. The crest factor is the ratio of the peak value of the signal to its rms value. Thus

$$F_c = \frac{s(t)_{peak}}{\{s(t)\}_{rms}} \,. \tag{6.12}$$

There are two fundamental methods used to obtain the rms value: one is to electronically square the signal, the other is to convert the signal to heat. The heat conversion method has the advantage of handling high crest factors, but the disadvantage of having a fixed averaging time because of the thermal characteristics of the method. Some rms detectors can measure the ac plus the dc component of the signal, others only the ac portion. Which should be used depends on the type of signal; for example, a square wave has no dc component (recall Eq.(2.12)) whereas a pulse train does (recall Eq.(2.17)).

The rms value of several common waveforms will now be determined.

1. Sine Wave. In this case Eqs.(6.11) and (6.2) yield

$$\{s(t)\}_{rms} = \sqrt{\frac{A^2}{T} \int_0^T \sin^2(2\pi t/T)dt} = \frac{A}{\sqrt{2}} = 0.707\,A\,. \tag{6.13}$$

which has already been obtained in Eq.(2.38). Using Eqs.(6.3) and (6.13) in Eq.(6.12) it is found that the crest factor of a sine wave is $\sqrt{2}$ (3 dB).

2. Sine Wave plus Third Harmonic. Using Eqs.(6.4) and (6.11) yields

$$\{s(t)\}_{rms} = \left[\frac{A^2}{T} \int_0^T (\sin\omega t + b\sin 3\omega t)^2 dt\right]^{1/2} = \frac{A}{\sqrt{2}}\sqrt{1+b^2} \tag{6.14}$$

which could also have been obtained from Eqs.(2.9b) and (2.10). It is seen that the rms value is independent of the phase of the third harmonic (or any other harmonic).

3. Square Wave. In this case Eq.(6.11) gives

$$\{s(t)\}_{rms} = \left[\frac{A^2}{T} \int_0^{T/2} dt + \frac{A^2}{T} \int_{T/2}^T dt\right]^{1/2} = A \tag{6.15}$$

which agrees with Eq.(2.15). It should be realized that the actual circuit that measures this value requires a sufficiently broad bandwidth. Recall Example 1 of Chapter 2 where it was found that

$$\{s(t)\}_{rms} = \sqrt{\frac{8A^2}{\pi^2} \sum_{n=1,3,5}^{\infty} \frac{1}{n^2}}.$$

If this result is numerically evaluated as a function of harmonic number n it is found that the bandwidth $B_n = nf_o$ required for an error of less than 1.0% is $n=21$; for 0.5%, $n=41$; and for 0.1%, $n=201$. Thus to correctly measure a 1 kHz square wave to 3 decimal places (0.1%) the rms detector must have a bandwidth of at least 201 kHz.

4. Gaussian Noise. If again the noise has zero mean and a standard deviation σ_o, then it can be shown that (Oliver and Cage [1971])

$$\{s(t)\}_{rms} = \sigma_o \qquad (6.16)$$

which is the definition of σ_o.

The results of these examples for the peak, rectified average and rms detectors are summarized in Table 6.1. It is common practice for some commercial detectors to correctly display the measurement of the voltage of a single frequency sine wave in terms of its rms value irrespective of the actual type of detector used. For those instruments that use rms detection the instrument is labelled a *true* rms device. Therefore Table 6.1 depicts what an rms, a rectified average and a peak detecting instrument would read when calibrated to indicate the correct rms value of a sine wave. Examination of Table 6.1 suggests a way one can determine the type of detector being used. Simply put a square wave of known amplitude into the detector. If the amplitude of the square wave is displayed by the device it is a true

Table 6.1 Comparison of rms, rectified average and peak detection

Waveform	True rms instrument indicates	Rectified average instrument calibrated in rms of a sine wave indicates	Peak instrument calibrated in rms of a sine wave indicates
Sine wave	$0.707A_o$	$0.707A_o$	$0.707A_o$
Sine wave plus third harmonic			
$b=0.5$ (in phase)	$0.79A_o$	$0.824A_o$	$0.761A_o$
$b=-0.5$ (out of phase)	$0.79A_o$	$1.461A_o$	$1.06A_o$
Square wave	A_o	$1.11A_o$	$0.707A_o$
Gaussian noise	σ_o	$0.886\sigma_o$	—

rms detector, if it reads 11% too high it is a rectified average detector, and if it reads 30% too low it is a peak detector.

6.1.2 Digital Voltmeter

A digital voltmeter (DVM) converts a slowly varying dc voltage of unknown magnitude into an observable digital number. Methods for performing the conversion have been discussed in Sections 5.4.1 to 5.4.3. In this section we shall investigate the major sources of error in DVMs.

Resolution
Digital voltmeters specify their resolution by an integer number of digits plus a fraction, usually ½; e.g., 3½, 4½, etc. The integer number specifies the resolution of the DVM: 3 corresponds to 1 part in 1000 or 0.1% of full scale range, 4 to 0.01%, and so on. The fraction indicates that a most significant digit of either a 0, 1 or 2 (and sometimes a 3) can be displayed. This provides the DVM the ability to display an overrange. To eliminate the ambiguity of what the fractional value means the integer plus ½ designation is qualified one of two equivalent ways: either by stating explicitly what the maximum reading of the DVM is on each range or with an additional integer known as the number of counts (or the scale length). For example, on a 1 V full scale range a 4½ digit meter could display from 0.0001 to 1.9999 with a resolution of 0.1 mV. In this particular example the DVM provides a 100% percent overrange. On each scale range, therefore, the maximum value that can be read is the scale range multiplied by 1.9999. On the other hand one could state that this DVM has a number of counts equal to 19999. The number of count designation is frequently used as part of the accuracy statement as discussed next.

Accuracy
The accuracy of a DVM is specified in one of two ways:

$$\epsilon_m = \pm \left(\epsilon_{reading} + \epsilon_{FS} \frac{V_{FS}}{V_R} \right) \quad \% \qquad (6.17a)$$

or

$$\epsilon_m = \pm \left(\epsilon_{reading} + 100 \frac{N_C}{N_{TC}} \frac{V_{FS}}{V_R} \right) \quad \% \qquad (6.17b)$$

where ϵ_m is the percentage total measurement uncertainty, $\epsilon_{reading}$ is the percentage uncertainty of the reading, ϵ_{FS} is the percentage uncertainty of full scale, N_C is the number of counts in error, N_{TC} is the total number of counts for the DVM, V_{FS} is the full scale range and V_R is the voltage reading. Both of these statements are usually accompanied by a temperature specification and a time period specification effective from the date of the DVM's last calibration. Equations (6.17) clearly show that the error is least when readings are taken as close to full scale as possible.

It should now be apparent why DVMs have such broad overranging capabilities. It permits their more frequent use near full scale.

Many DVMs have autoranging features, which automatically switch the voltmeter to a more sensitive (lower) full scale range when the input voltage drops below approximately quarter scale or to a higher full scale range when the overrange has been exceeded. However, autoranging takes additional time for the DVM to implement. Therefore if the DVM is being read with a computer, the computer program must allow extra time for this range changing possibility, otherwise an incorrect reading could be taken (see Table 7.2).

Sampling Speed
Some DVMs have the capability of user-selected resolution, typically ranging from 3½ digits to as high as 8½ digits. The main purpose for this variation is for those situations in which fast sampling speed is a consideration, such as in a multi-channel data acquisition system. The sampling speed is a nonlinear function of the number of digits. For example, one manufacturer specifies that 100 samples/s can be taken at 3½ digits, 10 samples/s at 5½ digits, but only one sample every 3.2 s for 7½ digits.

Sensitivity
The sensitivity is the smallest change the DVM can detect. Consequently the sensitivity is the resolution of the DVM on its most sensitive measurement range.

Impedance
The impedance of DVMs are typically greater than $1 \ G\Omega (10^9 \ ohms)$ for the dc voltage measurement, with some as high as 100 GΩ. These high values are usually for voltage ranges up to 10 V. Above 10 V the impedances are typically 100 times less. For measurement of ac voltages the impedances are typically $1 \ M\Omega (10^6$ ohms). To determine the significance of the input impedance of the DVM recall Figure 4.11. If e_i is the DVM's reading of the actual voltage e_g, then for the reading to be consistent with the resolution r_e of the DVM

$$\left| \frac{e_g}{e_i} - 1 \right| < r_e . \tag{6.18}$$

Using Eqs.(4.35) and (4.39), Eq.(6.18) can be written as

$$\frac{R_g}{R_i} < r_e$$

where R_i is the input impedance of the DVM and R_g is the impedance of the device producing e_g. If the DVM has N decimal digit resolution, then

$$R_g < R_i 10^{-N} . \tag{6.19}$$

Thus, if the input impedance of the DVM is 10 GΩ and the DVM is 6½ digits, then the source impedance R_g must be less than 10 kΩ $(= 10^{10} \times 10^{-6}$ ohms).

Example 1: Comparison of Two DVMs. Consider two DVMs with the following specifications at 20 °C.

<table>
<tr><td align="center">DVM-1
(4½ digits)</td><td align="center">DVM-2
(6½ digits)</td></tr>
<tr><td align="center">$\epsilon_{reading} = 0.2\%$</td><td align="center">$\epsilon_{reading} = 0.08\%$</td></tr>
<tr><td align="center">$\epsilon_{FS} = 0.1\%$</td><td align="center">$N_c = 50$</td></tr>
<tr><td align="center">$N_{TC} = 19999$</td><td align="center">$N_{TC} = 1999999$</td></tr>
<tr><td align="center">$V_{FS} = 10$ mV (lowest range)</td><td align="center">$V_{FS} = 1$ V (lowest range)</td></tr>
</table>

Both DVMs have the same resolution and sensitivity, $1\,\mu V$. Using Eqs.(6.17) the uncertainty of each DVM for a reading of 1 mV and 10 mV are, respectively:

<u>DVM-1</u>

$V_R = 1$ mV: $\epsilon_m = \pm(.2 + (.1)(10/1)) = \pm 1.2\%$

$V_R = 10$ mV: $\epsilon_m = \pm(.2 + (.1)(10/10)) = \pm 0.3\%$

<u>DVM-2</u>

$V_R = 1$ mV: $\epsilon_m = \pm(0.08 + 100(50/1999999)(1000/1)) = \pm 2.58\%$

$V_R = 10$ mV: $\epsilon_m = \pm(0.08 + 100(50/1999999)(1000/10)) = \pm 0.33\%$

For these DVM characteristics it is seen that DVM-1 has less uncertainty for both readings, even though it has two fewer digits resolution than DVM-2. However, when $V_R = 1$ V it is found that DVM-2 is better, 0.082% *vs* 0.3%.

6.1.3 Lock-in Amplifier (Phase Sensitive Detection)

A lock-in amplifier is a phase sensitive ac voltmeter that compares the input signal to a reference signal to produce a dc output voltage proportional to that part of the input signal with the same frequency and phase as the reference signal. A lock-in amplifier is used to measure low level periodic signals in the presence of noise. In its simplest form a lock-in amplifier consists of three parts: a signal conditioning channel, a reference channel and a phase sensitive detector. These are shown in Figure 6.1.

The signal conditioning channel is comprised of an input amplifier with a low noise figure and a filter to remove some of the noise from the input signal, which decreases the possibility of overloading the mixer in the phase sensitive detector that follows. The input to the reference channel is a periodic waveform whose fundamental frequency is the frequency of interest in the input to the signal channel. The reference channel electronics converts the reference signal to a square wave whose phase can be adjusted (shifted) between 0° and 360°. The

144

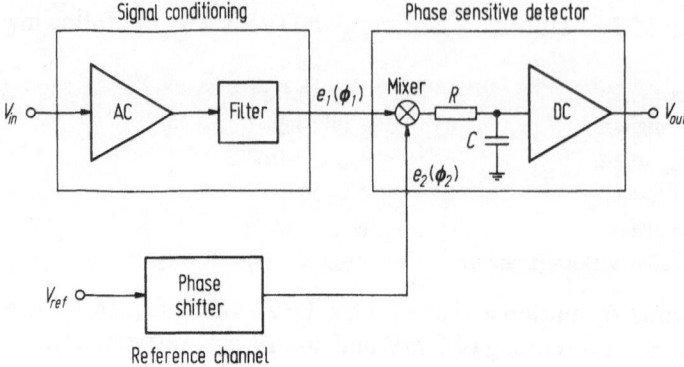

Figure 6.1 Basic components of a lock-in amplifier

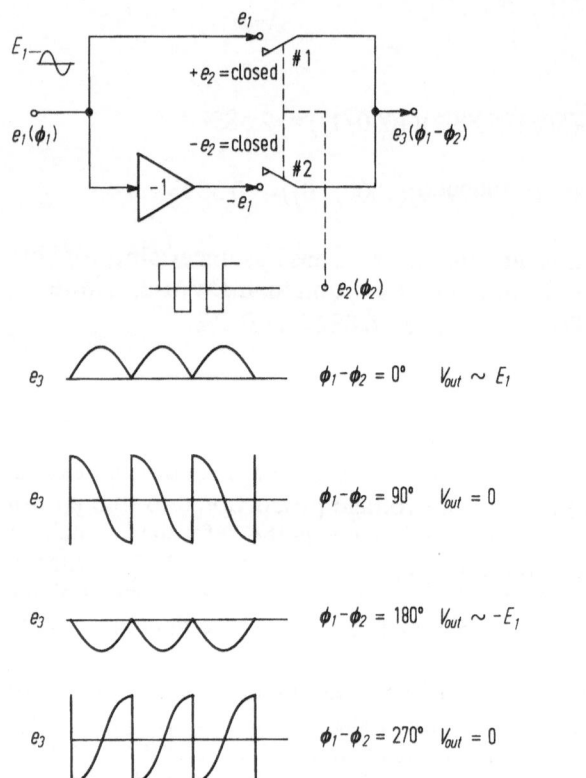

Figure 6.2 Mixer output as a function of the phase angle between the input signal $e_1(t)$ and the reference signal $e_2(t)$

phase controlled square wave is "mixed" with the filtered and amplified input signal. The operation of the mixer is shown in Figure 6.2 for a noise-free sine wave e_1 with a phase ϕ_1. The input signal is split into two paths. In the first path e_1 goes to switch #1 and in the second path e_1 is inverted so that $-e_1$ goes to switch #2. The reference channel signal is the square wave e_2 with phase ϕ_2. The signal e_2 closes switch #1 only when its amplitude is positive and closes switch #2 only when its amplitude is negative. The output of the mixer is e_3. Shown in Figure 6.2 are the waveforms of e_3 when $\phi_1 - \phi_2 = 0°$, $90°$, $180°$ and $270°$. It is seen that if ϕ_2 is adjusted so that $\phi_1 - \phi_2 = 0°$ then the mixer's output will be maximum. The signal e_3 is then filtered, amplified, scaled and converted to a digital number. For computer controlled experiments there are lock-in amplifiers that, under their own microprocessor control, automatically adjust ϕ_2 until a maximum value for the filtered output signal is obtained.

It should be noted that the original input periodic signal plus noise has been converted such that the amplitude of its synchronous (with respect to the reference signal) portion ends up at the output of the final filter. The noise that survives the mixer and the filtering can be greatly reduced by selecting the bandwidth of this last filter to be very narrow; that is, by selecting its RC time constant to be very large (recall Eq.(3.28)). The RC time constants typically vary from 1 ms to 10 s, with some units going as high as 300 s.

An application using a lock-in amplifier is described in Section 9.3.

6.2 Digital Counters: Time and Frequency Measurements

The digital electronic counter has become one of the most widely used instruments to measure time, frequency and phase of waveforms and events. A typical counter, illustrated in Figure 6.3, consists of several functional units: a time base, a gate, an input signal conditioner for each input, a decimal counting unit (DCU) and display and a control unit. Several of the components in this figure have not been interconnected, since their connection governs the type of measurement being made. The specific connections will be shown subsequently. The time base consists of a crystal controlled oscillator at frequency f_c and N decade (factor of 10) dividers, which generate N frequencies $f_n = f_c 10^{-n}$, $n = 1, 2, ..., N$. The gate is a switch that is opened and closed for a specific period of time by the control unit. The control unit also resets the counters and transfers the DCU results for display. The DCU counts the number of pulses it receives in a given time interval and converts it to a decimal number. The maximum number of digits displayed is N. The input signal conditioner converts the input signal into a pulse train having the proper characteristics for the electronic circuits that follow. This pulse shaping is sometimes determined by a user selected voltage level, called the trigger level, which can be either positive or negative. The trigger device has a voltage

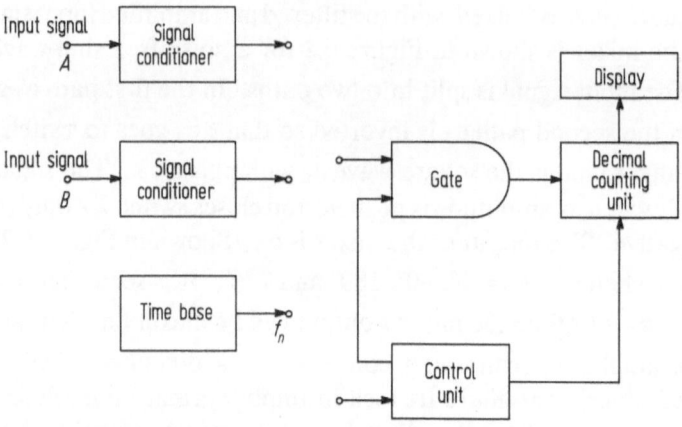

Figure 6.3 Major components of a digital counter

level V_L at which it goes from a low output to a high output voltage and another level V_H at which it goes from the high level to a low level. The difference $V_H - V_L$ is called the hysteresis and is shown in Figure 6.4. The choice of V_H and V_L determines the threshold of the counter, for the input signal must exceed V_H, and its insensitivity to some types of signal noise as shown in Figure 6.5. Depending on the mode of operation described below the accuracy of the counter is at least a function of the time base accuracy, the signal-to-noise ratio of the input signal and the ± 1 count uncertainty (Sanderson [1987]).

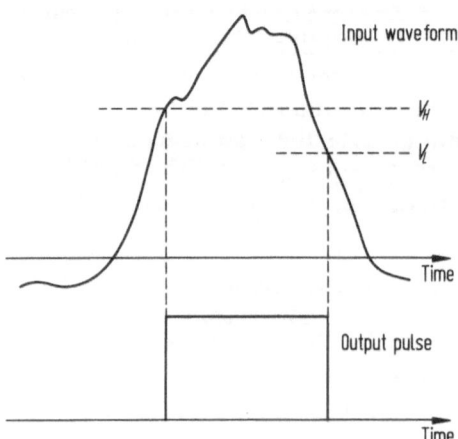

Figure 6.4 Hysteresis of the input waveform signal conditioner

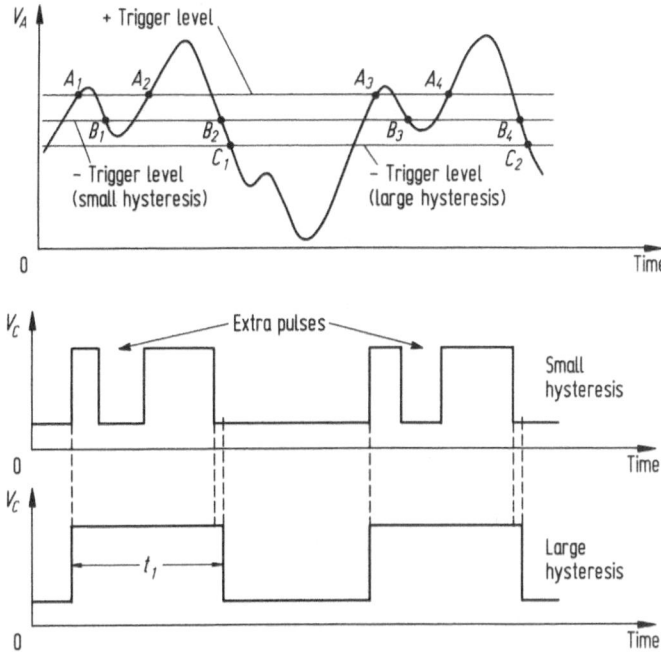

Figure 6.5 The use of hysteresis to reduce the effects of noise

Some of the types of measurements that can be made with a digital electronic counter will now be discussed.

Frequency

The frequency of the input signal is determined by measuring the number of pulses the input signal generates in a *known* period of time. The arrangement to measure frequency is shown in Figure 6.6a. The time base unit is used to create the known time interval. If the input signal has a frequency f_i and the time base frequency is f_b, then the number of pulses counted are $N_c = f_i / f_b$, where $f_i > f_b$. The error in this measurement is the ± 1 count uncertainty in turning the gate on and off. Therefore the frequency measured is $f_i / f_b \pm 1$. Thus in order to obtain the smallest error f_i should be much greater than f_b. There are, of course, instances where $f_i = 10 f_b$. In this case the ± 1 count error is 10%. To overcome this counters have another mode of operation called the period mode.

Period

The period is determined by measuring the number of pulses from the time base unit during one period of the input signal as shown in Figure 6.6b. From the time base unit the time interval between each pulse is $T_b = 1 / f_b$. If N_c is the total pulse count from the counter, then the period of the input signal is $T_i = N_c T_b$ or

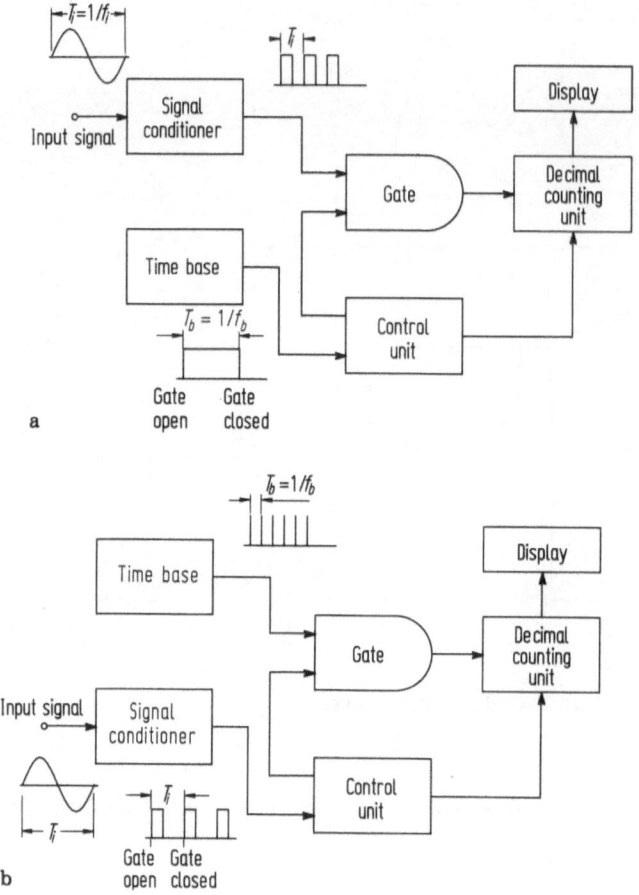

Figure 6.6 (a) Measurement of frequency (b) measurement of period

$f_i = f_b / N_c$. If $f_b / f_i = N_c \gg 1$, then the ± 1 count error is greatly reduced. For a given gate time and time base frequency there is a frequency below which it is more accurate to use the period measurement and above which it is more accurate to use the frequency mode. In some counters this is done automatically, in others the user has to change from one mode to the other.

Time Interval and Phase

Time interval measurement is the same as a period measurement except that two external signals open and close the gate instead of the signal from the time base. This is shown in Figure 6.7. When the input signals are periodic the configuration shown in Figure 6.7 can be used to measure the phase angle between two periodic signals [recall Eq.(3.10)]. To calibrate the counter for the time interval and phase measurements the same signal is fed into both inputs simultaneously and the trigger levels adjusted to obtain a display of zero.

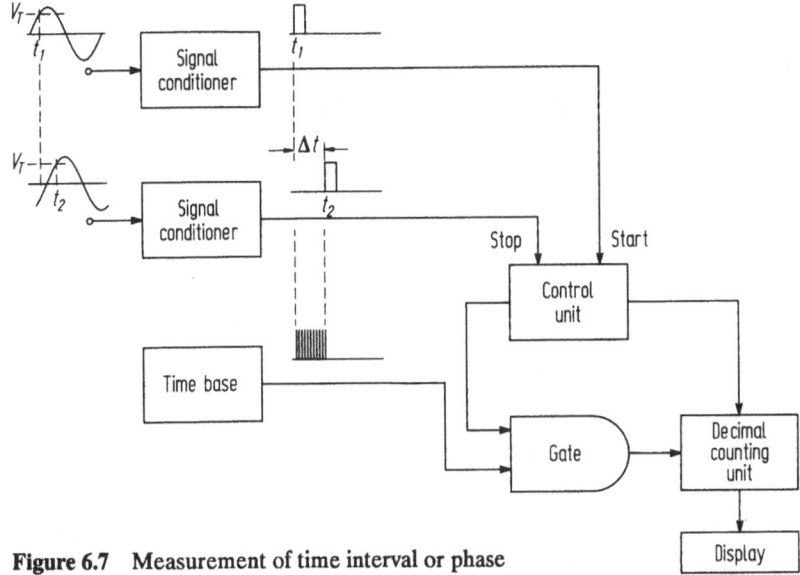

Figure 6.7 Measurement of time interval or phase

6.3 Power Supply

In common usage the term power supply means a source of constant (dc) voltage
or current. As illustrated in Figure 6.8 a power supply is comprised of several
distinct functional units. The full wave rectifier converts the line ac voltage into
positive-going half-waves, whose maximum amplitude is close to the maximum dc

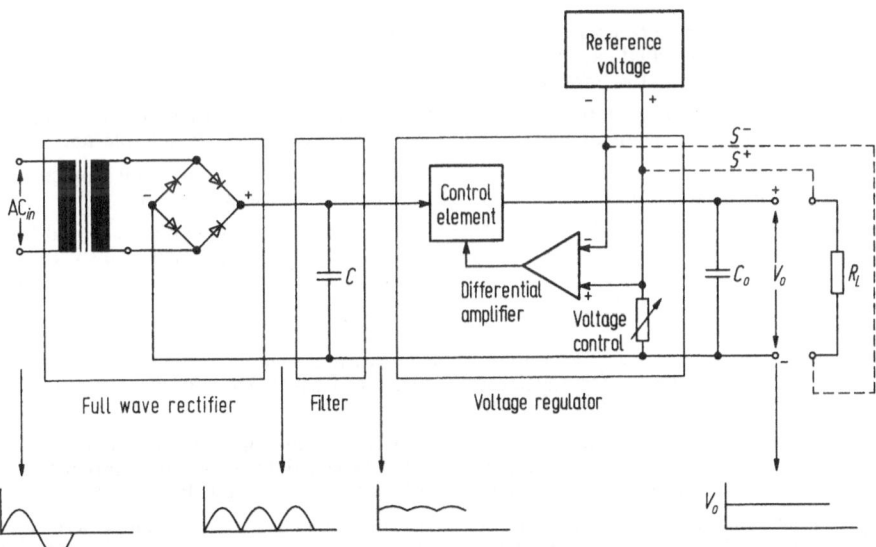

Figure 6.8 Basic features of a voltage regulated power supply

voltage rating of the power supply. This full wave rectified signal is filtered to remove much of the alternating portion of the rectified signal. What is not removed is termed ripple, which is defined, along with other terms, in Table 6.2. To further reduce the ripple a (voltage) regulator follows the filter section.

Table 6.2   Power supply terminology

Constant Current (CC) Power Supply. A power supply that maintains its output current within prescribed limits irrespective of changes in load, line voltage, ambient temperature, etc.

Constant Voltage (CV) Power Supply. A power supply that maintains its output voltage within prescribed limits irrespective of changes in load, line voltage, ambient temperature, etc.

Current Limiting. The action of limiting the output current to some maximum predetermined value and automatically restoring the power supply to normal operation when the overload condition is removed.

Drift. The maximum change of the output voltage or current during a specified period of time following the warm-up time with all other externally influencing parameters held constant. Drift is the low frequency (to dc) component of noise and is related to the internal temperature rise of the power supply.

Load Effect. The change in the steady-state value of the stabilized output voltage or current resulting from a full range producing change in the load resistance, with all other parameters constant.

Line Regulation. See Load Effect.

Output Impedance. The output impedance is the ratio of the output voltage to the change in the output current causing that change. It can be approximated as a resistance in series with an inductance.

Overvoltage Protection. A load protection circuit designed to prevent excessive output voltage. It rapidly (in microseconds) places a low resistance shunt (crowbar) across the output terminals of the power supply whenever a preset limit is exceeded. Such a power supply must also have current limiting.

PARD-Periodic and Random Deviation. Deviation in the regulated output voltage or current from its average value over a specified bandwidth with all influence and control quantities held constant. This term frequently replaces the terms ripple and noise.

Programming Speed. The maximum time required for the programmed output voltage or current to abruptly change from a specified initial value to another value within a specified tolerance.

Source Effect. The change in the steady-state value of the stabilized output voltage or current resulting from a change in the ac source voltage, with all other parameters constant.

Remote Sensing. A power supply feature in which which the load regulation point is shifted from the power supply's output terminals to an external location, usually the load itself. This is usually accomplished with an extra set of wires connected from two other power supply output terminals to the load. Since very little current flows in these extra wires, errors due to to voltage drop in the current-carrying wires are avoided.

Ripple. The deviation in the controlled output voltage or current that are harmonically related to the input line frequency and any internally generated switching frequencies. Usually stated as a rms or maximum peak-to-peak amplitude. See PARD.

Temperature Coefficient. The maximum change in the stabilized ouptut voltage or current due to a one degree change in the ambient operating temperature, provided the new temperature in within the power supply's operating temperature range. All other parameters are held constant.

Transient Recovery Time. The time interval between the beginning of an abrupt change in the load voltage or current until the stabilized output voltage or current returns to within a specified percentage of its previous output.

The regulator monitors the output voltage V_o at the power supply's output terminals. It is not uncommon for the power supply regulation to be 0.01% with some as low as 0.001%. However, when the load undergoes sudden changes of ΔA_o amperes, there is an additional voltage drop ΔV_w due to the resistance R_w of the connecting wires of $\Delta V_w = \Delta A_o R_w$. The values of R_w can be determined from Table 6.3. For example, consider a 40 V output voltage of a power supply with 0.005% regulation undergoing changes of 4 A. If the load is connected with 10 ft. of 20 AWG wire, then the total resistance of the wire is $R_w = (0.01016)(10) = 0.1016$ ohms. The voltage error at the load, therefore, is $\Delta V = (0.1016)(4) = 406$ mV, as opposed to 0.2 mV (0.00005x40) at the power supply's output terminals.

Table 6.3 Resistivity of wire as a function of size

Wire Size (AWG)	Resistivity (ohm/1000 ft.)
22	16.15
20	10.16
18	6.39
16	4.018
14	2.526
12	1.589
10	0.9994
8	0.6285
6	0.3953
4	0.2486
2	0.1564
0	0.09832

To compensate for this voltage drop most power supplies have remote sense lines. When an additional pair of wires, shown with dotted lines in Figure 6.8, are connected directly to the load, the voltage regulator corrects for the voltage sensed at the load. Since these wires are not carrying the current to the load they are not experiencing the voltage drop ΔV_w. The current in the sense wires are usually on the order of 10 mA. It should be realized that a power supply can only provide a maximum output voltage V_{max}. Therefore, if $V_o = V_{max}$ and V_{max} is required at the load and the sense lines find the voltage less than V_{max} at the load, then the power supply will not be able to provide V_{max} at the load. Instead it will remain $V_{max} - \Delta V_w$.

The sense lines are susceptible to electromagnetic pickup and it therefore recommended that shielded twisted cables be used for the sense lines. The load-end of the shield should be left unterminated and the power supply end connected to the ground terminal. The sense wires do have a side effect, they increase the output impedance of the power supply at medium and high frequencies. When this causes difficulty the capacitor C_o in Figure 6.8 is removed and connected instead across the load.

There is a family of power supplies designed to operate as either a constant voltage (CV) or a constant current (CC) source. See Table 6.2 for their definitions. When operating as a CV supply the output current changes as the load resistance changes, until the supply's maximum output current is reached. The maximum can be either a preset value or the supply's maximum output voltage. When operating as a CC supply the output voltage changes until the supply's maximum voltage is reached. Since these two modes cannot operate simultaneously, for any given load the power supply must operate as either a CC or CV supply. Consider Figure 6.9 which shows the operating curve of a CV-CC power supply. With no load attached $R_L = \infty$, $I_{out} = 0$ and the output voltage equals the predetermined voltage V_{set}. When R_L becomes finite I_{out} increases while V_{set} remains constant. Point A in Figure 6.9 represents a typical CV operating point. Further decreases in R_L increases I_{out}, with V_{set} still remaining constant, until I_{set} is reached ($R_L = R_c$). At this point the power supply changes its mode of operation to CC. Further decreases in R_L cause the power supply to decrease V_{out} while I_{out} remains at I_{set}. Point B represents a typical CC operating point. When $R_L = 0$, $V_{out} = 0$ but $I_{out} = I_{set}$. Notice that full overload protection is an inherent part of a CV-CC power supply since no combination of operating conditions will cause I_{set} and V_{set} to be exceeded.

Some microprocessor controlled CV-CC power supplies are autoranging, such that within given bounds the power supply can operate slightly beyond the nominal power rating. Consider the operating curve for a power supply shown in Figure 6.10, which has a power rating P_R W. It is seen that $0 \le V_{out} \le V_{max}$ and $0 \le I_{out} \le I_{max}$; however, $P_R \le V_{max} I_{max}$. If $I_{set} \le P_R / V_{max}$ and $V_{set} \le P_R / I_{max}$,

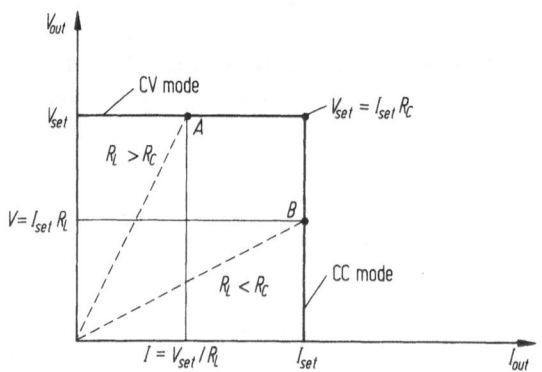

Figure 6.9 Typical operating curve of a CC/CV power supply

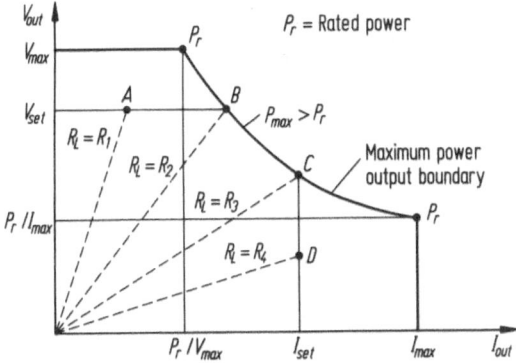

Figure 6.10 Typical operating curve of an autoranging CC/CV power supply

then the power supply acts as shown in Figure 6.9. However, when $I_{out} = I_{set}$ the power supply can deliver constant current for resistive load $R_L = R_4$ (point D) up until $R_L = R_3$ (point C). If R_L were to decrease beyond R_3 the power supply would be beyond its operating range where its performance would be seriously degraded. These power supplies indicate, usually by front panel indicators or error messages on their communications bus, that it has exceeded its rating. For a CV supply of $V_{out} = V_{set}$ the power supply can maintain constant voltage for $R_L = R_1$ (point A) until $R_L = R_2$ (point B). If R_L decreases beyond R_2 the power supply is forced to operate outside of its range and again degraded performance occurs.

6.4 The Wheatstone Bridge

The Wheatstone bridge shown in Figure 6.11 is a very useful circuit for converting a change in resistance to an output voltage, and is almost always used with strain gages (Sections 8.5.1 and 8.5.2) and frequently with resistance temperature detectors (Section 8.7.4). The output voltage from this bridge is

$$e_o = \frac{V_o(R_1 R_3 - R_2 R_4)}{(R_1 + R_2)(R_3 + R_4)(1 + R_e/R_m)} \tag{6.20}$$

where R_m is the input resistance of the device connected to terminals BD,

$$R_e = \frac{R_3 R_4}{R_3 + R_4} + \frac{R_1 R_2}{R_1 + R_2} \tag{6.21}$$

and it has been assumed that the resistance associated with the voltage source V_o is much less than R_m. In many strain gage applications $R_1 = R_2 = R_3 = R_4 = R$ and

154

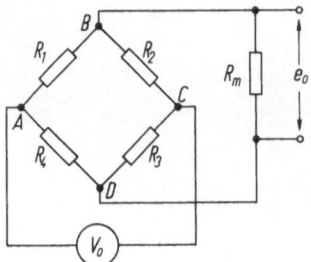

Figure 6.11 Wheatstone bridge

therefore, from Eq.(6.21) $R_o = R$. If we make $R/R_m < 0.001$ [recall Eq.(4.39)], we can ignore this term and Eq.(6.20) simplifies to

$$e_o = \frac{V_o(R_1 R_3 - R_2 R_4)}{(R_1 + R_2)(R_3 + R_4)}. \tag{6.22}$$

The bridge is balanced when

$$\frac{R_2}{R_1} = \frac{R_3}{R_4}. \tag{6.23}$$

It becomes unbalanced and produces an output voltage when each R_j undergoes a change $R_j + \Delta R_j$, $j = 1, 2, 3, 4$. Then Eq.(6.22) becomes

$$\Delta e_o = \frac{V_o \Delta N}{\Delta D} \tag{6.24}$$

where

$$\Delta N = (R_1 + \Delta R_1)(R_3 + \Delta R_3) - (R_2 + \Delta R_2)(R_4 + \Delta R_4)$$

$$\Delta D = (R_1 + \Delta R_1 + R_4 + \Delta R_4)(R_2 + \Delta R_2 + R_3 + \Delta R_3).$$

Equation (6.24) can be simplified, if it is assumed that each $\Delta R_j / R_j \ll 1$ and that Eq.(6.23) holds, to

$$\Delta e_o = \frac{r(1 - \eta)}{(1 + r)^2} \left(\frac{\Delta R_1}{R_1} - \frac{\Delta R_2}{R_2} + \frac{\Delta R_3}{R_3} - \frac{\Delta R_4}{R_4} \right) V_o \tag{6.25}$$

where $r = R_2 / R_1$

$$\eta = \frac{x}{1 + r + x} \tag{6.26}$$

and

$$x = \frac{\Delta R_1}{R_1} + \frac{\Delta R_4}{R_4} + r \left(\frac{\Delta R_2}{R_2} + \frac{\Delta R_3}{R_3} \right).$$

The quantity η gives a measure of the nonlinearity of the result. As mentioned previously a common situation is for $R_1 = R_2 = R_3 = R_4 = R$. Then Eqs.(6.25) and (6.26) become, respectively,

$$\Delta e_o = \frac{V_o}{4R}(\Delta R_1 - \Delta R_2 + \Delta R_3 - \Delta R_4)(1 - \eta) \qquad (6.27)$$

and

$$\eta = \frac{\sum_{j=1}^{4} \Delta R_j / R}{2 + \sum_{j=1}^{4} \Delta R_j / R}. \qquad (6.28)$$

If $\Delta R_j / R < 0.01$, then $\eta < 0.02$ and Eq.(6.27) is within 2%. In many applications η can be ignored.

The output voltage Δe_o is usually fed into an amplifier for amplification and to obtain a high input impedance to eliminate the potential loading effects on the bridge ($R_o / R_m \ll 1$). Two configurations can be used. Referring to Figure 6.11, if a single input amplifier is used terminal B is the signal line and terminal D the ground. If a differential amplifier is used terminals B and D are fed to its positive and negative inputs, respectively. However, as discussed in Section 4.4.2 the CMRR for the differential configuration must be large, for even when $\Delta e_o = 0$ and the bridge is balanced the common mode voltage is approximately $V_o / 2$.

For detailed discussions of many other uses of the Wheatstone bridge see Harris [1952].

References

1. Diefenderfer, A. J., *Principles of Electronic Instrumentation*, 2nd Ed., W. B. Saunders Co., Philadelphia, Pennsylvania, 1978.

2. Frenzel, L. E., *Digital Counter Handbook*, Howard W. Sams and Company., Indianapolis, Indiana, 1981.

3. Harris, F. K., *Electrical Measurements*, John Wiley and Sons, New York, 1952.

4. Higgins, R. J., *Electronics with Digital and Analog Integrated Circuits*, Prentice-Hall, Englewood Cliffs, New Jersey, 1983.

5. Magrab, E. B., and Blomquist, D. S., *The Measurement of Time-Varying Phenomena: Fundamentals and Applications*, John Wiley and Sons, New York, 1971.

6. Munroe, D. M., "Signal-to-Noise Ratio Improvement", in *Handbook of Measurement Science, Vol 1*, P. H. Sydenham, Ed., John Wiley and Sons, New York, 1982.

7. Oliver, B. M., and Cage, J. M., Eds., *Electronic Measurements and Instrumentation*. McGraw-Hill Book Co., New York, 1971.

156

8. Sanderson, M. L., "Electrical Measurements", in *Jone's Instrument Technology, Vol 3, Electrical and Radiation Measurements*, 4th Ed., B. E. Noltingk, Ed., Butterworths, London, 1987.

9. Wobschall, D., *Circuit Design for Electronic Instrumentation,* 2nd Ed., McGraw-Hill, New York, 1987.

10. "A Lock-in Primer", EG&G Princeton Applied Research, Princeton, New Jersey, 1986.

Exercises

1. Determine the peak, rectified average and true rms value of the triangular waveform shown in Figure 2.1b.

2. A DVM is to have a total measurement uncertainty of 0.35%. If the percentage uncertainty of its reading is 0.1% and that at full scale 0.05%, what is the lowest reading that one can make with respect to the full scale range.

3. A regulated power supply can provide up to 500 mV compensation with its remote sensing lines. If the range of current changes in the load is 6 A, will this power supply be able to provide remote compensation through 20 ft of 18 AWG wire. If not, what changes can be made so that the power supply will be able to remotely regulate its voltage to the load.

4. A Wheatstone bridge has its four resistors equal to 120 ohms and its bridge voltage set to 15 V. All the resistors except R_1 remain fixed and the bridge is initially balanced. The bridge's output voltage, caused by its unbalancing due to a ΔR_1, is connected to a differential amplifier that has a CMRR of 110 dB. What is the minimum value of ΔR_1 that can be measured so that the amplifier's output voltage is four times greater than the common mode voltage.

PART II

Systems Integration

7 Computer Integrated Experimentation

7.1 Introduction

Consider a typical computer-integrated instrumentation system shown in Figure 7.1. The process is being monitored by a transducer whose analog output voltage goes to a signal conditioner. The signal conditioner performs any combination of the following functions: amplification, filtering, impedance conversion, lineari-

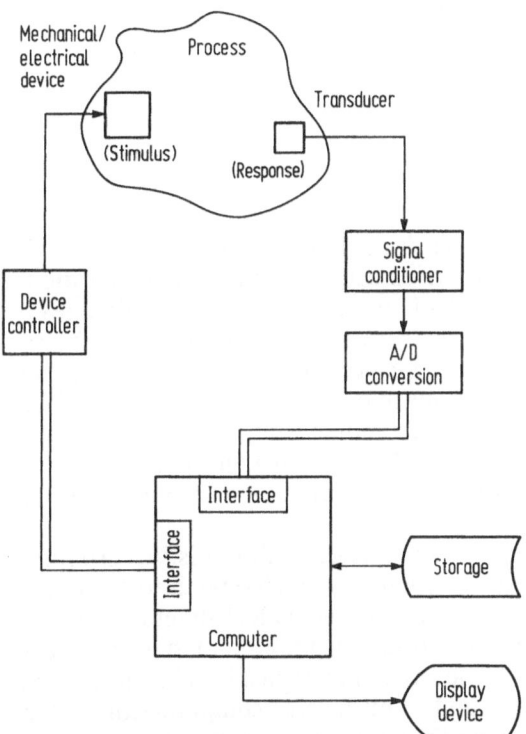

Figure 7.1   Typical components of a computer integrated instrumentation system

zation, rectification (conversion to a dc value), and/or a transformation (demo-dulation, voltage to frequency, etc.). The output from the signal conditioner is then digitized and sent to a computer for analysis and interpretation. Depending on the application the result can be displayed, stored, and/or fedback to the process as either an analog or digital signal. The display device can be the computer screen, a printer, a plotter or a special device such as a digital display, a light or a loud-speaker. The storage of the result can be in the computer's memory, on a magnetic tape or on a disk. On the other hand the type of device the feedback signal can go to is almost without limit. Some examples are step motors, relays, valves, heating elements, frequency generators, etc.

Integrating an experiment or a process with a computer can be a challenging and rewarding task. It does, however, introduce additional facets to the overall procedure that have to be addressed early in the experiment's design. Consider-ations such as instrument compatibility with the computer (e.g., availability of interfaces) and instrument acquisition speed and data transfer rates often are of great importance. As in every instrumentation system the degree of capability and flexibility of the resulting computer integrated configuration is a compromise among the availability of the appropriate instruments and transducers, cost, apparatus design and construction and the allowable time for the completion of the task. In this chapter some of these facets will be discussed in detail. They will be illustrated with examples in the subsequent chapters.

7.2 Equipment Block Diagrams

Clearly articulated objectives are required prior to any experimental design. Two techniques that help insure that the objectives can be met are the generation at the very outset of (a) an equipment block diagram, which clearly delineates the instrument interconnections and implies their order of operations and (b) a pro-gram flow chart, which clearly delineates the programming logic and its rela-tionship to the order in which the instruments and devices are to function. These two preliminary steps lead to a clearer understanding, and provide for a sharper focus, of how the experimental objectives will be met. In addition they often can uncover flaws that may exist in either the experimental design or in the objectives themselves, or both.

A computer integrated experiment generally consists of three distinct groupings: (i) the instruments and transducers, (ii) the computer and its standard peripherals and (iii) the actuated/controlled devices. The equipment block diagram facilitates the examination of the details within these groups, in addition to their interaction with each other and the computer. The attributes typically of concern when specifying and selecting instruments and transducers, computer-controllable devices and computers are listed below. Specific examples of equipment block diagrams are given in Chapters 8 and 9.

I) Instruments and Transducers

1) Functional Requirements

a) sensitivity
b) dynamic range
c) noise
d) bandwidth (response time)
e) accuracy (linearity)
f) reliability
g) environment
  i) temperature
  ii) humidity
  iii) radioactivity
  iv) noise (mechanical/electrical)

2) Digital Requirements

a) interface
  i) type (RS 232, IEEE 488, etc.)
  ii) transmission rates (transmit/receive)
  iii) synchronization/handshaking
  iv) protocol
  v) cables (connectors and lengths)
  vi) interrupts
  vii) electrical
      logic levels and type (TTL, CMOS)
      current requirements and power
b) data sampling/acquisition rates
c) filtering (anti-aliasing)
d) instruments' control languages

3) Calibration

4) Connectivity (cables, connectors, lengths)

a) noise
b) bandwidth
c) signal levels
d) impedance considerations

II) Computer and Peripherals

1) Programming

a) language
b) portability
c) process model
d) operating system
e) data format/structure
f) standardized input/output
g) documentation

2) Interfaces

a) program(s) to run them
b) programming language to access them
c) type (RS 232, IEEE 488, etc.)
d) cables lengths and connectors

3) Speed

a) Input/output
  i) data acquisition and memory storage

ii) device control
iii) disk storage
   b) computation (analysis and decision-making)

4) Distributed Control/Processing and Networking

   a) network languages and protocols
   b) multitasking
      i) interrupt (event) driven
            priority: hardware, software
      ii) polling
      iii) periodically in time

5) Storage Media

   a) type (computer, disk, tape)
   b) compatibility with other systems

6) Display of Results

   a) graphical
   b) tabular
   c) CRT (monochrome, color)

7) Backup Power

8) Data

   a) memory requirements
   b) data structure
   c) data sharing
      i) asynchronous/sequential access

III) Computer Actuated/Controlled Devices

See part I) above.

7.3 Consideration of Timing Requirements

The ease with which a computer can be integrated with various digitally control-lable devices and instruments oftentimes masks the need for careful scrutiny of the timing requirements of each individual component comprising the system. The core of the problem lies in the fact that each physical process, each transducer, each instrument, each actuated device, and the computer itself has its own time scale. This, in general, produces a very asynchronous system whose individual components must be properly coordinated in order for the final result to be meaningful, predictable and repeatable. To attain this coordination requires an understanding of the time scales of each component in the chain.

Consider the following three typical functional lines of program code:

```
200 ...
210 OUTPUT TO CAUSE STIMULUS
220 WAIT T_w
230 READ INSTRUMENT
240 ...
```

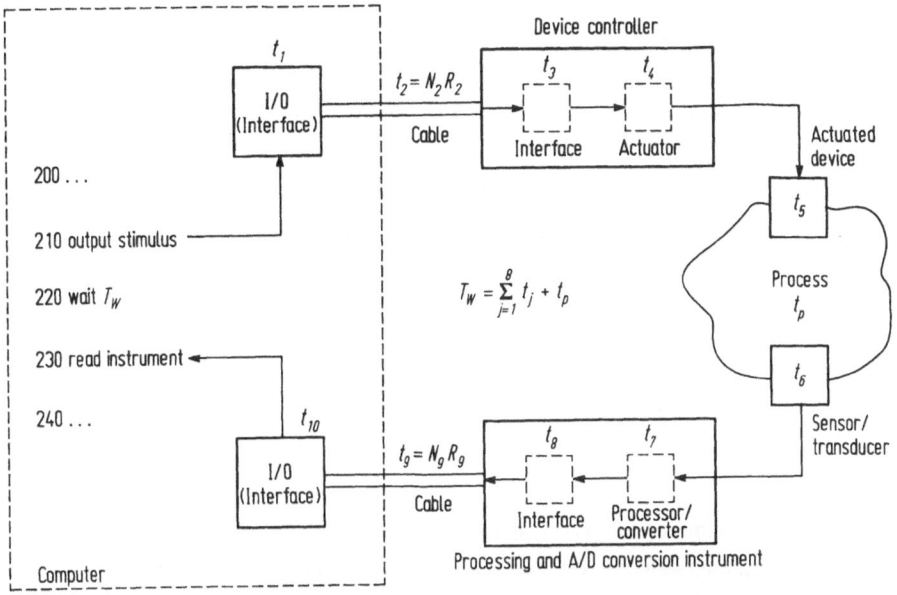

Figure 7.2 Computer program statements and their relationship to the physical devices' timing requirements

These three lines of code are shown in Figure 7.2 in conjunction with typical generic physical components. The definitions of the t_j appearing in this figure are listed in Table 7.1. The minimum wait time T_w is the sum of the times t_1 through t_8 and t_p.

In many applications t_1, t_2, t_3 and t_{10} are often small compared to the other times and can be ignored. The transmission time t_9 of the processed/recorded voltages can be on the order of several seconds when large amounts of data are required. Although this doesn't affect T_w, it does affect the amount of time that will elapse before the next line of code (line 240) can be executed.

The time it takes for the device controller to produce the desired change in voltage or current t_4 and the time it takes for the actuated device to respond to this change t_5 are each proportional to their respective bandwidths. The process time t_p is the time it takes for the process to reach its steady-state value after the actuated device causes the change. This time is not just related to the system's response time (bandwidth), but also to other factors of the process. For example, in a pulse propagation measurement t_p must also include the time required to travel the distance between the source and the receiver. Consequently, $t_p = t_{1p} + t_{2p}$, where t_{1p} is the response time of the process and t_{2p} is the time it

Table 7.1 Definitions of t_j

t_1	Time required to execute program line and set up and access interface.
t_2, t_9	Time required to transmit N_2 and N_9 bytes of data at the corresponding data rates of C_2 and C_9 bytes/sec, respectively. Thus, $t_2 = N_2/C_2$ and $t_9 = N_9/C_9$.
t_3	Time required for controller to execute instructions.
t_4	Time it takes electronics to produce final output voltage or current.
t_5	Time it takes device to reach requested state or perform requested function.
t_6	Time it takes transducer to reach rated accuracy.
t_7	Time it takes for instrument to process and convert analog signal to a digital number at rated accuracy.
t_8	Time required to present updated value(s) at its interface.
t_{10}	Time required to execute program line, set up and access interface and store data in memory.
t_p	Time required for process to reach requested state or perform requested action.
$T_w = \sum_{j=1}^{j=8} t_j + t_p$	Minimum wait (delay) between program lines to insure that meaningful quantity has been read.

takes to complete the process. Examples of this are given in Sections 8.2.10 and 8.3.2.

The response time of the transducer t_6 is a very critical element in the instrumentation chain. It is very important that t_6 be less than t_{1p} and, in particular, for transient measurements it should typically be less than $t_{1p}/10$. The instrument's processing and A/D conversion time t_7 has several facets to consider. It takes the instrument a certain amount of time to convert the analog signal to a digital number. This conversion time t_c is independent of the response time of the instrument t_{ins}, which still must be less than the process response time t_{1p} that the transducer is monitoring. At the end of each conversion an updated digital number is available at the instrument's interface every t_c irrespective of whether or not the instrument itself has reached its steady-state response.

The simplest example of such a situation is when a dc digital voltmeter is used to read a dc voltage that has abruptly changed from V_1 to V_2. Consider a DVM that has two types of range settings: preselected (fixed) and autoranging. For each type of range setting $V_1 = -10$ V and -1 V and $V_2 = 10$ V. Prior to abruptly changing the input voltage to the DVM the DVM's range setting mode was selected, a T_w of 2 s was used and the value of V_1 was read and recorded. The input voltage was then abruptly changed and seven reading were recorded at the DVM's maximum conversion rate, which for this DVM was 12.5 samples/s (every 80 ms). The results

Table 7.2 DVM response to a sudden change in its input voltage for two types of range settings and two initial voltages

| Initial Voltage = -10.005 | | | | Initial Voltage = -1.004 | | | |
| Range Preselected | | Autoranging | | Range Preselected | | Autoranging | |
Time (s)	Reading (V)	Time (s)	Reading (V)	Time (s)	Reading (V)	Time (s)	Reading (V)
0.02	-10.004	0.02	-10.005	0.02	-1.003	0.01	-1.000
0.08	0.864	0.09	4.997	0.08	6.631	1.23	9.993
0.16	9.982	0.17	9.986	0.17	9.990	1.31	9.995
0.24	9.993	0.25	9.994	0.24	9.995	1.39	9.996
0.32	9.994	0.33	9.995	0.32	9.995	1.47	9.995
0.40	9.995	0.41	9.995	0.40	9.996	1.55	9.995
0.48	9.995	0.49	9.995	0.48	9.996	1.63	9.996

are summarized in Table 7.2. When going from -10 V to 10 V the DVM's response time was approximately 0.3 s, whether autoranging or range preselection was used. This is because in going from -10 V to 10 V the DVM didn't have to change ranges. However, when the input voltage went from -1 V to 10 V the autoranging mode in the DVM took 1 s longer than in the preselected range mode. Thus, for these values of V_1 and V_2, if the range is preselected $T_w = 0.3$ s whereas if autoranging is used $T_w = 1.3$ s.

Regarding the availability of updated data from the digital instrument it should be realized that if one were to make multiple readings, then to take advantage of the instrument's processing speed the sum of the transmission time and the access and data storage time $t_9 + t_{10}$ must be less than t_c. However, for the converted voltages to be meaningful t_{ins} must still be less than t_{1p}.

The time it takes for the computer to read its interface and store the number in its memory can be done faster using direct memory access (DMA) in those systems/languages that provide this capability. DMA is a means whereby the computer system's overhead associated with placing a number into a memory location is greatly reduced by having the data at the interface go directly to specified memory locations without going through the usual system operations.

7.4 Special Interfaces

7.4.1 RS 232[1]

The RS 232 specification was originally conceived in 1969 as an "Interface Between Data Terminal Equipment and Data Communication Equipment Employing

---

[1] Lang [1987].

Serial Binary Data Interchange". This was a single-minded application of specifying Data Communication Equipment (DCE) and Data Terminal Equipment
(DTE). RS 232 refers to a means of transferring information serially. This permits,
in its simplest form, unidirectional transmission over two wires, a signal conductor
and a ground. Bilateral communications, therefore, only requires one wire to send,
one to receive and ground.

This simplistic three wire bilateral communications is asynchronous. That is, a
common clock signal, which would require another conductor, is not used. Instead
each end of the system (receiver/transmitter) generates its own clock pulses at a
mutually agreed-upon frequency termed the baud rate, or number of bits/s. Since
the clocks at each end are neither synchronized nor at the exact same frequency,
each receiver uses digital signal processing on the transmitted baud rate to lock
in the communication. The electronic device that does this is a universal
asynchronous receiver/transmitter (UART) chip which also converts the data to
and from the serial and parallel forms, the parallel forms being used internally by
the computer.

A typical RS 232 connector consists of 25 pins. Of these 25 pins only 8 are in
common use for connection with a computer's interactive devices. These are shown
in Figure 7.3. In particular, pins 2, 3 and 7 are the minimum number required to
conduct bilateral communications. It should be noted that the Transmitted Data
(TxD) is an output of a DTE, and that Received Data (RxD) is an input of the

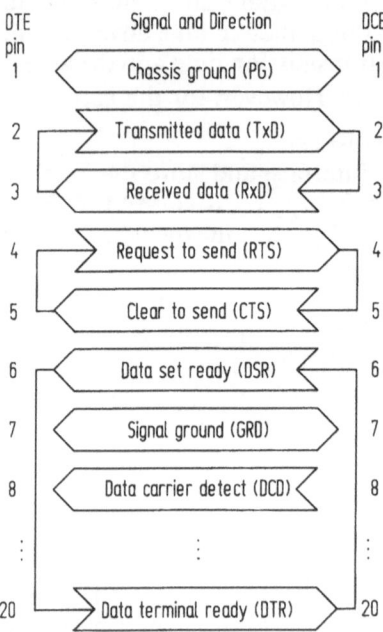

Figure 7.3 Signal designations for nine of the twenty five RS 232 pins

DTE and an output of the DCE. To overcome these semantics the following convention is used: each named signal has the same pin assignment at either a DTE or DCE. A device that transmits on pin 2 is of DTE gender, whereas one that transmits on pin 3 is of DCE gender. While RS 232 specifies the interconnection of a DTE to a DCE, it is, unfortunately, possible to interconnect two devices of the same gender. When this occurs one has to use a gender changing connection that simply cross-connects pins 2 and 3 with pins 3 and 2, respectively. Similar situations exist with pins 4 and 5 and pins 6 and 20 and are resolved in the same manner.

To obtain control, or "handshaking", of the communications pins 4, 5, 6 and 20 are used. Pin 4 is the Request to Send (RTS) and pin 5 the Clear to Send (CTS). When the CTS signal on pin 5 is idle the data on pin 2 (TxD) cross the line at the agreed-upon baud rate from the DTE to the DCE. If the DCE cannot handle all the input it activates the CTS line causing a halt to the TxD transmission until the CTS line returns to idle. When the RTS signal on pin 4 is idle the data on pin 3 (RxD) cross the line at the discretion of the DCE. When the DTE activates the RTS the data flow is halted until the RTS goes idle. The Data Set Ready (DSR) on pin 6 and the Data Terminal Ready (DTR) on pin 20 perform in the same manner as the CTS and the RTS, respectively. They are most frequently used in conjunction with auto-dial and auto-answer modems. Lastly, the Data Carrier Detect (DCD) on pin 8 is used only by a modem to transmit from the DCE to the DTE. The DCE then starts exchanging data.

The actual data transmitted as either TxD or RxD consists of a frame of B_o bits. The first bit is always the start bit. The next are the data bits, which form either a 7- or 8-bit ASCII (American Standard Code for Information Interchange) character. If it is 7 bits then a 128 character set can be transmitted/received; if 8 bits are used then the character set is 256. The data bits are followed immediately by an optional parity bit and then by a stop bit. The total of these bits is equal to the frame size B_o. The total number of ASCII characters that are transmitted per second are, therefore, (baud rate)/B_o. Baud rates are standardized at the following values: 50, 75, 110, 134, 150, 200, 300, 600, 1200, 1800, 2400, 3600, 4800, 7200, 9600, and 19,200. In most common use is a B_o of 10, with an 8-bit ASCII character, no parity and one stop bit. From this discussion it should be apparent that communications over RS 232 lines require *a priori* knowledge of the frame size, number of data bits, the number of stop bits and parity.

Virtually all computer controlled RS 232 interfaces can be set up with software commands to run at any baud rate and to match any frame size and all possible combinations within the frame. However, all the RS 232 standard really does is specify an array of 25 pins, it does not specify nor imply a communication protocol; that is an agreed-upon set of ASCII characters that convey specific meaning. For this one has to use the IEEE 488 interface discussed in the next section. However, because of its simplicity, its inexpensive cabling and its ability to communicate over long distances, the RS 232 has gained wide use.

The lack of a standard protocol for the RS 232 permits variability in the characters used to indicate the termination of a message or a data string. Typically they are the carriage return and the line feed. Just as frequently, however, the software in the instrument/device controller expects to receive only the line feed or only the carriage return (or sometimes neither). Similarly for the transmission of data from the devices. Consequently, one must use software with this interface that is flexible enough to permit non-typical termination character sequences to be sent and received.

7.4.2 IEEE 488

The IEEE 488 interface is one in which the communications are governed by a set of rules as set forth in a 1978 IEEE standard: IEEE 488. The interface consists of eight signal lines for the parallel transmission of one byte of data, and eight control lines, three of which are for communication control, called handshaking. This is shown in Figure 7.4. Up to thirty-one devices may be connected to the IEEE 488 bus at one time. Each device is given its own address from 0-30, which has to be manually set on each device with a series of switches. One of the devices, usually a computer, acts as the system controller. Other devices on the bus, if they have the capability, can be temporarily assigned the role of controller. During this time the device is designated the active controller. There is, however, only one system or active controller at a time.

Each device on the bus is addressed as either a talker or a listener. A listener is a device designated to receive data or instructions and a talker is a device designated to send (to the active controller) data or instructions. Most devices are talker/listeners; some devices, such as a function generator, are only listeners, others such as a counter are primarily talkers. One important rule governing the rate of transmission of data on the bus is that the data rate is determined by the *slowest* active listener. Therefore, preceding every data transmission the active

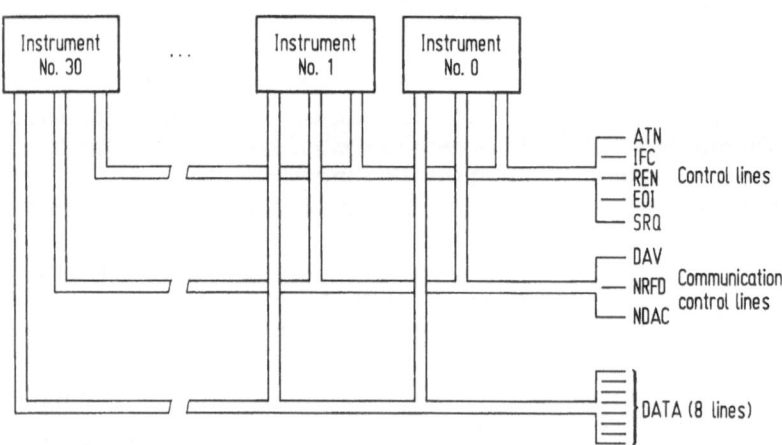

Figure 7.4   IEEE 488 interface lines and their functions

controller sends an *unlisten* command, which causes all devices connected to the bus to become unresponsive to all subsequent data transmissions. Following the unlisten command the active controller sends its address (0-30). Then the listener address is sent, which causes only that device to become a listener. Finally the data or messages are sent. The handshaking, to be discussed subsequently, is then used to insure that each transmitted byte has been received.

Conversely, to get a device to transmit data or a message to the active controller the unlisten command is again sent, followed by the talker's address. Then the listener's address, which is the active controller, is sent. Lastly, the data or message from the device just addressed is sent to the active controller, again using the handshaking lines to insure that the transmission of each byte has been completed.

In addition to sending data on the eight signal lines, command messages, consisting of one or more bytes, can also be sent over seven of these lines. For the device to be able to distinguish between data and commands the ATN (Attention) line is used. Some of the types of commands that can be sent are:

Data- which is sent from the talker to the listener.

Trigger- which causes the listening device to initiate a device-dependent action.

Clear- which returns listening devices to their device-dependent "clear" state.

Local Lockout- which disables a device's front panel controls preventing manual intervention to the device.

The three handshake lines are: DAV- data valid; NRFD- not ready for data; and NDAC- not data accepted. All devices currently set as active listeners would indicate that they are ready for data by using the NRFD line. When all the NRFD lines indicate that they are all ready for data the talker places a byte of data on the data lines and then uses the DAV line to inform the listeners that they may now read the data lines. After each active listener receives the byte of data it activates the NDAC line. When all the active listener NDAC lines have been set the talker knows that all the listeners have read the byte of data and the process starts over again with the next byte of data.

The remaining four control lines (the ATN line has already been discussed) have the following functions:

IFC- Interface Clear. Unconditionally terminates all bus activity. Can only be invoked by the system controller.

REN- Remote Enable. Allows instruments on the bus to be programmed remotely by the active controller.

EOI- End or Identify. Used to send blocks of data which contain the line-feed character. This instructs the listener to ignore the line-feed character within the data block and wait for the EOI character instead.

SRQ- Service Request. Since the active controller is always in charge of the order in which events on the bus occur, a device can make its needs known by activating the SRQ line. Since this is a request, not a command, the active controller can service it or not. However, the device will keep this line active until it has been satisfied. What the SRQ signifies is device-dependent.

The IEEE 488 interface specifies two transmission rate ranges: normal and high. The maximum value for the normal range is 500 kbytes/s, whereas for the high range it is 1 Mbyte/s. Many computer and instrument interfaces run at 40-55 kbytes/s. The implementation of DMA can increase these rates by 5 to 8 times. Typical overhead for this interface is around 5 ms. Although the IEEE 488 standard calls for a maximum of 31 devices to be placed on the bus, many interfaces only support (because of electrical considerations) one-half to one-quarter of this maximum. In addition, the cable lengths used to connect the devices are specified in the standard to be, on the average, 1 m, with the total length not exceeding 30 m. In many practical implementations this total length is typically 15-20 m.

7.4.3 Binary Coded Decimal (BCD)

The BCD interface is typically a 64 wire connection of which 40 lines are for the transmission of parallel data, 4 lines are for handshaking, 5 are special purpose, 8 are for output and the remainder are grounds. The BCD representation is the grouping of the forty lines (bits) into ten, four-bit sets to represent one of sixteen ASCII characters. Ten of the characters are the digits 0-9, one represents a line feed, one the letter E for the exponent and the remaining four are the plus and minus signs, the period and the comma. Nine of these characters are used to represent the digits of the number itself, eight for the mantissa and one for the exponent. The tenth digit is a function code whose interpretation is device-dependent, usually signifying the units of the measurement. Of the five special lines one is used to represent the sign of the number, another to represent the sign of the exponent and a third to indicate whether the device's input was overloaded.

7.4.4 CAMAC: IEEE 583[2]

CAMAC is a standardized means of connecting externally generated signals to a computer. It originated in the late 1960's as a European standard with the palindromic, but arbitrary, name CAMAC. Later it acquired the phrase Computer Automated Measurement and Control. It became an IEEE standard in 1982. This standard specifies (i) the mechanical and electrical characteristics of a crate, which houses plug-in modules and provides them with power, (ii) the protocol of a dataway, which is an internal 86-line data channel that provides the means by which a controller situated in the crate transfers data and other information to and from the other plug-in modules in the crate and (iii) the mechanical and electrical power specifications of the plug-in modules, which must be ≤25 per crate. The plug-in modules cover the wide range of digital and analog instrumentation found in digital control and digital data acquisition systems.

The CAMAC crate is connected to the computer by one of two methods. The first is to use a computer-specific CAMAC controller. This controller is, in turn,

---

2 Clout [1982].

connected to an interface card that is placed into one of the computer's input/output slots. When this is done the computer's operating system considers the CAMAC controller as just another device. Fortran and BASIC subroutines are usually provided by the controller's manufacturer to communicate with the crate's devices. A second method, which is more flexible and less expensive, is to use a data highway coupler. This coupler is an independent device that connects to the standard input/output ports of the computer, usually RS 232 or parallel. It receives instructions from the computer through either Fortran or BASIC language calls provided by the coupler's manufacturer. The other input/output port of the coupler connects to the crate's data channel and sends commands to, and receives data from, the crate's devices.

The 86 lines of the dataway are now described. Four lines, called subaddress lines, are used to access up to sixteen sections of one module and which are decoded by the module itself. Five function lines define 32 ($=2^5$) separate functions $F(0)$ to $F(31)$ that the module is to perform at each subaddress. Function $F(0) - F(7)$ are read functions, $F(16) - F(23)$ are write functions and the remaining functions are control and status. Two lines are strobe lines and are used to time the signals on the other lines so that transients and differential settling times do not cause erroneous results. There are 24 read lines and 24 write lines that are used in conjunction with the strobe signals to read data from, and to write data to, a module. There is one line to indicate that the dataway is busy, thus temporarily locking out other competing signals in the system. There is one status response line that may or may not be used by individual modules. There is one line each for the command accepted, initialize, inhibit and clear lines. These lines are used in conjunction with one or both of the strobe lines. The command accepted line indicates that the module recognizes the command as one it will be able to perform, even if it may not be able to perform it at that moment. The initialize signal resets all the modules' data and control registers to their initial state. The inhibit signal is module specific and will inhibit any function in the module to which it is connected. The clear signal clears all registers in the module to which it is connected. Of the remaining 21 lines 14 are for power and grounding and seven are free for use by the user for module-specific functions or interpretations.

7.5 Programming Considerations

Generating code for programs that are to be used to interact with instruments and devices requires special care, for numerical and logic errors in the program can have serious physical consequences to the experiment or process. To minimize these possibilities requires a carefully reasoned program and a thorough understanding of all the instruments and devices and the process under investigation (or control). The best procedure is to create a multilevel flow chart from which a structured computer program can be written. It is the purpose of this section to present suggestions concerning program attributes, procedures and techniques. These suggestions are implemented the examples in Chapter 8 and in the case studies in Chapter 9.

The creation of a multilevel flow chart is the key to writing successful code. The first level is one in which the major functions of the experiment, its control, its analysis, and its results are delineated in such a way that the order in which things are to be done is clearly shown. The next levels, if required, examine each of the functional groups in appropriate detail. Although there isn't any generic flow chart there are many attributes that these programs have in common. They are catalogued below.

General Attributes
When applicable the program should be written such that:

a) Small changes to the experimental procedures can be easily accommodated.

b) The program cannot start unless all equipment is working properly. If failures occur during program operation the program brings itself and the experiment/process to an orderly and safe halt.

c) Error detection and recovery from malfunctions are an integral part of the program design.

d) The program provides a window to the experiment/process by continually displaying appropriate information and status on the computer's CRT or printer.

e) The program continually monitors the integrity of the instruments, devices and process and the validity of the data read and transmitted.

f) All timing requirements are satisfied (recall Section 7.3).

g) The program is structured to insure its correct implementation; that is, unsafe or meaningless operations cannot take place.

h) The program is modular in structure so that each instrument and device can be verified and interacted with individually during the program's development. Modularity implies that each module (typically a subroutine) performs one task, which is satisfactorily completed before proceeding. In addition, each module should have only one entry point and one exit point, be independent of other program modules and pass information directly into and out of its module.

i) The program can read any data files it generates.

j) The program is well-documented, with detailed comments appearing throughout including: specific information regarding instrument/device specifications, timing and device-dependent language translations, memory allocations, overall programming logic, program's and programmer's names, date program was initially written and date of the last revision and/or version number.

Initialization
This portion of the program has several important functions:

a) It sets all the components used in the integrated experiment/process to a *known* state and checks that they are all functioning properly.

b) All the appropriate wait/delay times are set throughout the program, usually with the aid of COMMON statements. These values should appear in only one place in the program and, if possible, not be altered elsewhere in the program. An exception is when the process itself changes in such a way that these times have to be changed correspondingly.

c) It sets all the interfaces to their desired characteristics.

d) It informs the user of the status of all the devices and instruments prior to the actual running of the experiment. It should explicitly state the exact instrument/device and give a physical interpretation of the status.

e) Provides the user, where appropriate and/or feasible, with the choice of manually sequencing or checking some or all of the portions of the experiment/process prior to its fully automatic implementation. This feature greatly facilitates program development, verification and error detection and isolation.

f) Prints to the computer's CRT any special reminders or messages concerning the experiment/process.

g) Performs any transducer, instrument or process calibration procedures, either interactively or under complete computer control.

User Input
This portion of the program queries the user for experiment-specific information, such as the number of times the experiment is to be run, selection of operating parameters and ranges, file names for data storage, and so on. All user inputs should be scrutinized by the program to insure that each value is within valid ranges and that combinations of values are also within prescribed limits.

Instruments and Devices
Each computer-controllable instrument and device should be accessed from its own dedicated subroutine. Where appropriate each subroutine should contain two sections: one for setting the instrument/device to a desired state and one for reading its output. Within each subroutine great care should be exercised to insure that each value read is valid and that each value transmitted has been received by the device. Therefore, where provided for, one should always check each instrument's and each device controller's status and error indicators to insure that valid data have been transmitted/received. To detect communication link failures or instrument failures each access to an interface should be timed. If the specified time is exceeded then there is every reason to believe something unusual has occurred and the program should be halted and the operator notified.

Analysis of Data
In this portion of the program the data are analyzed. Depending on the experiment/process the data may be used either to make a decision regarding the next change in the experiment/process, to store the data in memory or on a disk, to convert the data to another form (transform (FFT), linearize, scale, etc.), to plot the data on the CRT or on a plotter, or any combination of these.

Storage of Data

It is good practice to keep a permanent record of any data acquired. This is usually done to a disk. At what point in the program the data are to be written to the disk depends on the timing requirements of the experiment/process. If there is sufficient time it is often best to save the data as it is acquired. This is especially helpful when there are premature terminations to the program execution, for then some information has been saved. It is important that each data file have sufficient header information so that it can be distinguished from similar files. To access these data files requires a program to read it in the same manner in which it was written. This capability should be part of the program that wrote the data, for it can then be used to verify that the data were initially written correctly, to further analyze or compare previously recorded data, or to display the data.

References

1. Barney, G. C., *Intelligent Instrumentation*, Prentice-Hall, Englewood Cliffs, New Jersey, 1985.

2. Brighouse, B., and Loveday, G., *Microprocessors in Engineering Analysis*, Pitman Publishing, London, 1987.

3. Clout, P., "A CAMAC Primer", Report No. LA-UR-82-2718, Los Alamos National Laboratory, Los Alamos, New Mexico, 1982 (Revision III: July 1986).

4. Gupta, S., and Gupta, J. P., Eds., *PC Interfacing for Laboratory Data Acquisition and Process Control*, Instrument Society of America, Research Triangle Park, North Carolina, 1989.

5. Lang, G. F., "Bits, Bytes, Baud, Bell, and Bull", S)V Sound and Vibration, Vol. 21, No. 9, pp. 10-17, September 1987.

6. Lawrence, P. D., *Real-Time Microprocessor System Design*, McGraw-Hill Book Co., New York, 1987.

7. Mellichamp, D. A., *Real-Time Computing with Applications to Data Acquisition and Control*, Van Nostrand Reinhold Co., New York, 1983.

8. Money, S. A., *Microprocessors in Instrumentation and Control*, Collins Professional and Technical Books, London, 1985.

9. Moore, J. A., *Digital Control Devices: Equipment and Applications*, Instrument Society of America, Research Triangle Park, North Carolina, 1986.

8 Transducers and Actuators

8.1 Introduction

Sensors are devices that convert physical quantities such as displacement, temperature, strain, etc., to an electrical change (voltage, resistance, capacitance, inductance or current) that can be measured. Transducers are considered the combination of the sensor plus any mechanical element plus any signal conditioning circuitry such that the output voltage is related to the physical quantity. Sensors are either self-generating (active) or passive. Self-generating sensors convert mechanical, thermal, chemical, or optical energy into electrical energy. Passive sensors on the other hand require electrical energy to be supplied in order for them to convert the physical energy into electrical energy. In addition sensors that convert the physical quantity directly into electrical energy are called direct converters. Examples of self-generating direct converters are thermocouples and piezoelectric materials. When another physical process intervenes prior to the electrical conversion the sensor is referred to as an indirect converter. Examples of passive direct converters are potentiometers, strain gages and differential transformers (LVDTs). Examples of passive indirect converters are capacitor microphones and load cells.

This chapter will introduce several of the more common sensors and transducers (with electrical output) used to measure physical quantities and to control processes. It is not intended to be a complete or detailed in its coverage; for this see Dally, et al. [1984], Doebelin [1990], Seippel [1983], Trietley [1986], Bentley [1988] and Neubert [1975]. It will also introduce several common actuators; namely, step motors, vibration exciters, loudspeakers and solenoids. The electrical impedance properties of most of these devices have already been discussed in detail in Section 4.5.

8.2 Displacement

8.2.1 Laser Interferometry

There are two laser interferometric techniques available to measure very small displacements with resolution of 160 nm (6 μin) to less than 10 nm (0.4 μin). Both techniques are such that they provide for their absolute calibration; that is all experimental scale factors and parameters can be determined from the experimental setup itself except for one, the wavelength of the laser light. The first technique uses a Michelson laser interferometer, which can be used to measure the displacement of both static and harmonically vibrating surfaces. This technique is used by the National Institute of Standards and Technology (formerly the National Bureau of Standards) to calibrate accelerometers from 1 to 20 kHz (Robinson, et al. [1987]). The second technique is commercially available and is called an ac interferometer. Despite its name, however, it only works for nominally static measurements.

Michelson Interferometer

Consider the setup shown in Figure 8.1. The laser beam first encounters a beam splitter which diverts a portion of the original beam to the "fixed" mirror and the remainder to the moving surface (mirror). The two split portions of the original beam reflect 180° from their respective surfaces and recombine at the beam splitter to be directed to the photodiode. The photodiode responds to the changes in the intensity of the recombined laser light as a function of the motion of the mirrors. (For the details of the properties and characteristics of photodiodes and other photosensitive detectors see Nunley and Bechtel [1987], Dereniak and Crowe [1984] and Keyes [1977].)

The total electric field strength impinging on the photodiode is

$$E(x,t) = E_F \cos(\omega t - kx_F) + E_M \cos(\omega t - kx_M) \tag{8.1}$$

where ω is the angular frequency of the laser light ($\cong 30 \times 10^{14}$ rad/s), k is its wave number ($= 2\pi/\lambda$; λ is the wavelength), E_F and E_M are the strengths of the individual beams, and

$$x_F = 2L_F + L_d, \qquad x_M = 2L_M + L_d$$

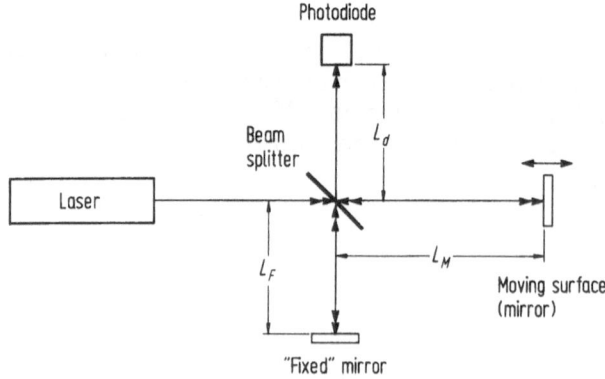

Figure 8.1 Michelson laser interferometer

are the total distances travelled by the beam in the fixed mirror and the moving mirror paths, respectively.

The photodiode's output voltage is proportional to the time average of the square of the amplitude of the total signal. Thus

$$E_p = \frac{1}{T} \int_0^T E^2(x,t)dt \qquad (8.2)$$

where T is the averaging time of the photodiode. Substituting Eq.(8.1) into Eq.(8.2), and noting that $\omega T \gg 1$ (typically on the order of 10^7), yields

$$E_p = \frac{E_F^2 + E_M^2}{2} + E_F E_M \cos 2k\Delta \qquad (8.3)$$

where

$$\Delta = L_F - L_M. \qquad (8.4)$$

Thus E_p is a constant unless Δ changes. When $2k\Delta = \pi$, $\cos 2k\Delta = -1$ and the output of the photodiode goes through a minimum. Consequently, this minimum corresponds to the moving mirror going through a displacement of

$$\Delta = \frac{\pi}{2k} = \frac{\lambda}{4}. \qquad (8.5)$$

Therefore we have obtained an absolute measurement of a relative displacement. It is seen that all that has to be known is the wavelength of the laser light. If we use a laser whose wavelength is 6328 °A (0.6328 μm), then $\Delta = 160$ nm (6.2 μin).

For single frequency vibration measurements we replace Δ with

$$\Delta = \Delta_o + d\cos\omega_o t$$

and Eq.(8.3) becomes

$$E'_p = \frac{E_F^2 + E_M^2}{2} + E_F E_M \cos[2k(\Delta_o + d\cos\omega_o t)] \qquad (8.6)$$

where Δ_o is the mean path length and d is the peak displacement amplitude of the single frequency sine wave at ω_o. Equation (8.6) can be expressed as (Schmidt et al. [1961], [1962])

$$E'_p = A + B\cos\theta \sum_{n=0}^{\infty} J_{2n}(z)\cos[2n\omega_o t] +$$

$$B\sin\theta \sum_{n=1}^{\infty} J_{2n-1}(z)\sin[(2n-1)\omega_o t] \qquad (8.7)$$

where $J_n(z)$ is the Bessel function of the first kind,

$$\theta = 2k\Delta_o = 4\pi\Delta_o/\lambda, \qquad z = 2kd = 4\pi d/\lambda$$

and A and B are unknown constants. Notice that the $n = 0$ term is independent of frequency and therefore can be combined with A.

Let us examine one way in which these results can be used. We pass the output voltage from the photodiode through a narrow bandpass filter centered at ω_0 $(n = 1)$. We now adjust the "fixed" mirror so that $\sin \theta = 1$ in Eq.(8.7). Then the filter's output voltage is

$$E_{p(n-1)} = BJ_1(z).$$

To calibrate the system, that is, to determine the constant B, we adjust the amplitude d from zero to d_{max}, which is that value for which $J_1(z)$ is a maximum. The corresponding output voltage is E_{pmax}. From a table of Bessel functions this maximum occurs when $J_1(z) = 0.5819$. Thus,

$$E_{p(n-1)} = \frac{E_{pmax}}{0.5819} J_1(4\pi d/\lambda)$$

and the system has been calibrated. Consequently for an arbitrary value of d we have

$$d = \frac{\lambda}{4\pi} J_1^{-1}\left(\frac{0.5819 E_{p(n-1)}}{E_{pmax}}\right). \tag{8.8}$$

One has to be careful, however, in using Eq.(8.8) because $J_1(z)$ is an oscillatory function. Therefore, the number of times $E_{p(n-1)}$ becomes zero must be recorded before using the Bessel function tables. Notice that this technique is useful only when one can control the type of signal and its amplitude.

AC Interferometer

The ac laser interferometer system makes use of the fact that light can be polarized; that is, since light consists of two transverse waves travelling at right angles to each other one (or both) waves can be filtered by a suitable polarizing lens. The ac interferometer takes advantage of this fact by using a laser that generates a laser light beam of two slightly different frequencies (wavelengths), each frequency being associated with a different polarization.

Consider the experimental setup shown in Figure 8.2. The two-frequency laser beam first encounters a polarized beam splitter, which diverts a portion of the incident beam containing both frequencies to photodiode P_1. The remainder of the beam passes through a second polarized beam splitter such that the laser beam of frequency f_2 is completely reflected to the fixed corner cube. The laser beam containing frequency f_1, however, passes through this polarized beam splitter and is reflected from the moving corner cube. The two independent beams recombine again at point "1" and their intensity is detected at photodiode P_2.

The analysis follows in a similar manner to that for the Michelson interferometer. The total electric field strength at photodiode P_1 is

$$E = E_1 \cos(\omega_1 t - k_1 x_1) + E_2 \cos(\omega_2 t - k_2 x_2) \tag{8.9}$$

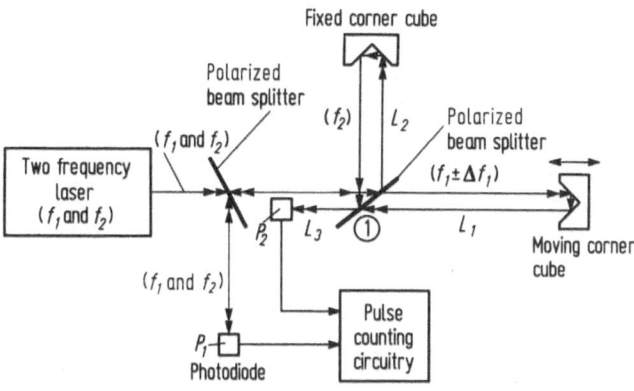

Figure 8.2 Two frequency ac laser interferometer

where ω_1 and ω_2 are the angular frequencies of the laser light and typically differ by several MHz, k_1 and k_2 are the corresponding wave numbers ($= 2\pi/\lambda_{1\,or\,2}$; λ is the wavelength), E_1 and E_2 are the strengths of the individual beams corresponding to each frequency, and

$$x_1 = 2L_1 + L_3, \qquad x_2 = 2L_2 + L_3$$

are the total distances travelled by the beam in the moving mirror and the fixed mirror paths, respectively.

Substituting Eq.(8.9) into Eq.(8.2) and again noting that $\omega_1 T \gg 1$ and $\omega_2 T \gg 1$ we find that

$$E_p = \frac{E_1^2 + E_2^2}{2} + \frac{2E_1E_2}{T\omega_d}\sin(\omega_d t + \Delta) \qquad (8.10)$$

where

$$\Delta = k_2 x_2 - k_1 x_1, \qquad \omega_d = \omega_1 - \omega_2.$$

Upon examining the argument of the sine function in Eq.(8.10) we see that ω_d and $k_2 x_2$ are constants: only $k_1 x_1 = \omega_1 x_1/c_1$ varies. This term can be considered the Doppler shift of the frequency ω_1, since x_1/c_1 changes when the moving mirror changes the value of x_1. Another way to state this is that the instantaneous frequency of Eq.(8.10) is

$$\frac{d}{dt}(\omega_d t + \Delta) = \omega_d - k_1\frac{dx_1}{dt}$$

where $k_1 dx_1/dt$ is the Doppler shift caused by the velocity dx_1/dt. Consequently, the argument of the sine function can be thought of as having a frequency comprised of the difference frequency ω_d and the shifted frequency $\Delta\omega_1$ brought about by the changes in x_1. To record this Doppler shifted frequency the outputs of both photodiodes are subtracted and the difference is proportional to the changes in x_1. Although the basic resolution of this technique is the same as the Michelson interferometer, special electronic techniques increase its resolution by a factor of 16 to 10 nm (0.4 μin).

Returning to Eq.(8.10) it is seen that by continually counting the changes in the light intensity from minimum to minimum the total relative displacement can be recorded. The maximum speed of the moving mirror is equal to the product of the minimum resolution and the number of counts/s of the counting unit. For example, if the count rate is 3 MHz and the resolution is 160 nm, then the maximum speed would be approximately 0.5 m/s.

An application using an ac interferometer is given in Section 8.2.10.

8.2.2 Linear Variable Differential Transformer (LVDT)

The linear variable differential transformer (LVDT) is a transducer used to measure nominally static displacements by converting the motion of a moveable core in a magnetic field into a voltage that is proportional to this movement. Referring to Figure 8.3 the core moves linearly inside a transformer consisting of a center primary coil and two outer secondary coils. The coil is energized with an ac voltage e_x of frequency ω, which induces secondary voltages that vary with the position of the core. The primary inductor has an inductance L_p, and two identical series-opposing secondary transformers each have inductances $L_s/2$. If the mutual inductances between each secondary and the primary are denoted M_1 and M_2 and the resistance associated with L_p is R_p and that with L_s is R_s, then the equivalent circuit is that shown in Figure 8.3. The resistance R_i represents the input resistance of the next instrument that gets connected to this sensor. Solving the circuit equations for this system in the frequency domain leads to the following

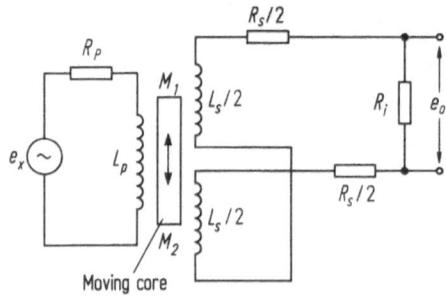

Figure 8.3   Schematic of a typical LVDT

transfer function (Doebelin [1983]):

$$\frac{e_o}{e_x} = \frac{j\overline{\omega}\Delta m}{1 - \overline{\omega}^2(c_o + (\Delta m)^2 R_p/R_i) + j\overline{\omega}(1 + c_o)} = A_o e^{j\phi} \qquad (8.11)$$

where

$$A_o = \frac{\overline{\omega}\Delta m}{\sqrt{\overline{\omega}^2(1 + c_o)^2 + [1 - \overline{\omega}^2(c_o + (\Delta m)^2 R_p/R_i)]^2}}$$

$$\phi = \tan^{-1}\frac{1 - \overline{\omega}^2(c_o + (\Delta m)^2 R_p/R_i)}{\overline{\omega}(1 + c_o)}$$

$$\overline{\omega} = \frac{\omega L_p}{R_p}, \qquad \Delta m = \frac{M_2 - M_1}{L_p}, \qquad c_o = \frac{R_p L_s}{R_i L_p} \qquad (8.12)$$

and we have assumed that $R_i/R_s \gg 1$. Equation (8.12) shows that the magnitude of e_o is proportional to frequency. Therefore, the higher the frequency the more sensitive the output voltage is to changes in its core position.

We now select our parameters such that when $\Delta m = 0$, $\overline{\omega} = 1/\sqrt{c_o}$ so that $\phi = 0$; therefore,

$$A_o = \frac{\Delta m/\sqrt{c_o}}{\sqrt{(1 + c_o)^2/c_o + [(\Delta m)^2(R_p/R_i)/c_o]^2}}. \qquad (8.13)$$

Thus the output voltage will be almost linearly proportional to Δm provided that for any value of Δm

$$\Delta m \ll \sqrt{(R_i/R_p)(1 + c_o)}\sqrt{c_o}.$$

We see from either Eq.(8.12) or Eq.(8.13) that when the core is centered the output voltage induced in the secondary is zero. When the core is moved from its center position the voltage in the secondary winding toward which the core is moved increases, whereas the voltage induced in the other secondary decreases. The resulting differential voltages, upon appropriate selection of parameters, is linear with the core position over some range of core travel. We also see from these equations that depending on the direction of travel A_o is either positive or negative. This is equivalent to adding to the phase angle ϕ either 0 or π, respectively.

The amplitude A_o modulates the excitation voltage e_x. Therefore to obtain an output voltage that is proportional to the amount of movement of the moveable core e_o has to be *demodulated* and its sense, that is, the direction of the core travel (+ or -) has to be determined. Demodulation is the process of multiplying the amplitude modulated signal e_o with another signal whose frequency is the same as that of e_x, and passing the resulting product through a low pass filter. Thus if

$$e_o(t) = e_m(t)\cos\omega_x t$$

where $e_m(t)$ is the voltage caused by the motion of the core, then the multiplication of $e_o(t)$ with another signal of frequency ω_x and amplitude b yields

$$e_d(t) = be_m(t)\cos(\omega_x t)\cos(\omega_x t) = \frac{b}{2}e_m(t)[1 + \cos(2\omega_x t)]$$

which, when passed through a low pass filter whose cutoff frequency is much less then ω_x, gives

$$e_d(t) = \frac{b}{2}e_m(t).$$

Thus, except for a scale factor, the output signal from the low pass filter is proportional to the voltage caused by the motion of the core. Note that since a low pass filter responds to dc, $e_m(t)$ does not have to be time dependent for the demodulation technique to work. We shall see that the modulation/demodulation process is common to many transducers.

There are several advantages to LVDTs. They give good accuracy, sensitivity and linearity over a wide range of travel; as small as 0.02 mm to as large as 64 cm. They have virtually frictionless operation with infinitely small resolution. They are rugged, very insensitive to cross axis motion and provide physical, electrical and environmental isolation. Some of their disadvantages are their physical size, their contacting moving mass (core), their susceptibility to stray ac magnetic fields, and their relatively complex circuitry to get their best performance.

Another form of the LVDT is the rotary variable differential transformer (RVDT), which is a device that produces an ac voltage that is linearly proportional to the angular position of its core (shaft). This core is a specially shaped magnetic rotor that typically gives a linear (±0.5%) output through ±60° of rotation.

8.2.3 Capacitance Gage

Consider a capacitor sensor that is formed by two parallel plates separated by an air gap d_o (inch) as shown in Figure 8.4. If the area of the sensing portion of one of the plates is A (in²) the capacitance formed by this gap is given by

$$C_d = \frac{0.225A}{d_o} = \frac{k_o}{d_o} \quad \text{pF} \tag{8.14}$$

which, unfortunately, varies nonlinearly with d_o. If the gap's distance changes by Δd, then C_d changes by ΔC_d and Eq.(8.14) becomes

$$C_d + \Delta C_d = \frac{k_o}{d_o + \Delta d}$$

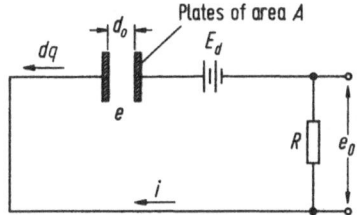

Figure 8.4 Schematic of a capacitance gage comprised of two parallel plates and its equivalent electrical circuit

or

$$\Delta C_d = -\frac{C_d \Delta d}{d_o + \Delta d}. \tag{8.15}$$

The sensitivity S_c of the capacitor is

$$S_c = \frac{\Delta C_d}{\Delta d} = -\frac{k_o}{d_o^2(1 + \Delta d/d_o)}. \tag{8.16}$$

Thus, the smaller d_o the more sensitive the gage becomes. However, if we rewrite Eq.(8.15) as

$$\frac{\Delta C_d}{C_d} = -\frac{\Delta d/d_o}{1 + \Delta d/d_o}$$

then we see that the percentage change in capacitance is always less than the percentage change around d_o.

Referring to Figure 8.4 we see that when the capacitor is stationary $e_o = E_d$. When the capacitor plate moves we have

$$e_o = e - E_d = iR = -\frac{dq}{dt}R. \tag{8.17}$$

But

$$q = C_d e = \frac{k_o e}{d} \tag{8.18}$$

where $d = d_o + \Delta d$. We now linearize Eq.(8.18) by taking a Taylor series expansion around d_o and E_d. Thus

$$q = \frac{k_o e}{d} - \frac{k_o e}{d^2}(d - d_o) + \frac{k_o}{d}(e - E_d) = \frac{k_o E_d}{d_o} - \frac{k_o E_d}{d_o^2}\Delta d + \frac{k_o e_o}{d_o}$$

which, upon differentiating with respect to time, gives

$$\frac{dq}{dt} = \frac{k_o E_d}{d_o^2}\frac{d(\Delta d)}{dt} + \frac{k_o}{d_o}\frac{de_o}{dt} = -\frac{e_o}{R}$$

where we have used Eq.(8.17) and noted that E_d is a constant. Taking the Fourier transform of this result yields

$$\frac{e_o}{\Delta d} = \frac{(E_d/d_o)j\omega\tau}{1+j\omega\tau}$$

(8.19)

where $\tau = RC_d$. We see that the capacitance gage forms a high pass filter [recall Eq.(3.35)] and as such does not respond to dc, or in this case to static or slowly varying changes in d_o. However, for $\omega\tau \gg 1$

$$e_o = E_d \frac{\Delta d}{d_o}$$

(8.20)

which is linear.

It would be better if we could use the capacitance gage at dc also. This can be accomplished by placing the gage in the feedback loop of an operational amplifier as shown in Figure 8.5. The voltage source e_x is an ac source at a fixed frequency. From Eq.(4.21) ff we find that

$$e_o = -\frac{C_1}{C_d}e_x = -\frac{C_1 e_x}{k_o}(d_o + \Delta d).$$

(8.21)

Thus, e_o is linearly proportional to Δd. However, it is amplitude modulated by e_x so that to recover the displacement we must again demodulate e_o [recall Section 8.2.2].

Notice that in both configurations considered the displacements are measured relative to a fixed frame. In addition, because the sensitivity is a function of both k_o and d_o these devices often have to be calibrated *in situ* in order to obtain reasonable accuracy. Advantages of capacitance gages are its broad frequency response and that it is non-contacting.

Applications using capacitance gages are given in Sections 8.2.11 and 9.3.

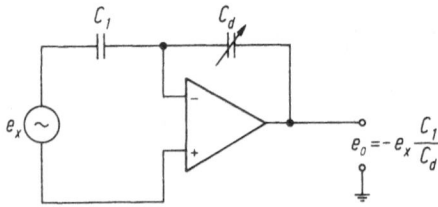

Figure 8.5 Capacitive gage placed in the feedback loop of an inverting amplifier to attain a response at dc

8.2.4 Eddy Current Sensors

An eddy current sensor measures the distance between the sensor and an electrically conducting surface (target). As illustrated in Figure 8.6 the sensor consists of two coils, one active and one passive, that form two arms of a bridge. The bridge is typically excited at a 1 MHz frequency and the active gage forms magnetic lines of flux to the conducting target, producing eddy currents. The eddy currents become stronger as the distance between the sensor and the target becomes smaller. This change in eddy currents changes the impedance of the active coil arm of the bridge, thus causing the bridge to become unbalanced and thereby producing an output voltage. This output voltage is then demodulated and filtered (recall Section 8.2.2) resulting in a signal whose amplitude is linearly proportional to the displacement. These transducers can have a frequency response as high as 100 kHz.

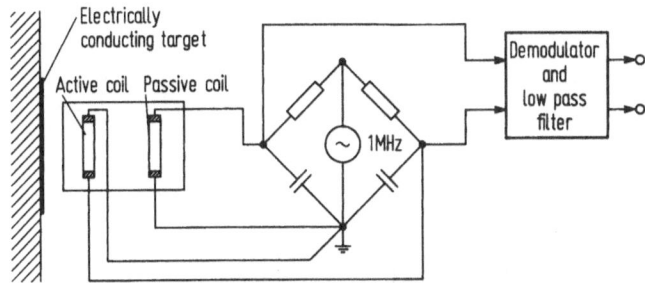

Figure 8.6 Eddy current gage with its associated bridge and demodulator

8.2.5 Potentiometers

A resistive potentiometer consists of a resistive element that is in contact with a moveable element (wiper). The contact motion can be either linear or rotational. The resistive element is either wire wound or continuous. The wire wound potentiometer can give resolutions as small as 0.025 mm. The continuous type have essentially infinitesimally small resolution. There are two common types of continuous potentiometers: cermet and conductive plastic film. The conductive plastic film potentiometers are made from plastics with conductive fillers and can be easily molded into a variety of shapes. Cermets consist of a thin metal film deposited on a ceramic substrate.

Wire wound potentiometers offer the best temperature and resistive stability, but have significantly poorer resolution and life than either cermet or conductive plastic film potentiometers. These latter potentiometers provide smooth output with position. Cermets offer the best power dissipation.

A typical potentiometer circuit, shown in Figure 8.7, has its output voltage related to its excitation voltage by

$$\frac{e_o}{V} = \frac{R_d}{R_T}\left[1 + \frac{R_d R_p}{R_T R_m}\right]^{-1} = \frac{d}{d_T}\left[1 + \frac{d}{d_T}\frac{R_p}{R_m}\right]^{-1} \tag{8.22}$$

where d/d_T is the fractional displacement of the wiper and R_m is the input resistance of the device reading the voltage. From Figure 8.7 we see that the maximum value of R_d is R_T. Therefore, from Figure 4.12 we see that for $R_T/R_m \ll 0.001$ the effects of R_m are negligible and Eq.(8.22) becomes

$$e_o = \frac{d}{d_T}V. \tag{8.23}$$

(Recall from Figure 4.5 and Eq.(4.14) that one way to obtain this high input resistance R_m is to use a voltage follower.)

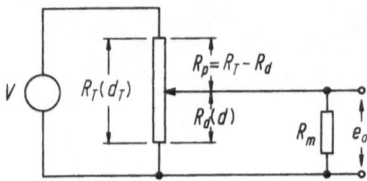

Figure 8.7 Typical potentiometer circuit

It appears from Eq.(8.23) that one can increase the sensitivity of the potentiometer by simply increasing V. This is true up to a limit, which is determined by the power the transducer can dissipate. Thus, since

$$P = \frac{V^2}{R_T}$$

then

$$V_{max} = \sqrt{P_{max} R_T}. \tag{8.24}$$

The advantages of potentiometers are that they are linear over a very large distances (64 cm) and can be made circular to measure rotation. Their disadvantages are their limited frequency response due to the mechanical wiper motion, their low service life (< 2 million cycles for the wire wound), and their mechanically elaborate construction.

Several applications of potentiometers are given in Section 9.4.

8.2.6 Fiber Optic Sensors

The most common type of fiber optic sensor is the intensity-modulated sensor, which detects the amount of reflected light from a target. It is, in essence, a displacement sensor. The sensor is usually comprised of either two bundles of fibers or a pair of single fibers. One bundle transmits light to a reflecting target, the other bundle traps the reflected light and transmits it to a detector. The intensity of the reflected light depends on the distance from the reflecting target to the end of the fibers. The relative merits of different bundle configurations are shown in Figure 8.8. It is seen that the front slope has a very linear portion over a relatively wide displacement range, whereas the back slope is highly nonlinear, with only certain restricted regions being somewhat linear.

Fiber optic sensors can be used in many different physical configurations to measure the presence or absence of an object, displacement of a surface (as described above), temperature, pressure, flow, liquid level, magnetic fields and rotation rates in gyroscopes (Krohn [1988]). Some of their major advantages are that they are nonelectrical, non-contacting, immune to radio frequency and electromagnetic interference, and because of their small mass less susceptible to vibrations.

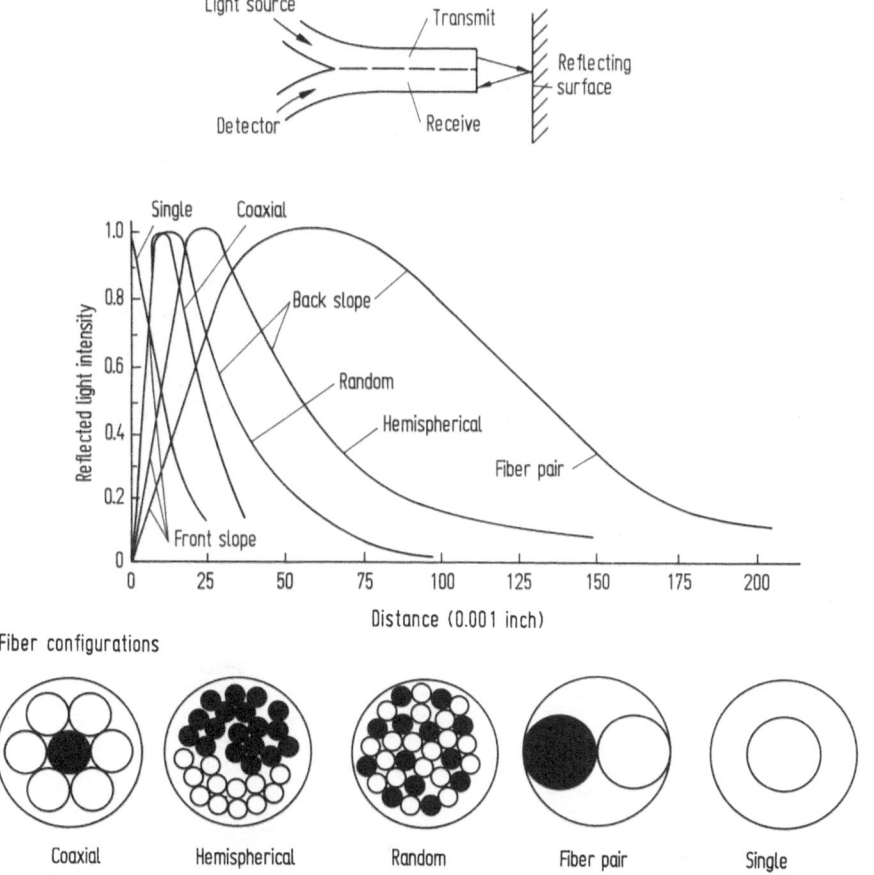

Figure 8.8 Reflective fiber optic sensor response for various fiber configurations

8.2.7 Proximity Sensors

There are many applications where the presence of a surface needs to be determined rather than its actual distance from some reference point. Although many of the previously described transducers would work very well, their cost and complexity are not commensurate with the stated objective. Instead one uses proximity gages, which are binary-type sensors that simply indicate that a certain predetermined displacement threshold has been exceeded. Three types of proximity sensors are commonly used: electrical contact switches, Hall effect switches and magnetic sensors.

The electrical contact proximity switch is a mechanical device in which a spring-loaded plunger protrudes from the switch. As a force (displacement) is applied to the plunger the spring resists until a point is reached where it snaps a mechanical connector into making electrical contact, thereby completing an electrical circuit.

The Hall effect sensor utilizes the Hall principle where a semiconductor supplied with a constant current produces an output voltage when a transverse magnetic field is applied. Either a permanent magnet is attached to the moving surface or the surface must be a ferrous material whose approach changes the reluctance of an internal magnetic circuit within the Hall sensor itself. These types of sensors usually give a constant output voltage after a certain magnetic field level has been sensed. The Hall effect sensors have the advantage of being non-contacting.

The magnetic gage is typically a passive device. It consists of a pole piece that is placed in a coil and is in contact with a permanent magnet as shown in Figure 8.9. When the pole piece is away from a magnetically conducting material the signal from the gage is zero. When a magnetically conducting material approaches the gage a change in magnetic field occurs, producing an output voltage. The amplitude and waveform of the output voltage are a function of the strength of the gage's magnet, the shape of the end of the pole piece, the number of turns of wire in the coil, the distance from the magnetic surface and the speed at which the surface passes in front of the pole piece.

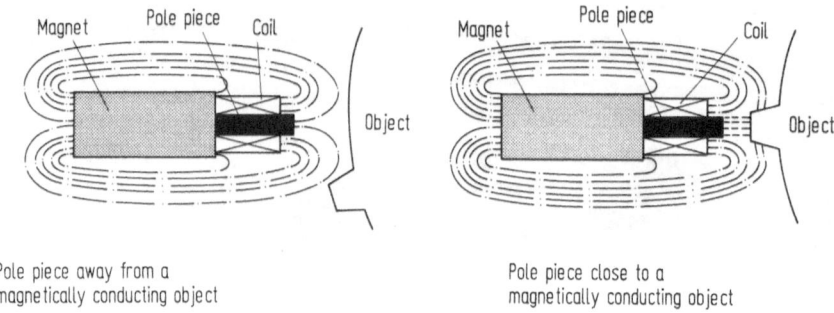

Pole piece away from a
magnetically conducting object

Pole piece close to a
magnetically conducting object

Figure 8.9 A typical magnetic gage and its magnetic field in the presence of a magnetically conducting material

Magnetic gages are typically used to detect rotational motion from gears, sprockets, cams and holes or depressions in flat surfaces. The device is electrically equivalent to an inductance in series with a resistor as described in Case 4 of Section 4.5.2.

Several applications of proximity gages are given in Section 9.4.

8.2.8 Translational and Angular Encoders

Translational and angular encoders are the only transducers with a parallel digital output. Since both types of encoders operate in the same manner only the angular encoder will be described.

Incremental Encoder

A rotary incremental optical encoder consists of a glass disk onto which an annular series of radially aligned marks are imprinted near the disk's perimeter. On one side of the marks is a light source and on the other side a photodetector. The number of marks determines the resolution of the encoder, which is limited by the manufacturing techniques used to place the marks on the disk. Many optical disks have one mark located radially interior to the encoder marks to provide a reference location. The output signal from the photodetector is converted a pulse. The counting of the pulses gives a number that is proportional to the amount of rotation of the disk. By monitoring the rate (frequency) of the pulses one can determine the rotational velocity of the disk. Many incremental encoders use quadrature encoding techniques, which employ a second light source and photodetector placed one-half a mark width apart from the original light source and detector. This technique greatly increases the systems immunity to electrical noise, allows detection of the direction of rotation (clockwise/counter clockwise) and provides a means of increasing the encoder resolution by a factor of four.

An application of an incremental encoder is given in Section 8.2.11.

Absolute Encoder

The most common type of absolute encoder uses optical means in which a common light source shines onto a disk having opaque and transparent tracks as shown in Figure 8.10. On the other side of the disk is an array of detectors, one for each track. The output of each detector indicates the absence or presence of the light and corresponds to one binary bit of information.

In Figure 8.10 two types of track arrangements are shown: one produces natural binary code, the other a gray code. (For the relationship between these two codes see the references cited in Chapter 5.) The reason the natural binary code is often not used in encoders is that in certain disk positions extremely close tolerances are required in the manufacture of the disk in order to properly record the change from, say, binary 0111 to 1000. As the number of bits increases the tolerances must be correspondingly better. With the gray code, however, this alignment problem is overcome, since with any incremental movement of the disk no two bits are ever changed simultaneously. It is seen that the output of an absolute encoder is unique, within its resolution, to only one angular position.

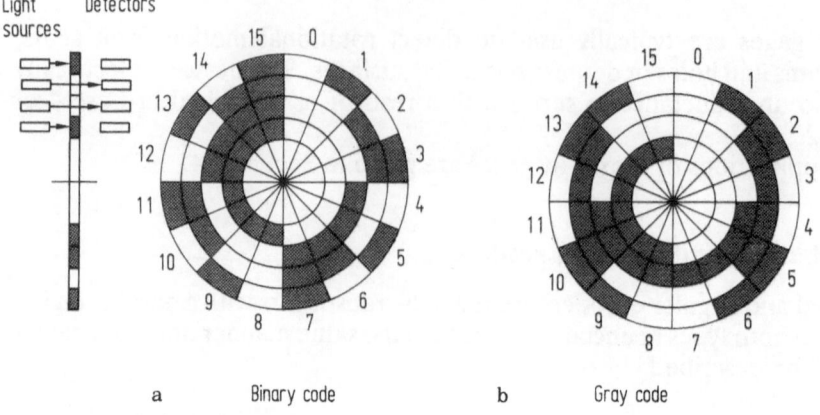

Figure 8.10 Absolute angular encoder: (a) natural binary (b) gray code

Practical considerations limit the number of encoder bits to around 14. In order to obtain additional resolution an extra track (the outermost track on angular encoders) is etched with a pattern that generates a sine wave of the same period as the LSB track. By placing two to four properly positioned sensors along this track positive and negative sine and cosine waveforms can be obtained. If each zero crossing of these waveforms is used to generate pulses, then suitable interpolation circuits can be employed to gain an additional 3 to 5 bits of resolution.

There are many advantages to an absolute encoder: there is no loss of position information on power down since it doesn't count pulses to know its position; it is inherently more immune to electrical noise; there is no need to locate a reference position, which is especially useful in those systems that can't reverse directions; and it can be used in high speed applications since the encoder does not need a continual observation of the encoder output.

Both types of encoders can be combined in a rack and pinion configuration to permit the rack to be used as a linear encoder.

8.2.9 Step Motors[1]

Angular
A step motor is an electro-mechanical device that converts electric pulses into discrete mechanical motion. One type of step motor consists of a rotor and a stator as shown in Figure 8.11. The rotor consists of a permanent magnet sandwiched between laminations having N teeth on its circumference ($N = 6$ in Figure 8.11). The windings for the rotor are contained in the stator, which results in a shaft with lower inertia. To increase the torque of the motor up to three of these laminated

---

1 Compumotor [1988].

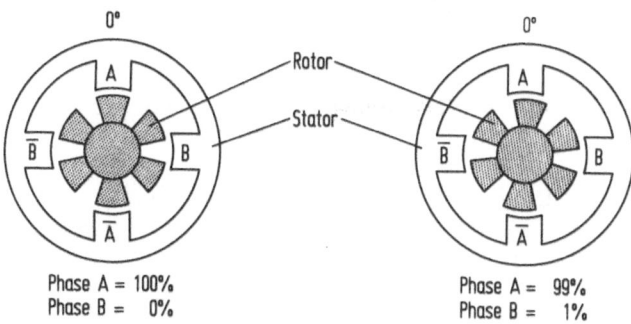

Figure 8.11 Angular step motor

rotors can be stacked. The torque increase is almost directly equal to the number of stacks.

The motor is stepped by placing properly phased sinusoidal currents in each of the stator windings. The sine wave currents, however, are not continuous, but are M discrete amplitudes from a microprocessor controlled D/A converter. As shown in Figure 8.11 the phasing in each winding creates a rotating electromagnet. The relative magnitude of the current controls the permanent magnet rotor (shaft) position. The frequency of the sine waves controls its rotational velocity. Reversing the current causes the rotor to reverse direction. Holding the current level maintains the rotor in its position with high rotational stiffness.

This four state sequence of phasing produces $4N$ full steps/revolution. Thus the resolution of the motor is $4NM$. Typical values are $N = 50$ (teeth) and $M = 125$ (discrete sine wave amplitudes per period). Therefore, the resolution is 25,000 steps/revolution or 0.0144 degrees/step. If this system were given 25,000 pulses it would rotate one revolution. If the frequency of these pulses was 25 kHz one complete revolution would occur in one second (60 rpm). By having the microprocessor control the frequency of the individual pulses, various velocity profiles as a function of time (accelerations) can be obtained.

An application of an angular step motor is given in Section 8.5.5.

Linear

A linear step motor operates on the same electromagnetic principles as a rotary step motor. Referring to Figure 8.12 the moving element is called a forcer and the stationary part the platen (stator). The platen is a passive, toothed steel bar extending over the entire length of travel. All permanent magnets, electromagnets and bearings are incorporated in the forcer. The forcer moves bi-directionally along the platen assuming discrete positions in response to the currents in its field windings.

Linear step motors are equipped with either mechanical roller bearings or air bearings. The advantages of mechanical bearings are their high stiffness to roll, pitch and yaw and their contribution to reducing settling times because of friction. The advantages of air bearings are their long life, even when operated at high

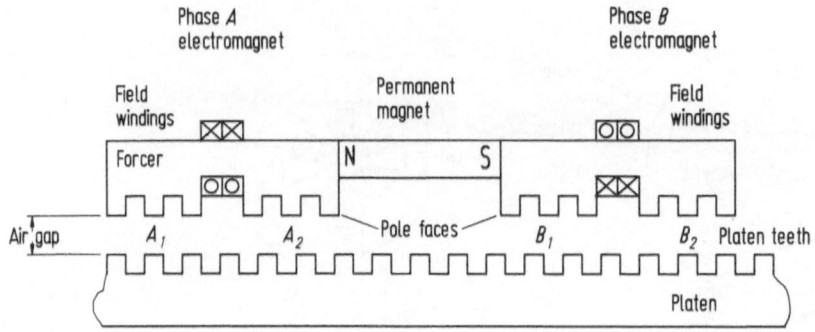

Figure 8.12 Construction of a linear step motor

speeds, higher force rating than a mechanical bearing of the same size, their continual cleaning of the platen of non-adhesive materials and the reduction of the forcer's operating temperature.

As shown in Figure 8.12 the forcer consists of two electromagnets, A and B, and a strong permanent magnet. The two pole faces of each electromagnet are toothed to concentrate the magnetic flux. Four sets of teeth are arranged on the forcer in spatial quadrature so that only one set of teeth at a time can be aligned with the

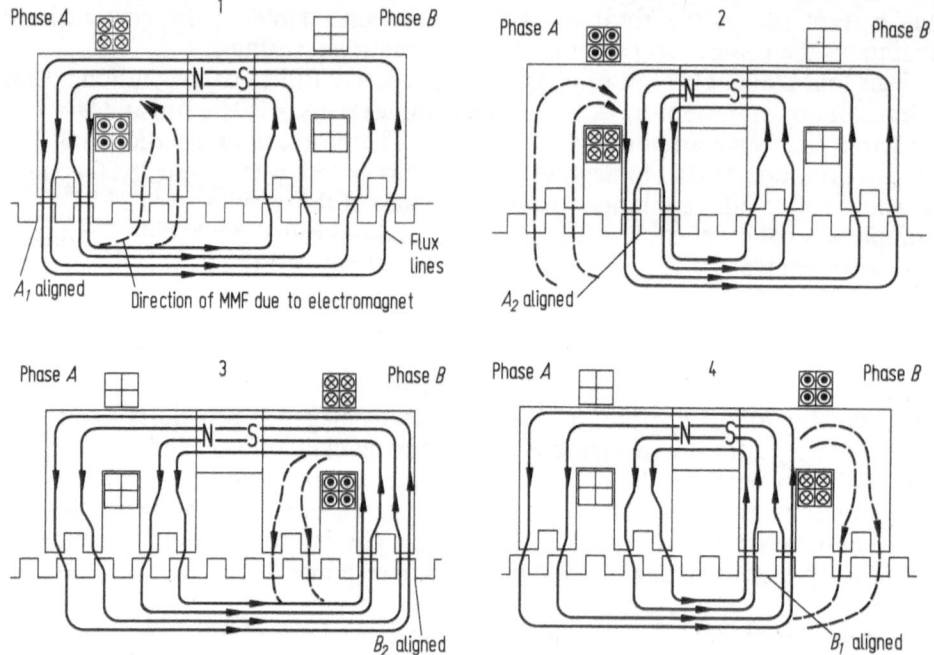

Figure 8.13 Four stages of field current directions to move a linear motor the distance of one tooth spacing

platen's teeth. The magnetic flux passing through the forcer and the platen results in a very strong normal attractive force between the two pieces. The purpose of the bearings is to maintain the precise clearance between the forcer and platen in the presence of this strong force.

When a current is placed in the field windings the resulting magnetic field tends to reinforce the permanent magnetic flux at one pole face and cancel it at another. By reversing the current the cancellation and reinforcement locations are reversed. Selectively applying current to A and B concentrates magnetic flux at any of the forcer's four pole faces. The face receiving the highest flux concentration will attempt to align its teeth with the platen. Figure 8.13 illustrates the four full steps of the forcer. These four steps result in the motion of one tooth interval to the right. Reversing the sequence moves the forcer to the left. Repeating the sequence will cause the forcer to continue movement. When the sequence is stopped the forcer stops with the appropriate tooth set aligned. At rest the forcer develops a restoring (holding) force that opposes any attempt to remove it from this equilibrium position. Too much of a disturbing force will cause the forcer to jump sharply and come to rest at an integer number of tooth intervals away from the original rest position. To obtain positions intermediate to the tooth interval a microprocessor controlled sine wave current is fed to the field windings as described for the angular step motors.

An application of a linear step motor is given in the next section.

8.2.10 Experiment 1   Calibration of an LVDT

Introduction
All transducers require calibration; that is, a measurement that relates their output quantity (usually voltage) to a *known* range of inputs of the appropriate physical quantity. It is generally accepted that the known input quantity must have an accuracy that is at least 10 times better than the resolution of the transducer being calibrated. If a displacement transducer is to be calibrated then, as noted in Section 8.2.1, a laser interferometer system would provide a means of determining displacement with an accuracy on the order of 20 to 50 nm (0.8 to 2.0 μin). It is the objective of this experiment to present a method for the automatic calibration of a LVDT using a laser interferometer and a linear step motor.

Experimental Apparatus and Instrumentation
Consider the experimental and equipment setup shown in Figure 8.14. The laser's retroreflector (the moving corner cube in Figure 8.2) was mounted to a plate that was itself mounted vertically to the forcer of the linear step motor. This vertical plate was the sensing surface for the LVDT. Straddling the linear motor's track was a stationary frame to which the LVDT was mounted. The frame, the track and the interferometer's beam splitter were mounted to a common platform. The forcer's motion was controlled by a set of instructions that were transmitted over a three-wire RS 232 connection to the linear motor's controller and amplifier. The forcer's resolution was specified as 10,000 steps/inch (0.0001 inches) and the

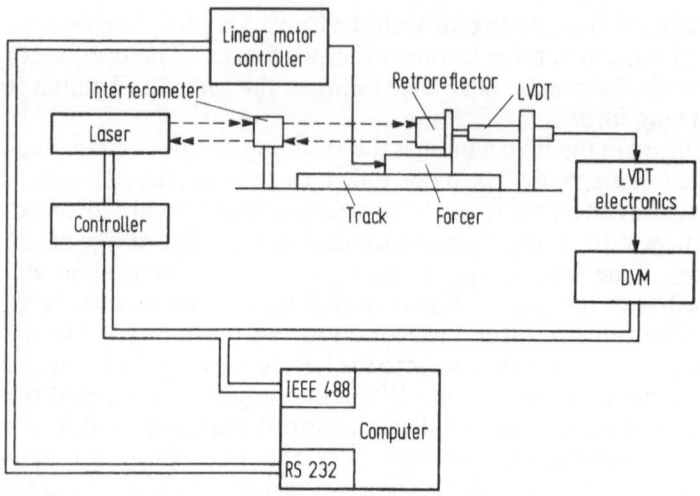

Figure 8.14 Experimental setup and instrumentation for the calibration of an LVDT using an ac laser interferometer

magnitude of the forcer's velocity and acceleration were programmable. The laser interferometer's measured and digitized displacements were transferred to the computer via an IEEE 488 interface. The output voltage from the LVDT's electronics was recorded by a 6½-digit DVM, which also communicated with the computer via an IEEE 488 interface. The particular LVDT that was tested had a full scale output voltage of approximately ±10 V over the corresponding displacement range of ±0.25 inches (±6 mm). The manufacturer's stated sensitivity for the LVDT was 39.812 V/inch ±0.7%.

Experimental Procedure
The sensing surface was positioned against the LVDT until its output was approximately -10 V. The forcer was then moved away from the LVDT in 20 equal steps of 0.025 inches each and then back towards the LVDT in the same manner. After each incremental step of the forcer the position of the vertical sensing surface and the output voltage of the LVDT were recorded by the computer. The amount of time that the program waited between the transfer of the command for the controller to move the forcer and for the DVM and the interferometer to have converted their inputs to a correct digital output was, at a minimum, the sum of the time of travel of the forcer (t_p in Table 7.1), which was a function of the velocity and acceleration selected, and the greater of the sampling speed of the DVM and the response time of the interferometer's digitizing process (t_7 in Table 7.1).

Using Eqs.(1.9) to (1.11) on the test results gave a sensitivity (slope) of 39.863 V/inch ±1.02% at the 95% confidence level, which compared favorably to that given by the manufacturer.

195

It is mentioned that since the laser optics were placed on the moving element of the step motor, it too could have been calibrated. However, this would have required a different set of step sizes, ranges and sequences.

8.2.11 Experiment 2   Measurement of Roundness

Introduction
In the manufacture of parts machined on lathes in which the roundness of the finished part is of major importance one must know the radial motion error of the machine's spindle. The magnitude of this radial error determines whether or not the machine has the ability to attain the desired roundness. Roundness is the deviation of the perimeter of the part from a perfect circle The objective of this experiment is to describe a procedure from which both the roundness of the part and the spindle's radial motion error can be determined.

Theory
Consider two almost concentric cylinders rotating about the axis of the bottom cylinder (spindle) as shown in Figure 8.15a. The measured deviation of the top cylinder's radial displacement is a function of (i) the eccentricity caused by the offset of the two cylinders' axes of rotation as defined by c and α, (ii) the roundness of the top cylinder $d_c(\theta)$ and (iii) the radial motion error of the spindle $d_s(\theta)$. If a displacement transducer measures only the radial motion of the perimeter of the cylinder $d(\theta)$, then if $c \ll R$ (Estler and Magrab [1985])

$$d(\theta) = d_c(\theta) + d_s(\theta) + c\cos(\theta - \alpha).$$

Now consider two configurations of the cylinder with respect to each other and with respect to the location of a displacement transducer as shown in Figures 8.15b and 8.15c. If it is assumed in Figure 8.15b that the deviation from the mean radius

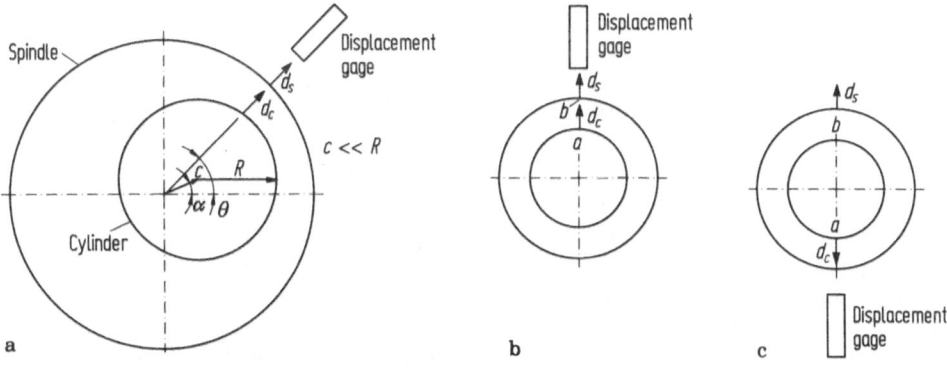

Figure 8.15   (a) Relation of the center of the cylinder to the center of the spindle   (b) displacement gage and cylinder in their original locations   (c) displacement gage and cylinder rotated 180° with respect to their original locations

at a point a on the surface of the cylinder is radially outwards and that the motion of the spindle is also radially outwards, then the displacement transducer measures

$$d_0(\theta) = d_c(\theta) + d_s(\theta) + c\cos(\theta - \alpha). \qquad (8.25)$$

If now the top cylinder and the displacement transducer are rotated 180° with respect to the location b on the spindle, then since the radial displacements of the cylinder and the spindle are still the same, each outwards as shown in Figure 8.15c,

$$d_{180}(\theta) = d_c(\theta) - d_s(\theta) + c'\cos(\theta - \alpha') \qquad (8.26)$$

where c' and α' define the location of the cylinder's new offset with respect to the spindle's axis.

If in each of these orientations (0° and 180°) the radial displacements are recorded at N nominally equally spaced positions $\theta_i, i = 1, 2, \ldots, N$, where $0° \le \theta_i < 360°$, then using an appropriate nonlinear curve fitting algorithm to $d_0(\theta)$ and $d_{180}(\theta)$, respectively, c, c', α and α' can be determined. Then Eqs.(8.25) and (8.26) can be rewritten as

$$e_0(\theta_i) = d_0(\theta_i) - c\cos(\theta_i - \alpha) = d_c(\theta_i) + d_s(\theta_i)$$

and

$$e_{180}(\theta_i) = d_{180}(\theta_i) - c'\cos(\theta_i - \alpha') = d_c(\theta_i) - d_s(\theta_i).$$

Solving for $d_c(\theta_i)$ and $d_s(\theta_i)$ yields

$$d_c = [e_0(\theta_i) + e_{180}(\theta_i)]/2$$

$$d_s = [e_0(\theta_i) - e_{180}(\theta_i)]/2$$

which are the roundness of the cylinder and the radial motion error of the spindle, respectively, at each θ_i.

Experimental Apparatus and Instrumentation
A horizontal lathe had an incremental encoder (recall Section 8.2.8) attached to its spindle as shown in Figure 8.16. Centered in the middle of the spindle was a right circular cylinder. The roundness of this cylindrical artifact was on the same order of magnitude as that expected for the radial error of the machine's spindle. The radial displacement of the cylinder with respect to an arbitrarily chosen reference point was measured by a capacitance gage whose sensing area was relatively small compared to the curvature of the cylinder and whose response went to dc (0 Hz; recall Eq.(8.21)). The sensitivity of the gage at an offset of 0.002 inches was approximately 2.5 mV/μin. The output of the capacitance gage was connected to a digital oscilloscope, which was an instrument comprised of a differential input amplifier, a 12-bit (without sign) A/D converter, an 8000-word memory for the digitized data and a CRT to display the digitized data. It did not have anti-aliasing filters.

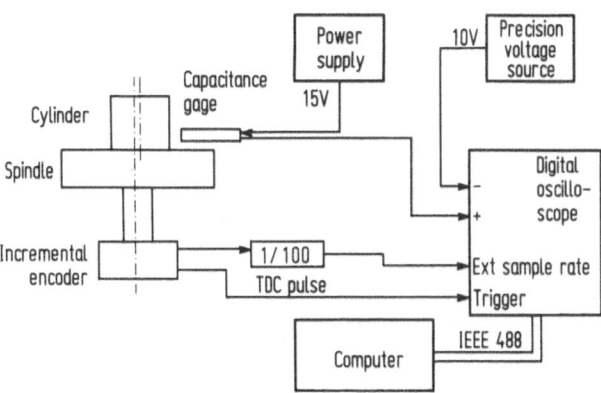

Figure 8.16 Experimental setup and instrumentation for the measurement of the roundness of the spindle

Experimental Procedure

As shown in Eq.(8.21) the output voltage of the capacitance gage is comprised of two parts. The first is a constant voltage that is proportional to the offset distance, 0.002 inches in this case. The second is the portion due to the deviation $d(\theta)$. The magnitude of this deviation was expected to be less than 20 μin. Thus the maximum change in the output voltage of the capacitance gage was 50 mV (20 μin $\times$ 2.5 mV/μin). At the offset of 0.002 inches the output of the capacitance gage was approximately 10.0 V. If the input range of the digital oscilloscope's amplifier was set to 10 V full scale then the resolution of the digitized input voltage would have been 2.44 mV (10 V $\div 2^{12}$), which corresponded to a resolution of approximately 1 μin. To improve the resolution a very stable, low noise and ripple 10 V dc voltage was placed at the negative input of the digital oscilloscope's differential input amplifier. This cancelled most of the portion of the capacitance gage's output that was due to the offset distance. (Recall from Section 4.4.2 that the actual amount of cancellation is a function of the amplifier's CMRR at dc.) The input amplifier's full range setting was then set to 1 V, providing a resolution of 0.1 μin.

The output pulses of the incremental encoder were used to synchronize the capacitance gage's output with the cylinder's angular position. The A/D converter in the digital oscilloscope was triggered by the encoder's top-dead-center (TDC) pulse. The digital oscilloscope was set up so that it responded only to the first TDC pulse of each run. The encoder provided 36,000 pulses per revolution, which was decreased to 360 pulses per revolution with a divide-by-100 digital circuit. This provided a displacement reading every 1°. The digital oscilloscope digitized its input signal until its 8000-word memory was full, which corresponded to 22 complete revolutions (8000/360). These data were then transferred to the computer, averaged at each θ_i and analyzed according to the equations above. A typical spindle error plot is shown in Figure 8.17.

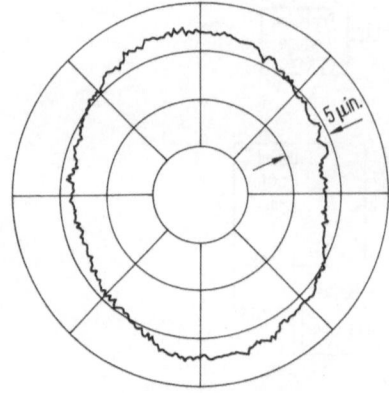

Figure 8.17 Typical spindle motion error for a high-accuracy lathe

8.3 Sound

8.3.1 The Capacitor Microphone

The capacitor microphone is another application of a capacitance gage. In this case, however, the capacitor is formed by two fixed surfaces, one rigid and the other flexible so that it deforms under the application of an external pressure over its surface. Consider the configuration of one type of capacitor microphone shown in Figure 8.18. The diaphragm of the microphone has an area of A_d and is subjected to a pressure p_i, which is not necessarily uniformly distributed over its area. The diaphragm is a distance d_o from the backplate that is interior to the microphone. This backplate has a sensing area A. It is supported by a disk that electrically isolates it from the microphone structure. In addition, the backplate has holes through it which provide damping B for the air mass M that is enclosed between the insulating disk and the diaphragm. The stiffness of the entrapped air is denoted K_s. In order for the gap distance to remain constant with changes in atmospheric pressure a very small capillary hole is drilled in the side of the microphone's casing. It is of length L, volume V and diameter d_t. The viscosity of the air is μ and its adiabatic bulk modulus is E_a.

For this configuration it has been shown that the relationship between e_o, p_i and Δd in the frequency domain is (Doebelin [1990])

$$\left(-M\omega^2 + j\omega B + K_s - C_d \frac{E_d^2}{d_o^2} \right) \Delta d + C_d \frac{E_d}{d_o} e_o = \frac{j\omega A_d \tau_l}{1 + j\omega \tau_l} p_i \quad (8.27)$$

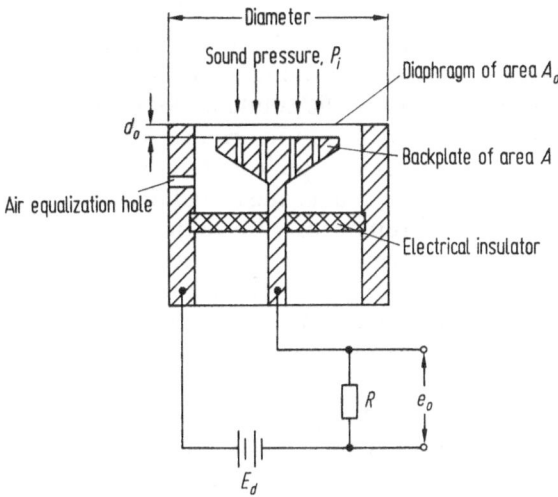

Figure 8.18 One type of capacitor microphone configuration

where C_d is the capacitance of the microphone, Δd is the change from d_o and

$$\tau_l \cong \frac{128\mu LV}{\pi E_a d_t^4} \quad \text{s}.$$

Substituting Eq.(8.19) into Eq.(8.27) yields

$$\frac{e_o}{p_i} = \frac{E_d}{d_o} A_d \frac{j\omega\tau_l}{(1+j\omega\tau_l)} \frac{j\omega\tau}{(a+jb)} \tag{8.28}$$

where $\tau = RC_d$ and

$$a = K_s - \omega^2(M + B\tau) - C_d E_d^2/d_o^2$$

$$b = \omega(B + K_s\tau) - M\tau\omega^3.$$

We see that the low frequency response is governed by two time constants, τ and τ_l. The circuit time constant is determined by the amplifier's input impedance [recall Eqs.(3.35) and (4.42)] and the value of the capacitance itself, C_d. On the other hand τ_l is only a function of the physical properties of the air equalization hole. The high frequency response is determined by the amount of damping B that is deliberately introduced to suppress the effects of the internal resonance caused by the mass of the entrapped air M.

Unfortunately microphone usage is not governed only by Eq.(8.28) but also by several other aspects (Magrab [1975]). If a microphone is placed in a sound field consisting of a plane wave that has a pressure p_o, the sound wave will be partly reflected from the diaphragm, causing an increase in the sound pressure. The magnitude of this increase depends on the wavelength of the sound, the dimensions of the microphone and the direction of travel of the sound waves with respect to the diaphragm. Often the objective is to measure the sound pressure that exists in a sound field before the microphone was placed in it. Hence we define the *free-field* response as the ratio of the rms output voltage from the microphone to the rms sound pressure of the undisturbed field; that is, the sound pressure at the microphone location with the microphone removed. We define the *pressure* response of a microphone as the ratio of the rms value of the output voltage of the microphone to the sound pressure uniformly distributed over the microphone diaphragm. Then the free-field response is equal to the pressure response plus the pressure increase or decrease caused by the microphone. When the pressure response is known, the problem of finding the free-field response is reduced to that of finding the pressure correction, which is a function of the shape and size of the diaphragm, the frequency of the sound wave and the sensitivity distribution over the diaphragm. It is therefore necessary for each microphone design to obtain the correction, which is then added to the microphone's pressure response. The correction becomes important when the ratio of the microphone diameter to the wavelength of sound becomes greater than 0.1. When the wavelength of sound is very small compared to the microphone's diameter and the sound waves are perpendicular to the diaphragm the sound pressure will increase 6 dB (double). When the sound waves are parallel to the diaphragm the output voltage will approach zero at high frequencies.

In the range between very low frequencies (where the pressure increase is negligible) and very high frequencies (where the pressure increase is 6 dB) a varying frequency response is obtained. A representative free-field correction is shown in Figure 8.19 for several directions of the incident sound. Shown in Figure 8.20 are the same data plotted as microphone directional sensitivity at several frequencies. These curves are obtained experimentally and should accompany each microphone. In order to use these correction curves one must know both the direction and frequency of the incident sound waves. Therefore one must Fourier transform the microphone's output voltage into the frequency domain prior to applying the corrections. In some applications the sound field is diffuse; that is, the sound's energy density is uniform and the mean acoustic power per unit area is the same in all directions. In this case one defines a microphone's *random incidence* response as the rms value of the free-field sensitivity for all angles of incidence. This response is usually obtained from an approximate formula that uses the microphone's experimentally determined free-field corrections. The various types of microphone usages are summarized in Figure 8.21.

Microphones are typically calibrated using battery operated devices into which the microphone is inserted. The microphone forms a closed cavity with the calibrating device and a known sound pressure for this enclosed volume is generated. The frequency of the sound is low, usually less than 1000 Hz (depending on the enclosed volume), and therefore the pressure distribution across the microphone's diaphragm is uniform. The calibration pressure is usually stated in dB referred to $20\,\mu Pa$ ($\cong 2.9 \times 10^{-9}$ psi).

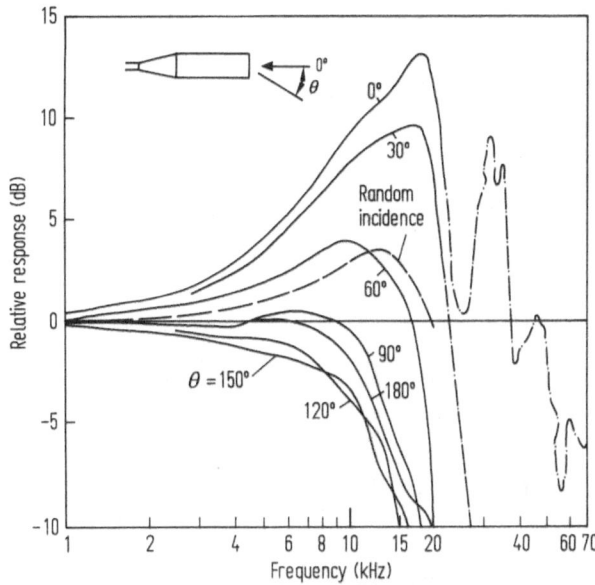

Figure 8.19 A set of microphone free-field correction curves for selected angles of incidence

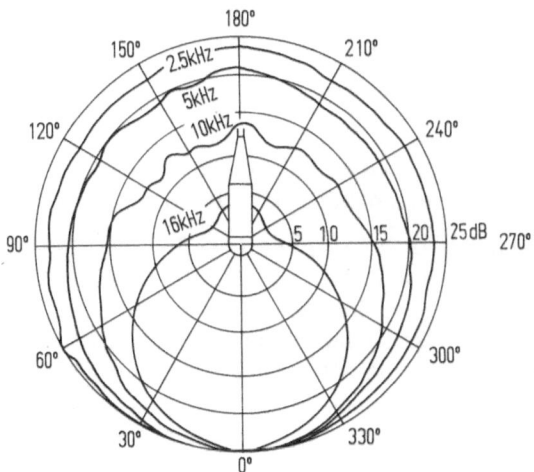

Figure 8.20 Figure 8.19 redrawn to obtain the microphone's directivity at several frequencies

Microphones used to measure sound in the environment are incorporated into battery operated devices called sound level meters. These devices have built into them the appropriate electronics and readout display from which sound pressure in dB *re* 20 μPa can be read directly.

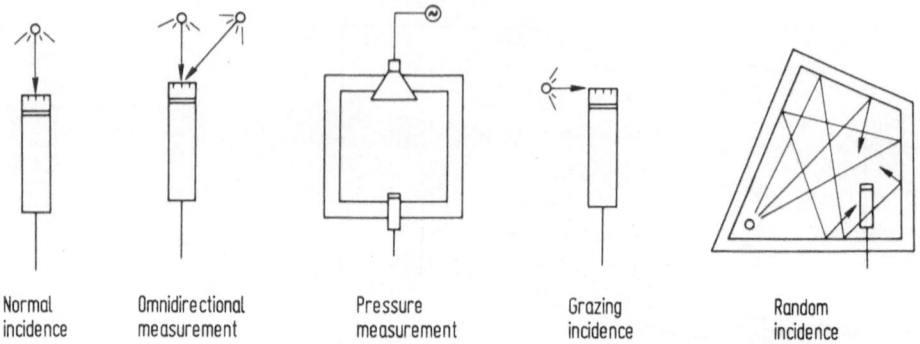

| Normal incidence | Omnidirectional measurement | Pressure measurement | Grazing incidence | Random incidence |

Figure 8.21   Microphone orientation with respect to several types of sound fields

8.3.2 Experiment 3   Measurement of the Speed of Sound in Air

Introduction
The measurement of the time of arrival of pulsed waveforms has many diverse applications. If the speed of propagation of the waveform in the medium is known, then the measurement of the time interval yields the distance travelled by the pulse. If, on the other hand, the distance is known then the time interval yields the speed of propagation in the medium. The objective of this experiment is to determine the speed of sound in air. The techniques that are used to make this measurement are also common to many other applications using time-of-arrival information.

Theory
The speed of wave propagation a medium can be determined by measuring the time t_o it takes a pulsed waveform to travel a distance d_o. In air this wave speed is called the speed of sound and is given by $c_o = d_o/t_o$. This method works best (i) with signals that do not change over their course of travel and (ii) in those environments (geometries) in which either reflections of the pulse from nearby boundaries do not interfere with the direct signal or the pulse itself is spatially focussed so that reflections from boundaries do not reach the measurement point until after the time interval for the direct path has been measured.

The speed of sound in air, when air is assumed to act as an ideal gas, is obtained from the relation

$$c_o = 20.05\sqrt{T} \quad \text{m/s}$$

where T is the air temperature in °K. At 20 °C (293 °K), $c_o = 343$ m/s.

It was shown in Eq.(3.12) that in order to maintain the original pulse shape of the signal both the signal's amplitude and phase information have to be preserved. Frequently sound waves in air are generated by loudspeakers which, because of their electro-mechanical and acoustical properties, do not replicate pulsed waveforms well. These difficulties can be overcome (i) by using a tone burst for the input signal which, as shown in Figure 2.10, requires a narrower bandwidth for the transmission of the pulse while still retaining enough information to make a time interval measurement and (ii) by determining the time of arrival of the pulse from the cross-correlation of the pulse measured at one location to that measured at the second location. The time-of-travel interval, therefore, is the correlation time at which these two measured pulses have the highest degree of similarity; that is, when the cross-correlation function is a maximum.

Consider the sine wave tone burst waveform of N periods and frequency f_o given by Eq.(2.69). The duration of the burst t_d is, therefore, $t_d = N/f_o$. If the speed of sound is c_o, then the spatial length x_p of this pulse as it propagates through the medium is

$$x_p = c_o t_d = N c_o / f_o .$$

Thus at 20 °C for example, if $f_o = 2000$ Hz and $N = 8$, then $x_p = 1.4$ m. In certain applications one must make sure that x_p "fits" into the medium so that it can be measured before any reflections arrive at the measuring points.

Referring to Figure 8.22 the distance between microphones is d_o and that from the loudspeaker to the nearest microphone is d_s. The distance between microphones should be reasonably large so that small errors in the measurement of d_o will have negligible effect on the calculation of c_o. If the pulsed waveforms at the two microphone locations are to be digitized, then the minimum number of samples

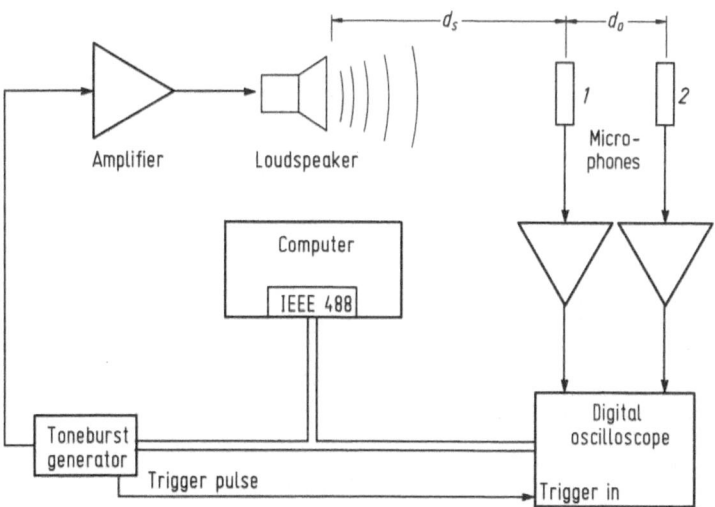

Figure 8.22 Experimental setup and instrumentation for the determination of the speed of sound in air

N_s from the time the waveform arrives at the first location until it is past the second microphone is

$$N_s = f_s \left(\frac{2N}{f_o} + \frac{d_o}{c_o} \right)$$

where $f_s = \beta f_o$, $\beta \gg 2$, is the sampling rate of the A/D conversion process and it has been assumed that $d_o \geq x_p$. If the digitization process is initiated from the moment the tone burst signal is generated then the minimum number of samples required is

$$N_s' = N_s + f_s d_s / c_o.$$

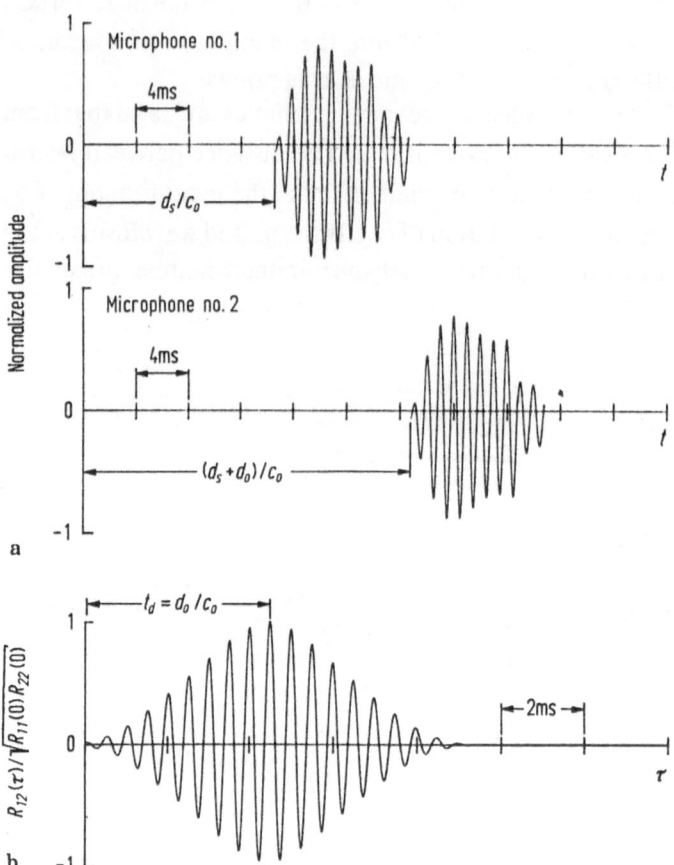

Figure 8.23 (a) Pulse waveforms received at microphone locations 1 and 2 (b) cross-correlation of the pulses in (a)

Since the resolution of the time-of-arrival measurement is $1/f_s$, f_s should be as high as practical.

Instrumentation and Experimental Procedure

The instrumentation and the experimental setup are shown in Figure 8.22. From a command issued by the computer the tone burst generator transmitted a single burst of eight periods of a 2000 Hz sine wave to the loudspeaker's power amplifier. The digital oscilloscope (see Section 8.2.11 for its description) was triggered by the tone burst generator to start acquiring data. The digital oscilloscope's sampling rate was set to 100,000 samples/s. The microphone spacing was d_s = 1.50 m and d_o = 2.50 m; thus $N_s' \cong 1966$. After the 8000-word memory of the digital oscilloscope was full the unit stopped taking data. The computer then transferred the data to its memory using the IEEE 488 interface, where it was analyzed using Eq.(2.62a). Typical results are shown in Figure 8.23, which, in the interests of clarity, have been edited to remove reflections. It was found that c_o=351 m/s.

8.4 Acceleration: The Accelerometer

A great majority of vibration measurements of acceleration, velocity and displacement must be made without reference to a fixed frame. This has lead to the development of seismic type accelerometers, which respond to the relative motion of the casing (housing) of the device and a mass located within it. A very commonly used accelerometer is the compression type that has a piezoelectric sensing element made of either quartz or PZT (lead zirconate titanate). A piezoelectric material is one that when mechanically deformed produces an electric charge. It also is a reversible sensor; that is, a voltage applied to it will cause it to mechanically deform. A device that uses this property is described in Section 9.3.

Consider a typical piezoelectric compression accelerometer construction shown in Figure 8.24. When the accelerometer is subjected to an acceleration, the mass will exert a force on the piezoelectric discs. The charge developed is proportional to the force, which in turn is proportional to the acceleration of the mass. As shall be shown, for frequencies much lower then the resonance frequency of the accelerometer assembly, the acceleration of the seismic mass is equal to the acceleration of its housing. We now model the accelerometer shown in Figure 8.24 as that shown in Figure 8.25. The seismic mass is M and the compression spring constant is k, which has damping c. The displacement of the base of the housing is x_i and that of the mass x_m. Referring to the free-body diagram in Figure 8.25 it is seen that the equation of motion is

$$M\frac{d^2x_o}{dt^2} + c\frac{dx_o}{dt} + kx_o = -M\frac{d^2x_i}{dt^2} \qquad (8.29)$$

where $x_o = x_m - x_i$ and d^2x_i/dt^2 is the acceleration of the base of the accelerometer. Taking the Fourier transform of Eq.(8.29) yields the transfer function

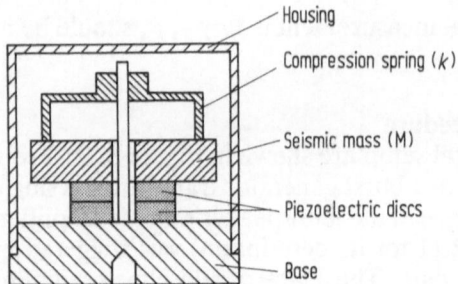

Figure 8.24 One configuration for a compression type piezoelectric accelerometer

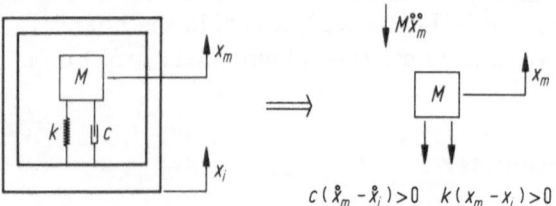

Figure 8.25 Forces and displacements on an accelerometer's components

[recall Eqs.(2.80) and (3.21)]

$$\frac{x_o}{a_i} = \frac{x_o}{\omega^2 x_i} = \frac{1}{\omega_n^2[\,1 - (\omega/\omega_n)^2 + 2j\xi(\omega/\omega_n)\,]} \tag{8.30}$$

where

$$2\xi = \frac{c}{\sqrt{kM}}, \qquad \omega_n = \sqrt{\frac{k}{M}}.$$

The quantity ω_n is the natural frequency of the accelerometer itself, assuming that its case is weightless. Since the accelerometer will always be mounted to an object that has mass, we re-examine the above model by including the mass of both the base and of the object to which the accelerometer is mounted. However, we restrict ourselves to finding only the new natural frequency. The new system is shown in Figure 8.26, from which it is seen that we get the two coupled equations

$$M\frac{d^2 x_m}{dt^2} + k(x_m - x_i) = 0$$

$$M_T\frac{d^2 x_i}{dt^2} - k(x_m - x_i) = 0$$

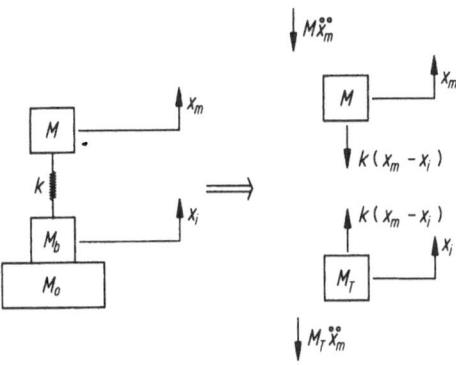

Figure 8.26 Forces and displacements on an accelerometer's components when the mass of the base and the mass of the object to which it is attached are included

which reduce to the single equation

$$\frac{d^2}{dt^2}\left[\frac{MM_T}{M+M_T}\frac{d^2x_i}{dt^2}+kx_i\right]=0$$

that has the non-zero natural frequency

$$\omega_n' = \omega_n\sqrt{1+\frac{M}{M_o+M_b}} \qquad (8.31)$$

where $M_T = M_o + M_b$, M_b is the mass of the accelerometer's base and M_o is the mass of the object to which the accelerometer is attached. When M equals M_T the maximum value of ω'_n is $\omega'_n = \omega_n\sqrt{2}$. This is referred to as the accelerometer's *unmounted resonance*. On the other hand when it is mounted to an object for which, say, $M_T = 10M$, $\omega_n' = \omega_n\sqrt{1.1} = 1.05\omega_n$. Thus the useable frequency range of the accelerometer varies with M_T. In general one should attempt to use the accelerometer in situations for which $M/M_T < 0.1$.

Returning to the piezoelectric element itself it has been found that its charge q is related to its deflection in one direction by

$$q = k_z x_o \quad coulombs \qquad (8.32)$$

where k_z is a constant in coulombs/unit of length, the same length units of the displacement x_o. The equivalent circuit for a piezoelectric accelerometer is shown in Figure 8.27, where it is seen that $E_i = q/C = k_z x_o/C$. The transfer function of the circuit is [recall Eq.(3.35)]

$$e_o = \frac{j\omega\tau}{1+j\omega\tau}E_i = \frac{j\omega\tau}{1+j\omega\tau}\left(\frac{k_z x_o}{C}\right) \qquad (8.33)$$

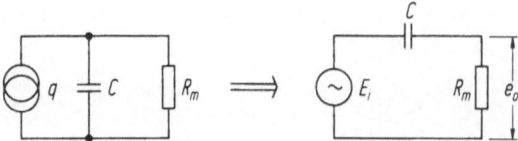

Figure 8.27 Equivalent circuits of a piezoelectric accelerometer

where $\tau = R_m C$. Combining Eqs.(8.30) and (8.33) yields the following expression that is a function of both the electrical (circuit) and mechanical properties of the system:

$$\frac{e_o}{\omega^2 x_i} = \left(\frac{k_z}{\omega'^2_n C}\right)\frac{j\omega\tau}{(1 + j\omega\tau)[1 - (\omega/\omega'_n)^2 + 2j\xi(\omega/\omega'_n)]} \quad (8.34)$$

where we have replaced ω_n by ω'_n in Eq.(8.30). As seen from Eq.(8.34) the lower frequency response behaves as a high pass filter shown in Figure 3.12b, whereas the high frequency response behaves as the second order system shown in Figure 3.5. Somewhere in between is the useable frequency range of the accelerometer. If the amplitude is to vary by less than, say, ±5% (±0.4 dB), then from Eqs.(3.36a) and (3.21) it is found that $3/\tau < \omega < 0.22\omega_n'$, provided $2\xi \ll 1$.

The sensitivity of the accelerometer depends on the type and size of the piezoelectric material and the value of the seismic mass. The larger the size (or number) of the piezoelectric material the greater the sensitivity, often at the expense of a lower natural frequency. Accelerometers also exhibit a sensitivity to accelerations applied at right angles to their main axis. This is called transverse sensitivity and is usually stated as some percentage of the accelerometer's main axis sensitivity. The transverse sensitivity also can vary with respect to the angular position in the plane perpendicular to the main (sensitive) axis. Accelerometer sensitivity is also affected by temperature and temperature changes and by the amount of strain of the base when mounted to the test object. The maximum acceleration at which a piezoelectric accelerometer should be used is typically one-third of its maximum shock rating.

The sensitivity of a piezoelectric sensor also depends on the type of amplifier to which it is connected, either a voltage amplifier or a charge amplifier. We re-examine Figure 8.27 and this time explicitly show the two different amplifier connections. Consider the simplified voltage amplifier illustrated in Figure 8.28a, where C_c is the cable capacitance and R_m is the input resistance of the amplifier [recall Figure 4.16]. The charge sensitivity S_q is related to the charge q by

$$q = S_q a \qquad \text{coulombs} \qquad (8.35a)$$

or

$$\frac{V_a}{a} = \frac{S_q}{C_a} \qquad (8.35b)$$

where α is the acceleration and C_a is the capacitance of the accelerometer. Referring to Figure 8.28a we see that the voltage sensitivity S_v is, after including the effects of the cable,

$$S_v = \frac{V_{ao}}{a} = \frac{C_a V_a}{C_T a} = \frac{S_q}{C_T} \quad \text{V/UA}$$

where $C_T = C_a + C_c$ and UA are the units of acceleration: m/s² or g($=9.8$ m/s²). Cable capacitance is not a constant, but a function of the length of cable used. Therefore, with a piezoelectric sensor the voltage sensitivity is a function of changes in C_c. Consequently voltage amplifiers are primarily used with relatively short cables (<1.5 m).

To overcome the cable capacitance problem the output from the piezoelectric element is fed into a charge amplifier, the configuration for which is shown in Figure 8.28b. Analysis of this circuit gives (Dally, et al. [1984])

$$\left[\frac{C_T}{A} + C_f\left(1 + \frac{1}{A}\right)\right]\frac{de_o}{dt} + \left(1 + \frac{1}{A}\right)\frac{e_o}{R_f} = \frac{1}{b}\frac{dq}{dt} \tag{8.36}$$

where A is the open loop gain of the first amplifier and $1/b$ is the gain of the second amplifier. Since A is usually greater than 10,000 Eq.(8.36) reduces to

$$C_f\frac{de_o}{dt} + \frac{e_o}{R_f} = \frac{1}{b}\frac{dq}{dt}. \tag{8.37}$$

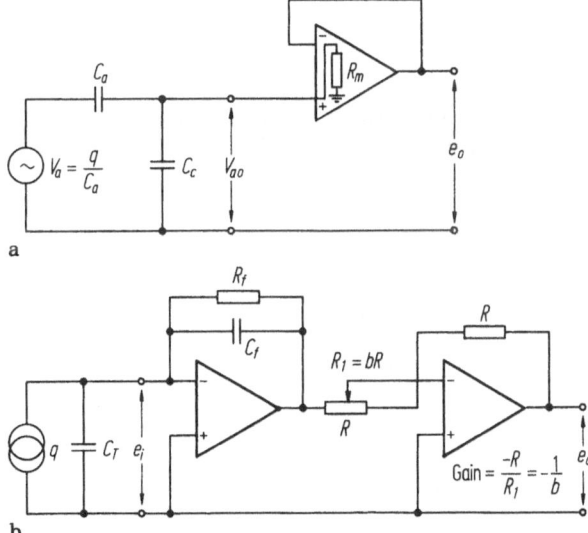

a

b

Figure 8.28 (a) Accelerometer connected to a voltage amplifier (b) accelerometer connected to a charge amplifier

Taking the Fourier transform of Eq.(8.37) gives

$$e_o = \frac{S_q a}{bC_f} \frac{j\omega R_f C_f}{1 + j\omega R_f C_f}$$

which is the response of a high pass filter whose cutoff frequency is given by [recall Eq.(3.35)]

$$f_c = \frac{1}{2\pi R_f C_f}. \tag{8.38}$$

It is seen that the voltage sensitivity is now independent of both the capacitance of the piezoelectric element and the cable. The combination of the gain $1/b$ and the feedback capacitance C_f are used to scale e_o to any convenient charge sensitivity. Also, by appropriately choosing the values of R_f and C_f the cutoff frequency can be made to approach zero. Commercially available charge amplifiers have an f_c as low as 2 μHz. Hence the charge amplifiers make it possible to use piezoelectric elements to measure both low frequencies and long duration events. Recall Figures 3.13 and 3.14.

Representative methods for the calibration of accelerometers can be found in Dally, et al. [1984] and Robinson, et al. [1987].

Various applications using accelerometers are given in Sections 8.6.2, 9.2 and 9.5.

8.5 Force, Torque and Pressure: The Strain Gage

8.5.1 Electrical Resistance Strain Gage

A strain gage converts an applied strain directly to a change in electrical resistance. The resistance R of a uniform metallic conductor of length L, cross-sectional area A and specific resistance ρ are related by

$$R = \frac{\rho L}{A}. \tag{8.39}$$

Taking the total differential of Eq.(8.39) we find that

$$\Delta R = \frac{L}{A}\Delta\rho + \frac{\rho}{A}\Delta L - \frac{\rho}{A^2}\Delta A$$

or, upon rearranging and using Eq.(8.39),

$$\frac{\Delta R}{R} = \frac{\Delta\rho}{\rho} + \frac{\Delta L}{L} - \frac{\Delta A}{A}. \tag{8.40}$$

The term ΔA represents the change in cross-sectional area of the conductor resulting from the applied axial load that causes the change in length ΔL. For this type of loading the axial strain of the conductor is

$$\epsilon_a = \frac{\Delta L}{L}$$

and the transverse strain is

$$\epsilon_t = -\nu \epsilon_a$$

where ν is Poisson's ratio of the conductor material. The change in area, therefore, is

$$\frac{\Delta A}{A} = \frac{A_f - A}{A} = \frac{A_f}{A} - 1 \qquad (8.41)$$

where A_f is the final cross-sectional area of the conductor. However, if the conductor is a wire of initial diameter d, then its final diameter d_f is

$$d_f = d(1 - \nu \epsilon_a). \qquad (8.42)$$

Therefore, substituting Eq.(8.42) into Eq.(8.41) gives, for circular cross-sections,

$$\frac{\Delta A}{A} = (1 - \nu \epsilon_a)^2 - 1 \cong -\nu \epsilon_a. \qquad (8.43)$$

Using Eq.(8.43) in Eq.(8.40) gives the final result

$$\frac{\Delta R}{R} = (1 + 2\nu)\epsilon_a + \frac{\Delta \rho}{\rho}. \qquad (8.44)$$

The sensitivity S_A is defined as

$$S_A = \frac{\Delta R / R}{\epsilon_a} = 1 + 2\nu + \frac{\Delta \rho / \rho}{\epsilon_a}. \qquad (8.45)$$

The first two terms on the right-hand side of Eq.(8.45) are due to the geometric changes in the wire: the first term is due to the change in the length of the wire and the second term to the change in its area. The third term is due to the variations in the number of free electrons and their increased mobility with applied strain. It has been experimentally shown that for the type of materials used for strain gages $2 < S_A < 4$. For this same group of materials $1 + 2\nu$ is around 1.6; therefore, the contribution due to the change in specific resistance ranges from 0.4 to 2.4.

Most resistance strain gages are the metal foil type, approximately 5 μm (200 μin) thick. This foil is photoetched onto a thin plastic film, which also serves as an electrical insulator between the gage and the object to which it is mounted. Since the gages are so thin, in order to get a workable nominal resistance they are made (for the measurement of the strain in one direction) in the form shown in Figure 8.29. The smallest size is approximately 0.2 mm and the longest 100 mm. Resistance values of these gages are mostly either 120 or 350 ohms.

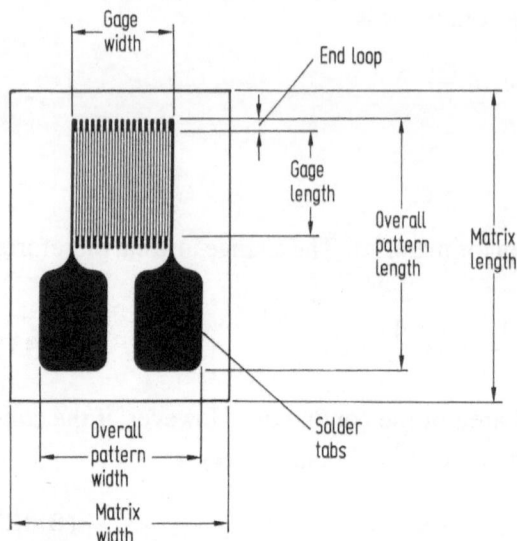

Figure 8.29 Typical strain gage configuration

In actual use a strain gage's change in resistance in the direction of its "grid" lines is related to the strain by

$$\frac{\Delta R}{R} = S_g \epsilon_a \qquad (8.46)$$

where S_g is the gage factor, or calibration constant, for the gage and is provided by the gage's manufacturer. Also, S_g is always less than S_A because the grid configurations used are always less responsive to strain than a straight uniform wire conductor.

Strain gages are temperature dependent; that is, $\Delta R / R$ is not necessarily zero for changes in temperature. The gage factor S_g changes linearly with temperature as $\alpha_a \Delta T$, where ΔT is the temperature change and α_a is the temperature coefficient of expansion for the gage material. On the other hand as the gage's grid elongates by $\alpha \Delta T$, the object elongates by $\beta \Delta T$ and the resistivity changes by $\gamma \Delta T$. Therefore,

$$\frac{\Delta R}{R} = [S_g(\alpha - \beta) + \gamma]\Delta T . \qquad (8.47)$$

In order for $\Delta R / R$ to equal zero for a modest range of ΔT,

$$\gamma = S_g(\beta - \alpha) . \qquad (8.48)$$

Thus, if we can choose a gage material in which γ is approximately equal to zero over a range of ΔT, and we if can select α that closely matches β, the temperature effects can be greatly minimized.

Strain gages are used with the Wheatstone bridge described in Section 6.4. Consider the case of one active strain gage. Equation (6.27) becomes, with $\eta = 0$,

$$\Delta e_0 = \frac{r}{(1+r)^2}\frac{\Delta R_1}{R_1}V_o = \frac{r}{(1+r)^2}S_g\epsilon_1 V_o. \tag{8.49}$$

The circuit sensitivity is

$$S_c = \frac{\Delta e_o}{\Delta R_1/R_1} = \frac{rV_o}{(1+r)^2}. \tag{8.50}$$

Again it appears that we can increase the sensitivity by increasing V_o. However, as with potentiometers [recall Eq.(8.24)] the power dissipated by the transducer, $P_T = I_{max}^2 R_1$, determines its upper limit. Therefore,

$$V_{max} = I_{max}(R_1 + R_2) = (1+r)\sqrt{P_T R_1}$$

and Eq.(8.50) becomes

$$S_c = \frac{r}{1+r}\sqrt{P_T R_1}.$$

Although the circuit sensitivity can be increased by increasing any combination of r, R_1 and P_T, it is best to increase the product $P_T R_1$.

The lower limiting voltage is a function of both the noise in the amplifier and the noise generated by the strain gage resistance itself [recall Eq.(4.10)]. The latter effect can be significant when the gage is used to measure small strains over a large bandwidth.

Lastly there are two effects that can be compensated for: changes in resistance of the gage due to temperature and the effects of lead wire and changes in its resistance due to temperature. The preferred way to deal with these potential sources of error for one active gage is shown in Figure 8.30, in which three lead wires and one dummy (non-strained) gage are used. In this case Eq.(6.25) becomes, with $\eta = 0$,

$$\Delta e_o = \frac{rV_o}{(1+r)^2}\left[\left(\frac{\Delta R_g}{R_g + R_L}\right)_\epsilon + \left(\frac{\Delta R_g}{R_g + R_L}\right)_{\Delta T} - \left(\frac{\Delta R_g}{R_g + R_L}\right)_{\Delta T}\right]$$

which simplifies to

$$\Delta e_o = \frac{rV_o}{(1+r)^2}\left(\frac{\Delta R_g}{R_g + R_L}\right)_\epsilon.$$

Thus, the effects of any temperature changes that induce changes in resistance of either R_g or R_L are cancelled.

For a detailed discussion of the theory and use of strain gages see Dally and Riley [1978].

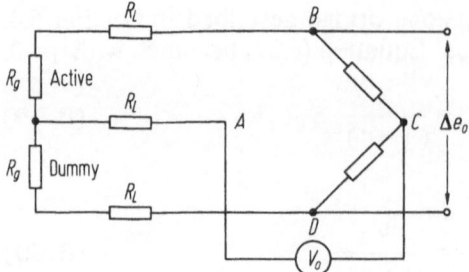

Figure 8.30 Single active strain gage and a bridge configuration used to cancel the effects of temperature on the strain gage and the lead wires

8.5.2 Strain Gage Transducers

Strain gages can be combined very effectively with elastic elements to form transducers that can measure force, torque, and pressure. The characteristics of the transducers are determined by the size and shape of its elastic member, the material used and the number and orientation of the strain gages. Three representative devices will now be examined. An application of a fourth configuration, called a load washer, is given in Section 9.4.

Axial Load Cell
Consider a bar of rectangular cross-section of area A that is subjected to an axial compressive or tensile force P as shown in Figure 8.31a. If the bar has a Young's

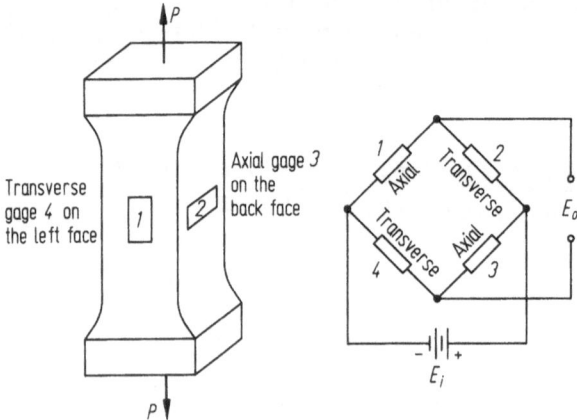

Figure 8.31 (a) Axial load cell strain gage orientations (b) corresponding placement in a Wheatstone bridge

modulus E and a Poisson's ratio ν, then the axial strain is

$$\epsilon_a = \frac{P}{AE} \tag{8.51a}$$

and the transverse strain

$$\epsilon_t = -\nu\epsilon_a . \tag{8.51b}$$

If four gages are attached to the rod as illustrated in Figure 8.31a and they are connected to a Wheatstone bridge as shown in Figure 8.31b, then from Eqs.(8.46) and (8.51)

$$\frac{\Delta R_1}{R_1} = \frac{\Delta R_3}{R_3} = S_g\epsilon_a = \frac{S_g P}{AE} \tag{8.52a}$$

and

$$\frac{\Delta R_2}{R_2} = \frac{\Delta R_4}{R_4} = S_g\epsilon_t = \frac{-\nu S_g P}{AE} . \tag{8.52b}$$

Substituting Eqs.(8.52) into Eq.(6.25), with $\eta = 0$ and $r = 1$, gives

$$\Delta e_o = \frac{P V_o}{C_o}$$

where

$$C_o = \frac{2AE}{S_g(1+\nu)} .$$

We see that the output voltage is linearly proportional to P. Load cells of this type are typically rated at full scale as

$$\frac{P_{max}}{C_o} = \left(\frac{\Delta e_o}{V_o} \right)_{FS} = a_o \ \ \text{mV/V} .$$

where a_o is typically on the order of 3. Then for any load P

$$P = \frac{\Delta e_o / V_o}{(\Delta e_o / V_o)_{FS}} P_{max}$$

where P_{max} and a_o are given by the manufacturer.

An application of this type of transducer is given in Section 8.5.6.

Torque Cell
A torque cell usually consists of a circular rod of diameter D that is subjected to a torque T. It has four strain gages positioned at 45° to the rod's axis as shown in Figure 8.32a. It can be shown (Dally, et al. [1984]) that the principal strains act in the direction of the gages as shown in the figure and are given by

$$\epsilon_1 = C_1 T , \qquad \epsilon_2 = -\epsilon_1 \tag{8.53}$$

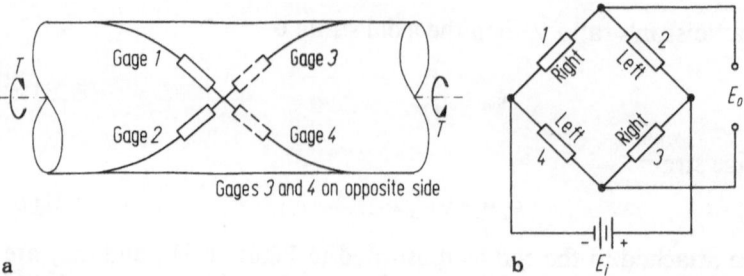

Figure 8.32 (a) Torque cell strain gage orientations (b) corresponding placement in a Wheatstone bridge

where

$$C_1 = \frac{16(1+v)}{\pi D^3 E}$$

and E and v are as previously defined. Therefore, from Figure 8.32b we see that

$$\frac{\Delta R_1}{R_1} = -\frac{\Delta R_2}{R_2} = \frac{\Delta R_3}{R_3} = -\frac{\Delta R_4}{R_4} = S_g \epsilon_1$$

and Eq.(6.25), again with $\eta = 0$ and $r = 1$, gives

$$T = \frac{1}{C_1 S_g} \frac{\Delta e_o}{V_o}.$$

Diaphragm Pressure Transducers

Consider a thin circular plate clamped along its outer edge. If the radius of the plate is R_o, its thickness h and it is subjected to a uniform pressure P_o over its surface, then the radial strain is given by (Dally, *et al.* [1984])

$$\epsilon_{rr} = \frac{3P_o(1-v^2)}{8Eh^2}\left(R_o^2 - 3\bar{r}^2\right)$$

and the tangential strain by

$$\epsilon_{\theta\theta} = \frac{3P_o(1-v^2)}{8Eh^2}\left(R_o^2 - \bar{r}^2\right)$$

where $\bar{r}$ is the radial distance from the plate's center and $0 \leq \bar{r} \leq R_o$. It is seen that the maximum value of $\epsilon_{\theta\theta}$ occurs at $\bar{r} = 0$, the center of the plate. The maximum (negative) value of ϵ_{rr} is at $\bar{r} = R_o$, its outer boundary. In order to take advantage of these maximum strains a commercially available strain gage consisting of four gages as shown in Figure 8.33 has been created. The two inner most gages are the

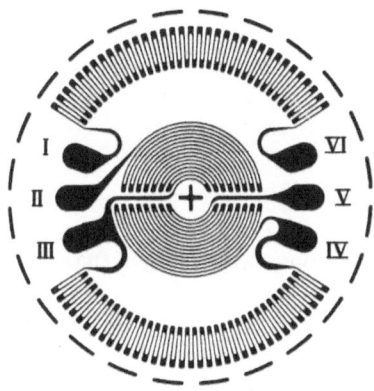

Figure 8.33 Specially constructed diaphragm pressure transducer strain gage

elements R_1 and R_3 while the two outermost gages are the elements R_2 and R_4, since they record negative strain. Notice that because of their physical size the gages cannot be located at the points where they give the maximum values of strain. Therefore the preceding equations cannot be used directly. However the manufacturer does provide the following formula from which the pressure can be determined:

$$P = \frac{1.22Eh^2}{R_o^2(1-\nu^2)}\left(\frac{\Delta e_o}{V_o}\right).$$

Applications of the diaphragm pressure transducers are given in Sections 8.5.5 and 8.8.3.

8.5.3 Dynamic Force Transducers

The strain gage transducers described in the previous section are primarily used to measure static and low frequency (<25 Hz) dynamic forces. This bandwidth limitation is not because of the strain gage or its subsequent electronics. It is limited by the useful (non-resonant) bandwidth of the elastic member to which it is attached. To measure force over a broad bandwidth (to 15 kHz) piezoelectric transducers are used. The application of a force to the piezoelectric material produces a measurable change in its charge q.

These transducers are typically used in two types of applications. In the first type they are placed directly in contact with the object whose force is to be measured, such as punch presses and bearing supports. In the second type of application the force gage is sandwiched under high preload between two structures, one "fixed" and one "floating". The floating structure is essentially a mass connected to the fixed structure through a spring, whose spring constant is the stiffness of the force

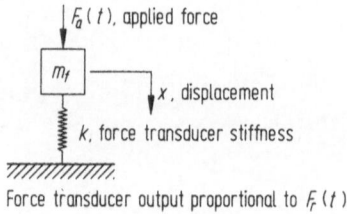

$F_a(t)$, applied force

m_f

x, displacement

k, force transducer stiffness

Force transducer output proportional to $F_r(t)$

Figure 8.34 Schematic representation of a force gage placed between a fixed support and a floating structure of mass m_f

gage as shown in Figure 8.34. Consequently, if $F_a(t)$ is the force applied to the center of m_f, the mass of the floating structure, and $F_r(t)$ is the force sensed by the transducer, then

$$F_a(t) = F_r(t) + m_f \frac{d^2 x}{dt^2}.$$ (8.54)

Since $F_r(t) = kx(t)$, where k is the stiffness of the force gage, the Fourier transform of Eq.(8.54) yields

$$F_a(\omega) = F_r(\omega)[1 - (\omega/\omega_n)^2]^{-1}$$

where $\omega_n^2 = k/m_f$. For the gage to be affected by the floating mass by less than, say, 5%, $\omega < 0.2\omega_n$ [recall Eq.(8.34) ff]. If the fixed side of the transducer is not fixed but attached flexibly, then the above analysis must be modified to include this additional degree of freedom.

Since piezoelectric materials are capacitive devices, dynamic force gages are typically used with charge amplifiers.

An application of this transducer is given in Section 9.5.

8.5.4 Pneumatic Actuator

A pneumatic actuator is a device that converts a control signal (current) into a large force to manipulate some mechanical element. The converter for such a system is a flapper/nozzle system shown in the top portion of Figure 8.35. Regulated low pressure air (usually less than 20 psig) passes through a fixed flow restriction into a tube, one end of which has a nozzle that vents to the atmosphere. The other end goes to the mechanical device that is to respond to the pressure in the pneumatic line. Placed in front of the nozzle is a pivoted metal rod called the flapper. The position of the rod with respect to the nozzle's orifice is controlled by the amount of current in the solenoid, which determines the magnitude of its attractive force on the flapper and, consequently, the size of the gap between the orifice and the flapper. When the gap is large the pressure in the line decreases,

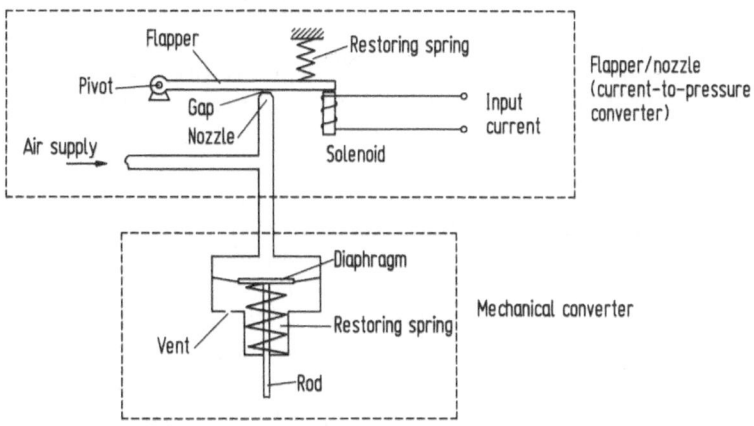

Figure 8.35   Pneumatic actuator

when it is small the pressure is large. Although the relationship between the gap size and the line pressure is nonlinear, there is a range in which the line pressure will vary linearly linearly with the gap distance. Typically, for an input air supply pressure of 20 psig the linear range is between 3 and 15 psig.

In the linear range of the gap's travel the resulting line pressure variation can be used to generate a relatively large linear translation as shown in the bottom of Figure 8.35. A rod is connected to a spring loaded diaphragm that has the varying line pressure on one side and is vented to the atmosphere on its other side. The diameter of the diaphragm determines the maximum force on the restoring spring. At the maximum force the rod's maximum travel is determined by the spring's stiffness. The response of this device is that of a second-order system (recall Section 3.3.1).

A typical application for this device in the automation of control valves, an example of which is given in Section 8.8.4.

8.5.5 Experiment 4   Determination of Leakage in Plastic Food Containers

Introduction
Changes in food container packaging and processing in recent years have lead to a shift in the type of containers being used: from rigid containers with metal lids that initially contain food at an elevated temperature to flexible containers with non-metallic lids that initially contain food at nominally room temperature. The food processes that use elevated temperature create, upon cooling, a partial vacuum within the sealed container, and most existing methods judge the presence of a leak by measuring the deflection of the metal lid. These methods, however, will not work with flexible containers because (i) no vacuum is created, (ii) the lid is non-metallic and (iii) due to the lid's flexibility it is never in a fixed position with

220

respect to any reference point on the container. Typically the lids of flexible containers are sealed with heat applied to its perimeter and are designed to be easily peeled open by the consumer. The objective, therefore, is to determine a way in which the integrity of the seal of the lids of flexible food containers can be measured.

Proposed Method
Consider a flexible container that has had its bottom and its four sides restrained from moving. If one of its sides is now compressed a fixed amount the air trapped between the food or liquid in the container and the container itself forces the flexible top of the container to expand outwards. If the top is prevented from expanding, then the pressure in the air pocket increases further. If the surface that is restraining the lid's expansion contains a transducer whose output is proportional to this restraining force, then this measured output force is proportional to the container's internal pressure. When the force remains constant there is no leak. If the force decreases with time (relaxes) there is a leak in the lid and/or its seal.

The variation of the internal pressure $p(t)$ of a pressurized vessel that discharges to the atmosphere through a small opening can be approximated by (Karditsas [1982])

$$\frac{p(t)}{p_o} = [Ak_3t\sqrt{T_o}/V + 1]^{-2k/(k-1)} \tag{8.55}$$

where p_o and T_o are the initial pressure and temperature (°K) of the entrapped air, respectively, A is the cross-sectional area of the opening (m²), V is the volume of the entrapped air (m³), and for air $k = 1.4$ and $k_3 = 3.312$ m/s/$\sqrt{°K}$. For a container with an air pocket volume $V = 32.8$ cm³ and a hole in the lid of diameter 0.20 mm, Eq.(8.55) shows that it takes approximately 1.8 s for the internal pressure to decrease to 50% of its initial pressure; that is, $p(1.8) \cong 0.5p_o$. Although Eq.(8.55) is for choked flow (sonic velocity at the hole) it does indicate the order of magnitude of the time in which one can expect the pressure to decrease 50%. Thus, if the internal pressure or, equivalently, the force exerted by the entrapped air on the container's lid, is measured by an appropriate transducer, it is feasible to detect this relaxation and, consequently, the presence of a hole within a relatively short period of time.

Experimental Apparatus
A schematic of the method's apparatus is shown in Figure 8.36. The flexible container fit snugly into a specially constructed well, which was open at the top. Since the walls of the well were rigid compared to the container's walls and its shape conformed very closely to that of the container, the container was prevented from expanding laterally when compressed. A small opening on one side of the well permitted a rounded end of a rod that was attached to a pneumatically actuated piston to partially compress one wall of the container. The magnitude of the displacement of the piston could be adjusted by varying the position of the piston with respect to the well. The piston itself was double-acting so that air pressure

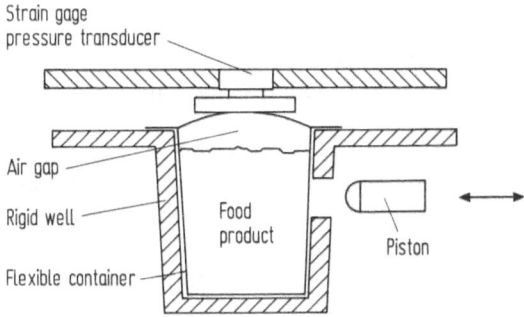

Figure 8.36 Apparatus to determine leakage in flexible food containers

was required to move it in both directions: in and out. The lateral compression of the container forced the container's lid to bulge against the circular sensing plate of the transducer, which was rigidly mounted above the container's lid as shown in Figure 8.36. The transducer's mounting fixture was removable to permit the placement of the container into and out of the well. The transducer was a strain gage pressure sensor whose sensing element was that shown in Figure 8.33. To the gage's sensing surface was bonded a disk as shown in Figure 8.36 which, upon compression of the container, came into contact with most of the lid.

Instrumentation
The equipment block diagram of the experimental setup used to test the proposed leakage detection method is shown in Figure 8.37. The transducer's output signal

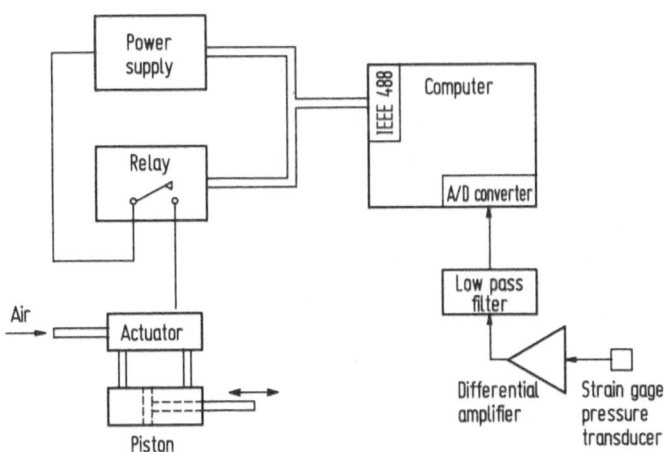

Figure 8.37 Experimental setup and instrumentation for the determination of leakage in flexible food containers

was amplified by a differential amplifier, which was powered by a 15 V power supply. The amplifier's output, in turn, was digitized with an A/D converter and the result stored in the computer's memory. Another power supply was used to power the piston's actuator, one lead of which was opened and closed with a computer controlled switch.

In order to remove some of the mechanical ringing from the transducer's signal a low pass RC filter was introduced between the output of the transducer's amplifier and the input of the A/D converter. The filter's cutoff frequency was set to 75 Hz, which was at least 10 times higher than the bandwidth of the pressure relaxation phenomenon (approximately 5 Hz for a hole with a diameter of 0.5 mm).

Experimental Procedure
The computer controlled test for leakage in the flexible container performed the following functions: (i) initialized the appropriate instruments and the A/D converter, (ii) retracted the piston (if already retracted nothing happened), (iii) set the values of the test duration (typically 1 s), the sampling rate (typically 200 samples/s) and the number of times the test was to be repeated, (iv) ran the test by closing the relay to the piston and starting the A/D converter, (v) plotted the variation of the transducer's output with time to both the screen and the plotter after the data were acquired and (vi) saved the data to a file. Typical shapes of the force-time (relaxation) responses as a function of hole size are shown in Figure 8.38. For a sealed container it is seen that after the filtered mechanical ringing died out the normalized force remain almost constant with time. The actual magnitude of the steady-state value of the force depended on the volume of

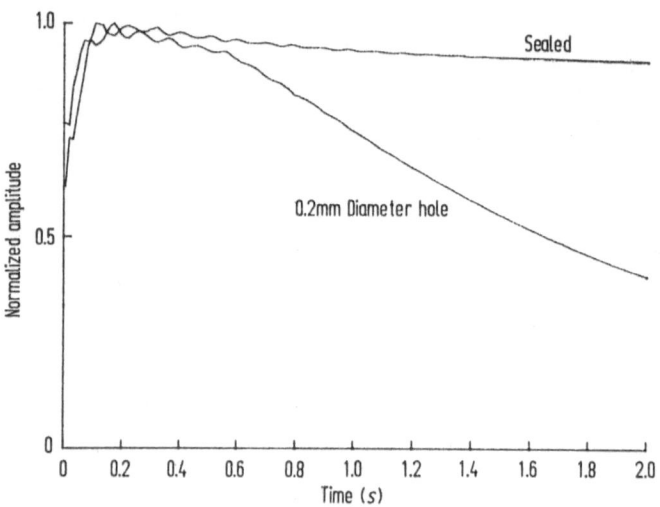

Figure 8.38 Typical relaxation response of a plastic food container with and without a leak in its lid

entrapped air in the container and the amount of displacement of the piston. For relatively large diameter holes (> 0.40 mm) the normalized force decreased quickly to a steady-state value of almost zero. For a 0.20 mm diameter hole the force decayed somewhat exponentially, with the time constant τ of the decay being related to the area of the hole by Eq.(8.55). To create these holes stainless steel wires of known diameter were embedded into the seal of the lid during the sealing operation. These wires were then removed prior to testing, presumably leaving a hole in the seal of equal diameter.

Based on an extensive series of tests, of which the results in Figure 8.38 are representative, the following method was used to determine from the data whether or not there was a leak in the lid's seal for holes greater than 0.20 mm in diameter. Two characteristic values of the transducer's output signal were required: (i) the time constant τ of the exponential decay and (ii) the mean value V_{ss} of the signal for the time interval 0.9 to 1.0 s. The values of the signal at ten equally spaced points in the time interval 0.3 to 0.8 s were used to determine the value of τ. The time constant was obtained by fitting a straight line to the natural logarithm of the amplitude at each point. The slope of the line was the estimate of $1/\tau$. If the value of τ was less than 10 s, then pressure relaxation was present and a leak had been detected. This case most often represented holes that ranged from 0.20 to 0.50 mm in diameter. For larger holes, however, the force signal usually settled in less than 0.3 s, thus resulting in a value of τ less than 10 s. In order to account for these cases the value of V_{ss} was examined and it was found that when V_{ss} was less than 0.5 V a relatively large hole was present. The total time it took to start the piston, digitize the data and then analyze the data to make a decision as to whether or not a leak was present was on the order of 1.5 s.

8.5.6 Experiment 5   Determination of Process Variables in Metal Forming

Introduction
A manual manufacturing process that was used to make resilient metallic ring seals of various diameters through several stages of forming was analyzed prior to redesigning it to be a semi-automated process. The final configuration of the ring seal is shown in Figure 8.39. The process variables that could be controlled during the forming of these ring seals were the forming force, the speed of the mating forms (rollers), the feed rate of the rollers, the number of revolutions required to

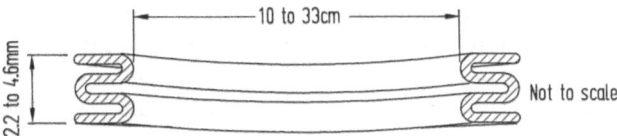

Figure 8.39   Cross-section of a ring seal

form the ring to its final shape and the deviation of the ring's plane with respect to the plane of the rollers. The effects of these variables on the resulting geometric properties of the final product were determined manually. The values of those process variables producing the best geometric features were considered the optimum ones and the subsequent design was based on them. However only the method used to acquire the data from which these optimum process variables were determined will be described.

Experimental Apparatus and Instrumentation

In order to achieve precise control and measurement of the input process parameters, an existing manually-operated forming machine was retrofitted so that the feed of the moveable roller was under computer control. As shown in Figure 8.40 this control was attained by replacing the manual drive of the screw that advanced the moving roller against the fixed roller with a computer controlled step motor. This step motor consisted of a servo motor, a controller that provided the pulses for the motor and a power amplifier. This automated system was able to position the moveable roller to within 0.001 inch and to control its feed rate as a function of position and time. To compensate for any misalignment of the motor's drive shaft and the roller's screw drive a flexible coupling was used to complete the connection. Although the roller's position control was open loop in the sense that the position of the roller was not fedback to the controller, the servo system itself used feedback. The servo motor was capable of reproducing the same position to within ±30 arc seconds. A proximity switch on the moving roller's carriage provided the signal necessary for defining a zero position for the motor. In operation the computer instructed the servo system to move away from the fixed roller to clear the proximity switch, then to move forwards until the proximity switch opened, at which point it stopped. By performing this sequence a very repeatable forward motion of the roller was obtained since backlash of the lead screw was eliminated. The controller communicated with the computer through a three wire RS 232 interface. The power for both the step motor controller and the proximity switch interlock signal were obtained from a power supply whose output voltage was controlled by the computer through an IEEE 488 interface.

The manually controlled inputs to the metal forming process were the roller speed, which was set using the existing manual system on the machine, and the ring's orientation with respect to the plane of the roller, which was fixed using a special fixture.

Several transducers were attached to the machine to provide computer access to the various states of the machine. The machine states were determined by the speed of the driving (fixed) and the driven (moving) rollers, the force between the rollers and the number of revolutions of the ring in each rolling position. The roller speed was measured with an optical encoder attached to the shaft of each roller. The pulses from the encoder were sent to a digital display and also transformed into an analog signal using a frequency-to-analog converter and amplifier. The voltages from the two amplifiers were smoothed with 22 μF capacitors as shown in Figure 8.40 and were connected to a sequentially scanned multi-channel A/D converter in the computer. The time constant of the analog roller speed

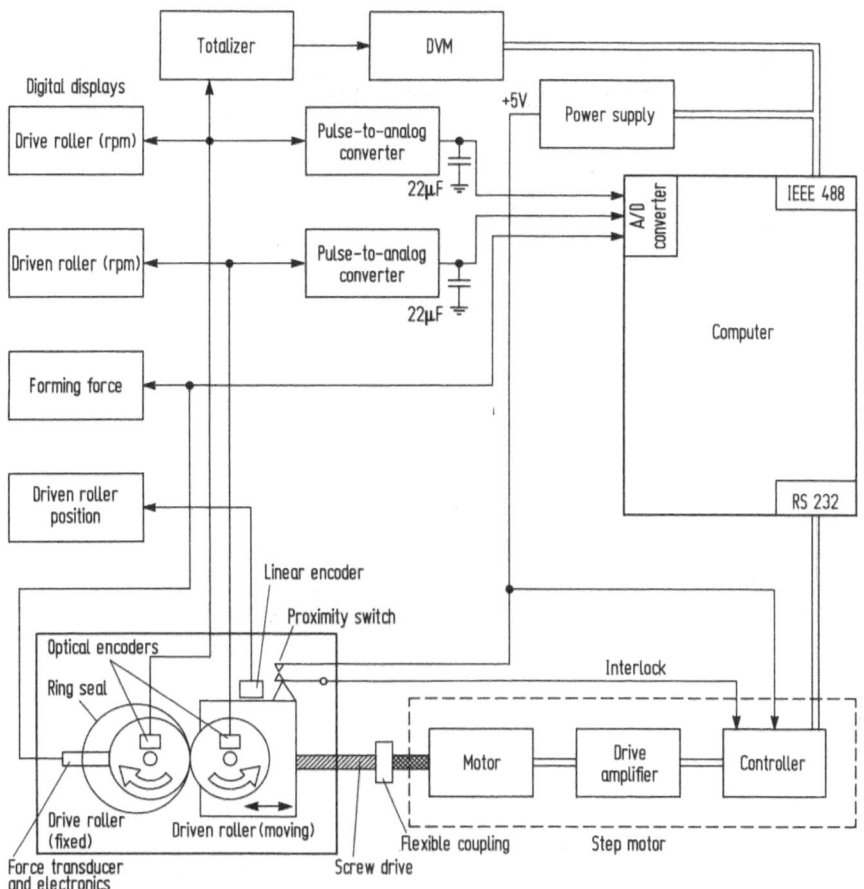

Figure 8.40 Experimental setup and instrumentation for the measurement and control of several process variables in the forming of metal ring seals

signals was approximately 250 ms. The force between the rollers was measured by a strain gage type load cell placed behind the fixed roller. The electronics used with the load cell provided for both a digital display and an analog output signal. The analog output voltage was also fed to the multi-channel A/D converter.

The computer controlled the number of revolutions of the ring for each fixed position of the moving roller by reading a signal measuring the number of revolutions of the roller prior to advancing the roller to its next position. The number of ring revolutions was measured by counting the number of revolutions of the driving roller. For each pair of mating rollers a certain number of pulses corresponded to one revolution of the ring. The pulses were taken from the optical encoder attached to the driving roller and sent to a totalizer (counter). The totalizer had push button keys on its front panel that controlled the display, reset the count and programmed the total count. This last feature provided a signal at

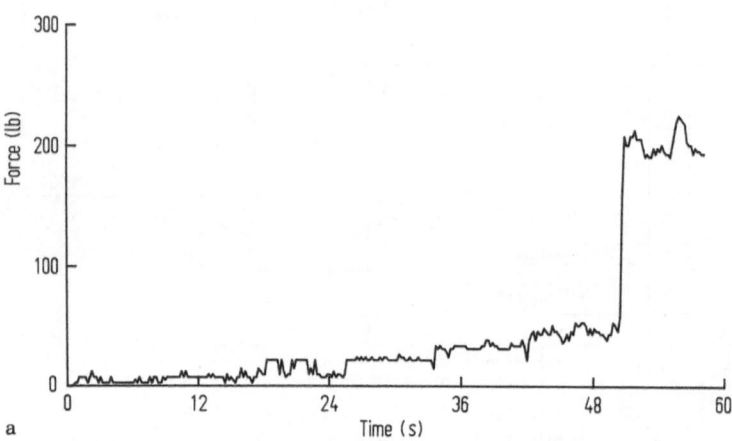

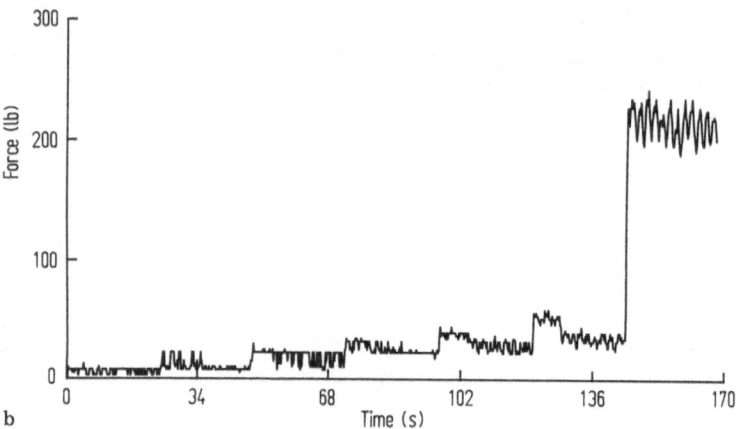

Figure 8.41 Typical values of the forming force as a function of time for (a) 7 steps with 1 revolution /step   (b) 7 steps with 3 revolutions /step

a certain total count that corresponded to N revolutions of the ring. When the totalizer reached the set count a switch closed and the signal from the totalizer was read by a digital voltmeter that communicated with the computer via the IEEE 488 interface. After reaching the total count, the totalizer reset itself to zero.

The location of the actual position of the moveable roller, although not necessary for the control of the process, was very helpful in monitoring its repeatability. It was therefore monitored visually by a linear encoder whose output was connected to a digital display.

Experimental Procedure
The controllable variables were the number of feed steps, the feed rate, the number of revolutions of the ring at each feed step and the orientation of the ring's plane with respect to the plane of the rollers. The roller speeds were controlled independently and manually by making the appropriate changes to the machine.

At the start of the experiment an identification number was given to the ring. The process parameters were then entered and the proper instructions were sent to start the servo motor system. Immediately thereafter the process state data consisting of the two roller speeds and the force between rollers were collected every 0.01 s. Typical force-time histories for two values of the number of steps per total travel of the moveable roller are shown in Figure 8.41. After the process was completed, the movable roller was retracted as described previously and the system was ready for the next ring.

The outcome of each rolling operation was determined by manually measuring certain dimensions of the formed ring. These dimensions were later entered from the keyboard and appended to the appropriate disk file for that ring identification number. Subsequent statistical analyses on all the stored data resulted in a set of design parameters for the semi-automatic machine.

8.6 Electrodynamic Actuators

8.6.1 Vibration Exciters and Loudspeakers

Electrodynamic devices can be used to generate displacement, force and sound. Consider a cross-sectional view of an electrodynamical system shown in Figure 8.42. The cross-hatched section of the frame and the center piece create the magnetic field in the air gap shown in the figure. The magnetic field is generated by a dc field coil in those devices requiring large forces. For small forces, and in loudspeakers, the dc field coils are not used and instead the cross-hatched section is a strong permanent magnet. A drive coil is suspended in the air gap by means of elastic supports (flexures) and has attached to it a surface. In vibration exciters the surface is a plate to which test objects can be attached. In a loudspeaker the surface is usually hemispherical and acts as the acoustic radiator. Placing a current i in the drive coil causes it to move with a force

$$F = Bli = \Gamma i \quad \text{N} \tag{8.56}$$

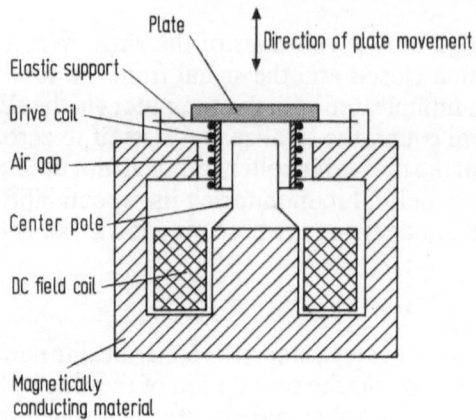

Plate

Direction of plate movement

Elastic support

Drive coil

Air gap

Center pole

DC field coil

Magnetically
conducting material

Figure 8.42 Electrodynamic system

where B is the magnetic flux density (Teslas) and l is the total length of the drive coil (m). If the situation is reversed and a force F were applied to this device to move the coil, then a current i would be generated. This is the basis of a linear velocity transducer.

The electrodynamic system is an electrical system coupled to a single-degree-of-freedom system as shown in Figure 8.43. If the oscillator has an output voltage V_a, an output impedance R_a and an output current i_a, and if the drive coil has a resistance R_c and an inductance L_c, then the electrical circuit equation for the circuit shown in Figure 8.43 is (Buzdugan, et al. [1986])

$$L_c \frac{di_a}{dt} + (R_c + R_a)i + \Gamma \frac{dx}{dt} = V_a(t) \qquad (8.57)$$

where dx/dt is the velocity of the drive coil and $\Gamma dx/dt$ is the back emf. The equation of motion for the coil and flexure assembly is

$$m \frac{d^2x}{dt^2} + c \frac{dx}{dt} + kx = \Gamma i \qquad (8.58)$$

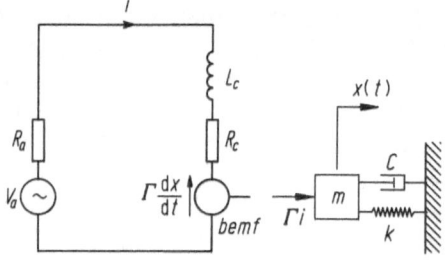

Figure 8.43 Electrodynamic system coupled to a single-degree-of-freedom system

where m is the mass of the coil, k the stiffness of the flexures and c the damping. Taking the Fourier transform of Eqs.(8.57) and (8.58) and solving yields the force

$$F(\omega) = V_a \Gamma \frac{H(\omega)}{D(\omega)}$$

and the displacement

$$x(\omega) = \frac{F(\omega)}{H(\omega)}$$

where

$$H(\omega) = (k - m\omega^2) + j\omega c$$

and

$$D(\omega) = (k - m\omega^2)(R_a + R_c) - \omega^2 c L_c +$$

$$j\omega[c(R_a + R_c) + L_c(k - m\omega^2) + \Gamma^2].$$

These results clearly show that applying a constant voltage to this system gives both a frequency dependent displacement and force, neither of which may be uniform (constant) over any operating frequency range of the device. It is also seen that the magnitudes of the force and the displacement are proportional to the strength of the magnetic field Γ.

When using this type of device as a vibration exciter the above results have to be modified to include another second-order system attached to the coil mass m (see Buzdugan et al. [1986]). This coupling effect is shown in Sections 8.6.2 and 9.2.

Solenoids
A special case of this electrodynamic device is its response at dc. Thus Eq.(8.56) gives

$$F_{dc} = \Gamma i_{dc} = k x_{dc} \quad N.$$

With considerably less elaborate construction such a dc responding device is called a solenoid. It is used extensively in the control of pneumatic and hydraulic valves. Applications of soleniods are given in Sections 8.5.4 and 9.4.

8.6.2 Experiment 6   Constant Acceleration Vibration Testing

Introduction
There are many environmental and qualification test standards and specifications that require a system or product to withstand a given vibration acceleration level over a certain frequency range for a stated period of time. The shape of the acceleration spectrum over the frequency range is usually selected to simulate the expected vibration environment into which the product is to be placed. Thus it is

necessary to have the means to create and sustain the desired spectral shape at the point of attachment of the object to the excitation source. If an electrodynamic vibration exciter is used then, as discussed in Section 8.6.1, the acceleration level at the vibration exciter's table is a function of both the exciter's acceleration frequency response and the interaction of the test object with the exciter. Described below is a method that gives a constant acceleration spectrum at the point where the test object is attached to the exciter.

Theory
Consider a bandlimited periodic signal (recall Eq.(2.8) ff)

$$f(t) = \sum_{k=1}^{M} A_k \sin(2\pi k f_o t + \phi_k) \qquad (8.59)$$

where the bandwidth of the periodic signal is $M f_o$, A_k are arbitrary, but specified, amplitudes at each $k f_o$, ϕ_k is a random phase angle given by

$$\phi_k = \frac{2\pi}{50} \text{INT}(100 r_k - 50)$$

where $\text{INT}(\ldots)$ indicates the integer value of its argument and r_k is the kth random number in the range $0 < r_k < 1$. If $f(t)$ is now sampled and digitized at N discrete time intervals t_s apart such that $T_o = 1/f_o = N t_s$, then Eq.(8.59) becomes

$$f(n t_s) = \sum_{k=1}^{M} A_k \sin(2\pi k n/N + \phi_k) \qquad 0 \le n \le N-1 \qquad (8.60)$$

where the spacing of the frequency components is $\Delta f = 1/T_o = 1/N t_s$ and the bandwidth $M/N t_s$. As discussed at the end of Section 2.5.2 the reason for selecting $T_o = N t_s$ was to create a periodic signal for which the DFT was exactly equal to its continuous (integral) Fourier transform. Equation (8.60) is a pseudorandom signal of period T_o, with deterministic amplitudes and random phase angles. To shape the spectrum of the signal over the frequency range $M/N t_s$ one has to simply adjust the amplitudes A_j accordingly. However, it can only adjust the amplitudes every Δf apart, which is the resolution of the method. The implementation of this equation is described subsequently for the case where $M = 400$, $N = 2048$ and $t_s = 0.2441$ ms; therefore, $\Delta f = 2$ Hz.

Instrumentation and Experimental Procedure
The experimental apparatus consisted of a small (45 N) electrodynamic vibration exciter to which a long thin steel beam was attached at its midpoint. The instrumentation used to control the excitation of the beam is shown in Figure 8.44. The arbitrary function generator (AFG) was a computer controlled waveform generator that had the ability to generate an arbitrarily shaped periodic waveform consisting of 2048 user-specified amplitudes spaced t_s apart, where t_s was also

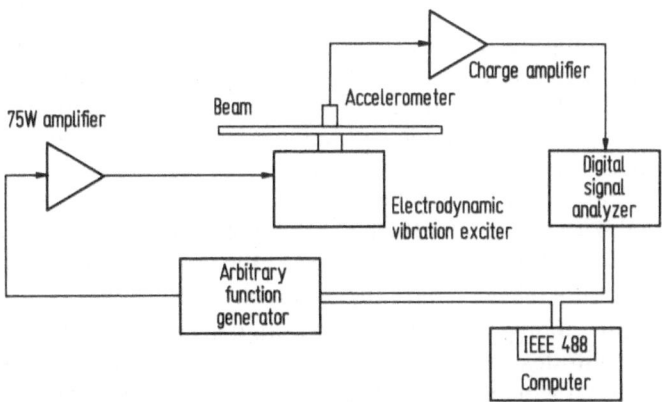

Figure 8.44 Experimental setup and instrumentation for the generation of constant acceleration over a frequency range

specified by the user. These 2048 digital amplitudes were converted to a periodic analog signal by continually recirculating them through a D/A converter. The digital signal analyzer (DSA) was an instrument that digitized an analog signal and, using the FFT algorithm, converted its input signal to an amplitude spectrum. The spectrum was then transferred to the computer for further analysis via an IEEE 488 interface.

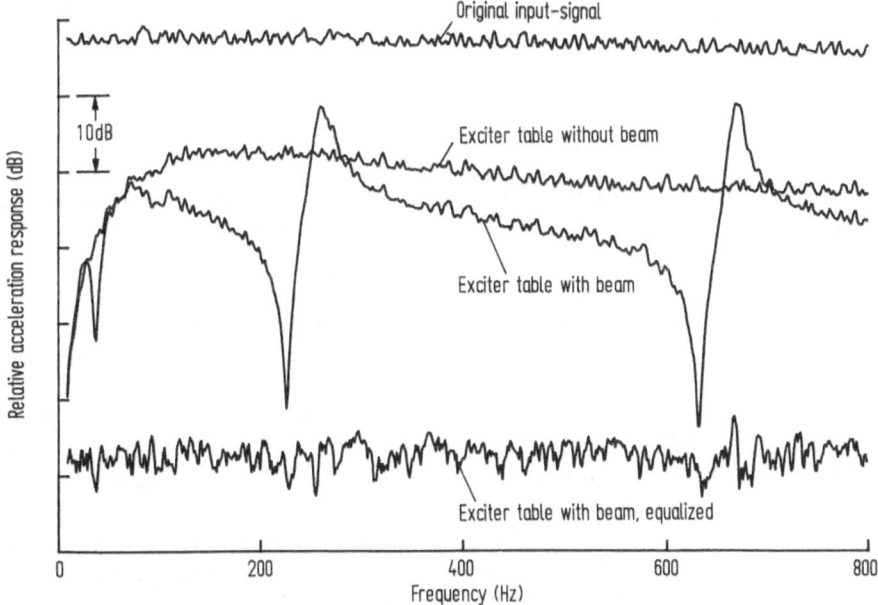

Figure 8.45 Spectra of the original periodic pseudorandom input signal and the acceleration response of an electrodynamic exciter's table under various conditions

The computer first generated a signal of constant amplitude from 2 to 800 Hz using Eq.(8.60). Then 2048 amplitude values of this signal were sent to the AFG. The AFG's output was connected to the vibration exciter's power amplifier. The amplitude spectrum of the AFG's original output signal is shown in Figure 8.45. The responses of the vibration exciter's table with and without the beam are also shown in Figure 8.45, where the effects of the coupling of the test object with the moving element of the vibration exciter are clearly seen. The output spectrum of the accelerometer's signal with the beam attached was then transferred from the DSA to the computer. This spectrum was normalized to the amplitude at 100 Hz and the reciprocal value obtained for each frequency band. The original A_j were then multiplied by the corresponding reciprocal value obtained from the measured spectrum. These new amplitudes were then used in Eq.(8.60) to obtain a new pseudorandom periodic signal which, in turn, was transferred to the AFG. This resulted in the equalized acceleration level shown in Figure 8.45, which was uniform within ±4 dB from 2 to 800 Hz.

8.7 Temperature

8.7.1 Introduction

Qualitatively, temperature can be considered the level of thermal energy of a body and the driving force for heat flow. The unit for measuring temperature, the degree, is arbitrary. There are four temperature scales: Celsius (°C), Kelvin (°K), Fahrenheit (°F), and Rankine (°R). The four scales are related by:

$$°C = (5/9)(°F - 32), \qquad °F = (9/5)°C + 32,$$

$$°K = 273.15 + °C, \qquad °R = 459.67 + °F.$$

Although a temperature scale is arbitrary a reference temperature is required. A reference temperature can be achieved by properly obtaining a *fixed point*, which is a physical phenomenon that occurs reproducibly at the same temperature; e.g., melting ice at a certain atmospheric pressure. There are many such reference temperatures ranging from the triple point of hydrogen at 13.81 °K to the freezing point of gold at 1337.58 °K as denoted by the International Practical Temperature Scale.

There are many methods that measure temperature and these can be found in Schooley [1986] and McGee [1988]. Here we shall briefly introduce only three of the most common types of temperature sensors: the thermistor, the thermocouple and the resistance temperature detector (RTD).

The temperature response $T(t)$ of these three sensors and the hot-wire anemometer that is discussed in Section 8.8.2 are approximately governed by the following heat conduction equation (Ozisik [1980]):

$$\tau_t \frac{dT}{dt} + T = T_a(t) \tag{8.61}$$

where

$$\tau_t = \frac{mc_p}{hA_s} \quad s$$

is the time constant of the heat transfer process, m is the mass of the temperature sensing material, c_p its specific heat, A_s its surface area, and h its convective heat transfer coefficient, which depends on the fluid into which the thermal device is placed and the fluid's flow velocity. If $T_a(t) = T_o u(t)$, where $u(t)$ is the unit step function given by Eq.(2.54), then the solution to Eq.(8.61) is

$$T(t) = T_o(1 - e^{-t/\tau_t}).$$

It is seen that this response is the same as that obtained for a low pass RC filter in Eq.(3.32). Consequently, the time and frequency response characteristics of these devices are the same as those discussed in Section 3.3.2.

8.7.2 Thermistors

A thermistor is a special type of semiconductor whose resistance varies in a predictable, but nonlinear, manner as its temperature changes. They are available in different sizes and shapes, such as rods, films, discs and washers. Thermistors are usually made of fired ceramic compounds to which platinum, silver or gold connecting wires have been inserted.

The resistance-temperature relationship for a thermistor is usually expressed as

$$R = R_o e^{\beta(1/T - 1/T_o)} \tag{8.62}$$

where R is the resistance of the thermistor at a temperature T (in °K) and R_o is its resistance at T_o (in °K). The constant β has a value between 3000 and 5000 °K. Thermistors are usually specified by their values of R_o and T_o. The sensitivity of a thermistor is

$$S_T = \frac{\Delta R / R}{\Delta T} = -\frac{\beta}{T^2}. \tag{8.63}$$

Thus

$$\frac{\Delta R}{\Delta T} = S_T R = -\frac{\beta R}{T^2}. \tag{8.64}$$

Hence, if $R = 2000$ ohms at $T = 300$ °K and $\beta = 4000$ °K, then from Eq.(8.64) $\Delta R / \Delta T = 89$ ohms/°K. This is a very large resistance change that can easily be read directly with an ohmmeter or can be converted to a voltage and read with a voltmeter. Several methods for reading thermistors and their corresponding circuits can be found in McGee [1988], Dally, et al. [1984] and Wobschall [1987].

The advantages of a thermistor are its high sensitivity to temperature change, its fast response time (time constants on the order of ms) and the fact that it requires only two wires for accurate measurements. Its disadvantages are its nonlinearity, its limited temperature range, its fragility and that it is self-heating.

An application using a thermistor is given in Section 9.3.

Example 1: Determination of β. A thermistor has an $R_o = 15,000$ ohms at $T_o = 298.15$ °K (25 °C). If we place the thermistor in a melting ice bath ($T = 273.15$ °K) it is found that its resistance is 45,000 ohms. We can now determine β from Eq.(8.62). Thus

$$45,000 = 15,000 e^{\beta(1/273.15 - 1/298.15)}$$

and, therefore, β=3,579 °K.

8.7.3 Thermocouples

A thermocouple is a temperature sensor consisting of two dissimilar wires in thermal contact as illustrated in Figure 8.46. The temperature difference between T_1 and T_2 causes a thermoelectric potential to be generated. (This is called the Seebeck effect after T. J. Seebeck who discovered it in 1821.) This potential can be determined from the empirical relation

$$E_o = C_1(T_1 - T_2) + C_2(T_1^2 - T_2^2) \qquad (8.65)$$

where C_1 and C_2 are thermoelectric constants that depend on the materials that form the junctions which are at temperatures T_1 and T_2, respectively. In practice junction "1" is the unknown temperature and junction "2" is at a known temperature, usually water and ice (0 °C). These constants must be determined individually for each material in each temperature range of interest. The rules governing thermocouple usage and be found in Benedict [1984], Dally et al. [1984] and McGee [1988]. Time constants for thermocouples vary according to their size at the junction, type of electrical grounding and whether they are in air or in water. The values range from 0.1 to 85 s.

The advantages of a thermocouple are its ruggedness and simplicity, that it is self-powered and inexpensive and its wide temperature ranges (-200 to 2300 °C). Its disadvantages are its nonlinearity, low voltage output and very low sensitivity

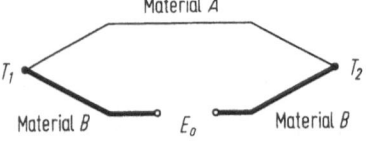

Figure 8.46 Schematic of a typical two junction thermocouple

[25 to 50 μV/°K (20 to 40 μV/°C)], its instability and that a reference temperature is required.

8.7.4 Resistance Temperature Detector (RTD)

A resistance temperature detector (RTD) is made of a metal with a positive temperature coefficient α, usually platinum. The variation of resistance is a factor of ten less than that for a thermistor, but its response is very nearly linear over a wide temperature range. The resistance R_T of a RTD changes with temperature as

$$R_T = R_o(1 + \alpha \Delta T) \qquad (8.66)$$

where ΔT is the temperature change from a reference temperature at which R_o is determined. For the platinum typically used in RTDs $\alpha = 0.00392$ ohm/ohm/°K. A more accurate relationship for R_T for platinum is

$$R_T = R_o(1 + AT - BT^2) \qquad (8.67)$$

where $A = 3.90 \times 10^{-3}$ and $B = 5.7 \times 10^{-7}$ for temperatures between 0 and 420 °C. This predicts R_T within 0.1 to 0.5 °C. The resistance R_o has been standardized at 100 ohm. For other temperature ranges (-220 to 850 °C), and for materials other than platinum, standard tables exist. (See Benedict [1984] and McGee [1988].) Typical time constants for RTDs are 0.4 s.

The advantages of an RTD are its stability, its high accuracy and its relatively high linearity. Its disadvantages are its cost, low sensitivity (α), self-heating and that a current source is required.

8.8 Flow Velocity

8.8.1 Hot-Wire Anemometers

To measure velocity in a flowing fluid it is desirable for the sensor/transducer to have a high frequency response to be able to accurately follow flow transients, be small enough to measure at a point with minimum flow disturbance, measure a wide range of velocities in only the direction of interest including flow reversals, work in a wide range of fluid temperatures, densities and compositions, have high resolution and accuracy and be easy to use. One technique that sufficiently satisfies most of these criteria is a hot-wire anemometer.

A simple constant temperature hot-wire anemometer and its corresponding bridge circuitry is shown in Figure 8.47. The resistance of the sensor is governed by Eq.(8.66). Thus

$$R_{HW} = R_r[1 + \alpha(T_m - T_r)] \qquad (8.68)$$

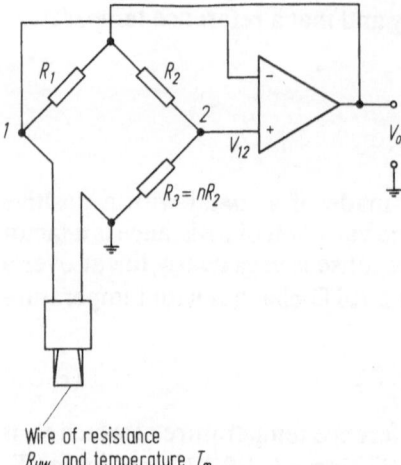

Wire of resistance
R_{HW} and temperature T_m

Figure 8.47 Basic elements of a constant temperature hot-wire anemometer

where R_{HW} is the resistance of the hot-wire, R_r is the reference resistance at T_r, T_m is the average sensor temperature and α is the temperature coefficient of resistance.

When there is an increase in the velocity across the hot-wire, the hot-wire cools, lowering T_m and, hence, R_{HW} and the voltage at point 1. The lowered voltage at the negative input to the amplifier causes the voltage V_{12} to increase, which in turn increases its output voltage V_o. This increased voltage on the bridge increases the current through the sensor, which heats the sensor. This heating decreases V_{12} until the system returns to equilibrium. Since V_o changes as fast as R_{HW} changes, V_o is proportional to the velocity cooling it.

The flow velocity is related to a convective heat transfer function ϕ from the hot-wire to the flowing fluid and a conductive heat transfer function K from the sensor to its supports by

$$V_o^2 = \frac{(R_{HW} + R_1)^2}{R_{HW}}(\phi + K).$$ (8.69)

Expressions for ϕ and K can be found in Fingerson and Freymuth [1983].

8.8.2 Turbine Flowmeter

A turbine flowmeter is a freely suspended, concentrically mounted axial turbine whose blades (propellers) are rotated as the liquid or gas flows past them. The rotational speed of the turbine is proportional (over a wide range of rotational speeds) to the velocity of the fluid. Since the cross-sectional area of the flowmeter is constant the turbine's rotational speed represents the volume of liquid or gas flowing past it. The rotational speed of the turbine is detected by a magnetic sensor mounted exterior to the flow. Each passage of a turbine blade gives a pulse, which is usually fed to a digital counter. With the counter in its frequency mode the "instantaneous" flow rate is measured. With the counter in its totalizing mode the total flow over a given time period is measured.

8.8.3 Experiment 7   Energy Balance of a Heat Pump

Introduction
Heat pumps are machines that pick up heat at a low temperature and, with the expenditure of energy, lift it to a higher temperature. This expenditure of energy can be either in the form of work, as in the case of vapor compression heat pumps or in the form of heat, as in the case of absorption heat pumps. When the desired product of the system is the high temperature heat the system is called a heat pump; when the desired product is the cooling effect at the low temperature the system is called an air conditioner or a refrigerator. An important quantity that describes the performance of the system is the Coefficient of Performance (COP). In this experiment the COP is determined, along with other quantities of interest, for one type of heat pump system.

Theory
Consider the heat pump system shown in Figure 8.48. The flow of the refrigerant is in the counter-clockwise direction. The compressor on its suction side lowers the pressure in the evaporator causing the refrigerant to boil (evaporate) absorbing latent heat from its environment. The cool gas flows to the compressor where its pressure and temperature are increased by the work of the compressor. The hot gas flows to the condenser where the temperature of that environment causes the gas to condense. Upon condensation it becomes liquid again and in doing so gives up the heat acquired in the evaporator and the compressor. The condensed liquid refrigerant is collected in the receiver and sent through the expansion valve into the evaporator and the cycle repeats.

The heat exchangers' fluids are a glycol (anti-freeze) and water mixture having a specific heat c_p. The mass flow rate at which the fluid moves through the condenser's heat exchanger is $\dot{m}_c$ and that through the evaporator $\dot{m}_e$. If Q_{cond} is the heat (enthalpy difference) transferred to the environment by the refrigerant in the condenser, Q_{evap} the heat absorbed by the refrigerant in the evaporator and W_{com} the work of the compressor, then according to the first law of thermodynamics

$$Q_{cond} = Q_{evap} + W_{com}. \tag{8.70}$$

In practice Eq.(8.70) is considered balanced if the ratio of its left and right hand sides is between 0.95 and 1.05. The COP for the heat pump is defined as

$$COP = \frac{Q_{cond}}{W_{com}}. \tag{8.71}$$

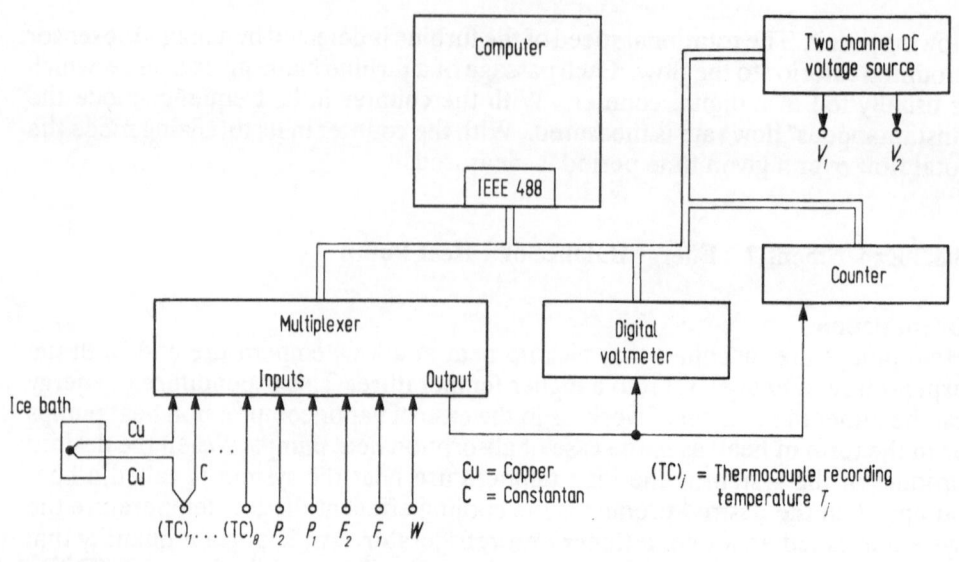

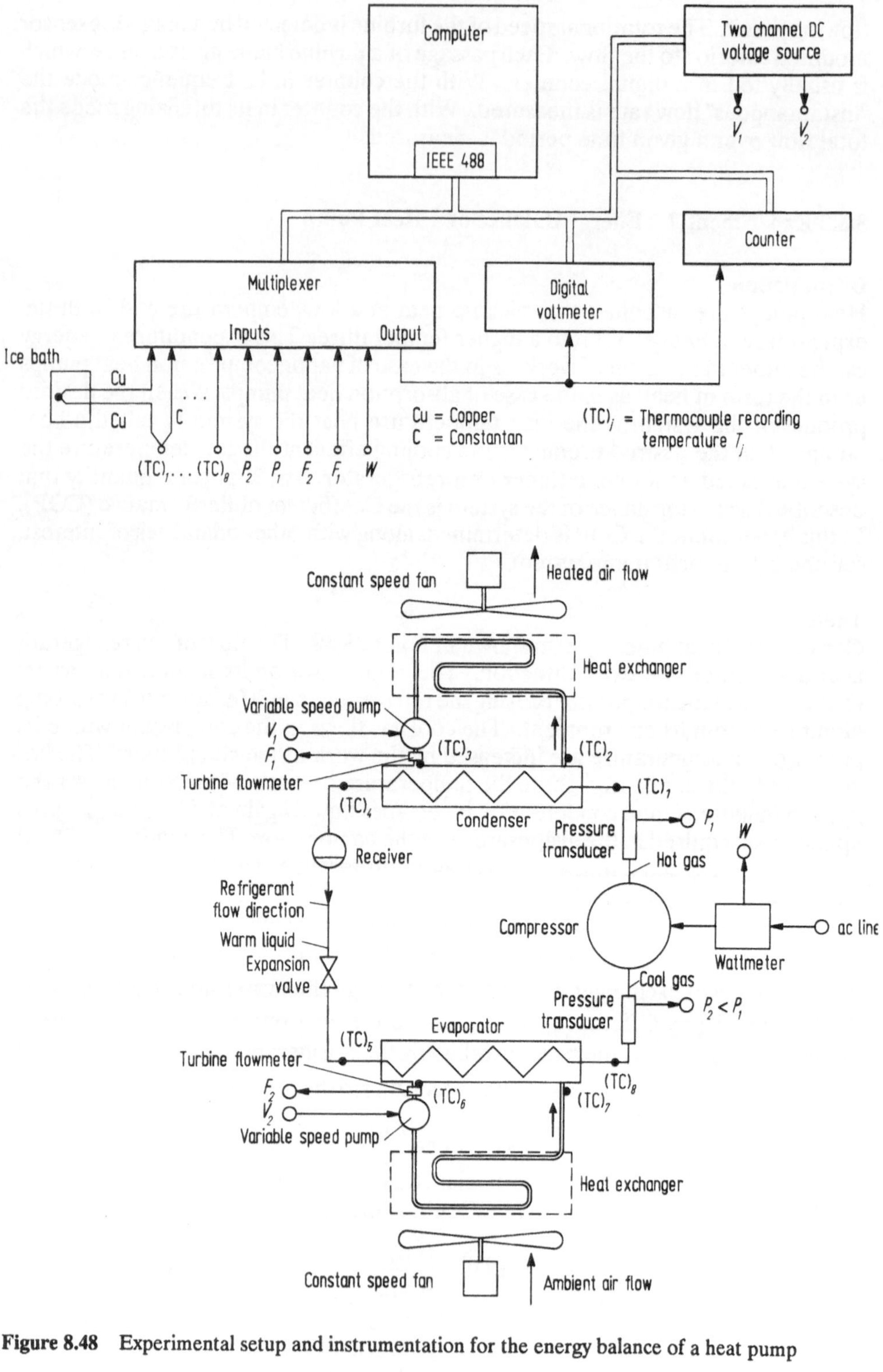

Figure 8.48 Experimental setup and instrumentation for the energy balance of a heat pump

(If the system is used as an air conditioner then $COP = Q_{evap}/W_{com}$.) The quantities Q_{cond} and Q_{evap} are given by

$$Q_{evap} = \dot{m}_e c_p (T_3 - T_2), \qquad Q_{cond} = \dot{m}_c c_p (T_6 - T_7)$$

where the T_j are the respective temperatures at the points indicated in Figure 8.49. Because most of the heat that is transferred from the evaporator to the condenser is the latent heat of phase change, more energy is delivered from the condenser then is applied by the compressor. Therefore the COP will always be greater than 1.

The COP depends on a number of parameters: the temperature levels in the evaporator and condenser, the heat exchangers' effectiveness (see below), the setting of the expansion valve, the pressure drop of the refrigerant in the heat exchangers and their connecting tubing and the heat loss to, and the heat gain from, the environment. Some of the conditions can be altered in the experiment and their effects on the COP determined. The heat exchanger effectiveness of the condenser ϵ_c and that of the evaporator ϵ_e are proportional to the ratio of the following temperature differences, respectively,

$$\epsilon_c \propto \frac{T_1 - T_4}{T_1 - T_3}, \qquad \epsilon_e \propto \frac{T_5 - T_8}{T_5 - T_7}$$

and vary between 0 and 1.

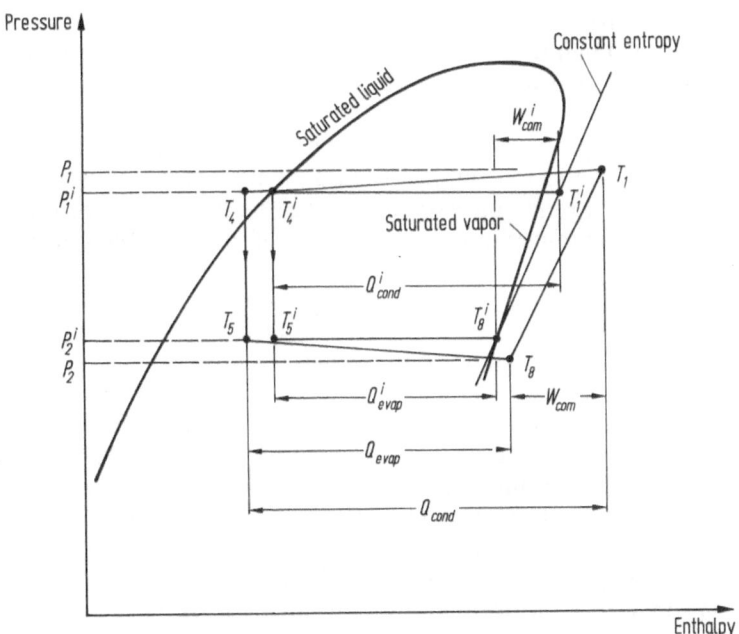

Figure 8.49 Ideal and actual heat pump processes for a given refrigerant

Another measure of the performance of the system is its closeness to an ideal reverse Rankine cycle. Referring to Figure 8.49 it is seen that for a given refrigerant an ideal cycle should perform as shown, where in the ideal case $Q_{cond}^i = Q_{cond}$, $Q_{evap}^i = Q_{evap}$ and $W_{com}^i = W$, where W is the power delivered to the compressor. Deviation from this ideal gives a measure of the performance of the individual heat exchange processes and the compressor.

Instrumentation and Experimental Procedure

The instrumentation is shown in Figure 8.48. The outputs from the eight thermocouples ($T_1, \ldots, T_8$), the two pressure gages (P_1, P_2), the two turbine flowmeters (F_1, F_2) and the wattmeter (W) were connected to a multiplexer. The output of the multiplexer was connected to both a 6½-digit autoranging digital voltmeter (DVM) and a digital counter. The digital counter was read only when the output of the multiplexer was from one of the flowmeters. The DVM was used to read all other outputs of the multiplexer. The flow rate of the glycol and water mixture in each heat exchanger was controlled independently by two dc voltages (V_1, V_2) that were inputs to the motors' speed controllers.

The speed range of these pumps was from 0 to 1750 rpm for $0 < V_j < 10$ V, $j = 1, 2$, which gave a maximum flow rate of approximately 10 gallons/minute (gpm) at 1725 rpm. The actual flow rate in each heat exchanger was monitored by the corresponding flowmeter. The turbine flowmeters had an equivalent sensitivity of approximately $0.01f$ gpm, where f was the measured frequency in Hz. The full scale range was 10 gpm. The pressure gages were the strain gage diaphragm type described in Section 8.5.2. Their range was from 0 to 300 psia, with a corresponding full scale voltage of 5 V. The wattmeter was rated at a 1000 W and had a sensitivity of 100 W/V.

Prior to the start of the experiment the thermocouples were calibrated by first placing them in the reference thermocouples' ice bath. Their output voltages were then checked to see that they read approximately $0.0\,\mu V$. Then they were placed in boiling water at a known atmospheric pressure and, after a suitable period of time, their output voltages were read. Assuming a linear relationship between output voltage and temperature rise each thermocouple's sensitivity was obtained.

At the start of the test the pumps were turned on to give a maximum flow rate in each heat exchanger of approximately 10 gpm. Then the compressor was turned on and all of the sensors were read every 30 s. When each temperature changed by less than ± 0.5 °C from the average of the preceding ten measurements and, similarly, the pressures changed by less than $\pm 1.0\%$ the process was considered to have reached steady state conditions. The next ten values from each transducer were then averaged and the results stored for further analysis, along with the time it took to reach steady-state. Then the flow rates were decreased to three-quarters of their maximum and the procedure repeated until all desired combinations of flow rates were investigated. Results from a typical run are shown in Figure 8.49. The superscript i on some of the quantities in the figure refers to the ideal process.

8.8.4 Experiment 8   Control of Air-To-Fuel Ratio in Multi-burner Combustion Systems

Introduction
One of the primary goals in combustion systems is to attain an appropriate air-to-fuel mixture so as to extract the maximum available energy in the fuel while minimizing air pollution. Conventional methods are often based on the analysis of the carbon monoxide and oxygen content of the flue gases. This method has serious shortcomings in those combustion processes that use more than one burner, since it cannot distinguish the operating conditions of the individual burners. Insufficient air (a rich mixture) leads to the excessive production of carbon monoxide, smoke and solids in addition to an incomplete release of available energy. Excess air, on the other hand, results in decreased efficiency due to the increased volume of hot waste gases passing through the stack. Additionally, in those fuels containing sulfur, excessive air in the mixture often leads to the formation of SO_3, which causes acid smut and corrosion in the heat exchangers and flue stack. It is the purpose of this experiment to demonstrate a method of controlling the combustion properties of a two-burner system.

Theory
An analysis of the spectral properties of flames indicates three distinct regions: region I, where the fuel reactants are relatively cool; region II, where the combustion reaction takes place; and region III, where the burned gases reside. In region I the presence of relatively cool air and fuel molecules absorb the infrared radiation and, consequently, very little infrared radiation occurs. In region II the combustion chemistry takes place which results in strong visible and ultraviolet emission spectra, although some of the radiation is chemilumnescent rather than thermal. In region III the burned gases, which are downstream of the combustion zone and are comprised mainly of carbon dioxide, water vapor, nitrogen and other trace gases, exhibit strong infrared emissions (or absorptions) in various frequency bands. Consequently, detection of infrared emissions from regions II or III can provide information regarding the burning process.

There are many factors that influence the actual combustion process: the air-to-fuel ratio and the volume of air flow, the properties of the fuel itself, the method of mixing the air and the fuel during combustion, the geometry of the burner and the location of the burner in the combustion chamber. For a given set of conditions, however, it has been found that near optimum combustion takes place at an air-to-fuel ratio where the infrared radiant heat emission is maximum (Penzias [1966]). A typical variation of the total radiation level as a function of the air-to-fuel ratio is shown in Figure 8.50. Therefore, if the air-to-fuel ratio is made richer, then a decrease in the radiation level means that the air-to-fuel ratio was initially rich, whereas an increase in the radiation level means that the air-to-fuel ratio was initially lean. If the air-to-fuel ratio is made leaner the converse would be true. It is this property of the combustion process that can be used for its optimization.

If a heat exchanger is placed in the combustion chamber, then the amount of heat the exchanger absorbs Q_{he} is given by

$$Q_{he} = U A \Delta T_m$$

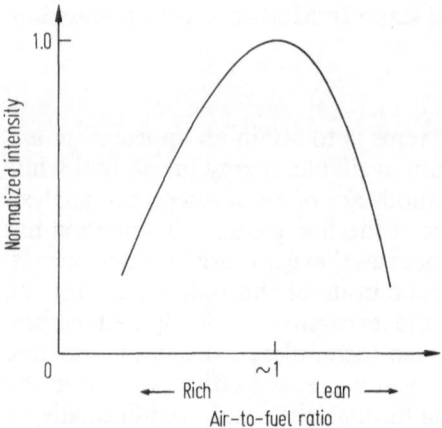

Figure 8.50 Normalized infrared intensity as a function of air-to-fuel ratio

where U is the overall heat transfer coefficient for the heat exchanger, A is the total area of the heat exchanger's outside surface and ΔT_m is the log mean temperature difference

$$\Delta T_m = \frac{(T_3 - T_1) - (T_4 - T_2)}{\ln[(T_3 - T_1)/(T_4 - T_2)]}$$

and T_i are the temperatures recorded by the thermocouples at the locations shown in Figure 8.51. The overall heat transfer coefficient U is an empirically determined constant that is a function of the velocity and temperature of the fluid or gas in the heat exchanger at its inlet conditions, the outer diameter of the heat exchanger's tubing and its material and a fouling, or cleanliness, factor. The heat balance of the process is then given by

$$Q_{fuel} = Q_{he} + Q_{loss}$$

where Q_{fuel} is estimated from the product of the calorific value of the fuel and the fuel's mass flow rate. The quantity Q_{loss} is the heat loss to the environment.

Instrumentation and Experimental Procedure
Consider the two burner combustion system shown in Figure 8.51. The intake combustion air flowed into two ducts, each of which fed one of the burners. The amount of air flow in each duct was governed by the orientation of the dampers, which were controlled by pneumatic actuators. The actual amount of air flowing in each duct was recorded by the turbine flowmeters. The amount of fuel fed to each burner was determined by the settings on one common metering pump, which is a variable displacement pump that delivers a known flow rate of fluid at a given pump speed and displacement. Its flow rate was controlled by the voltage V_m.

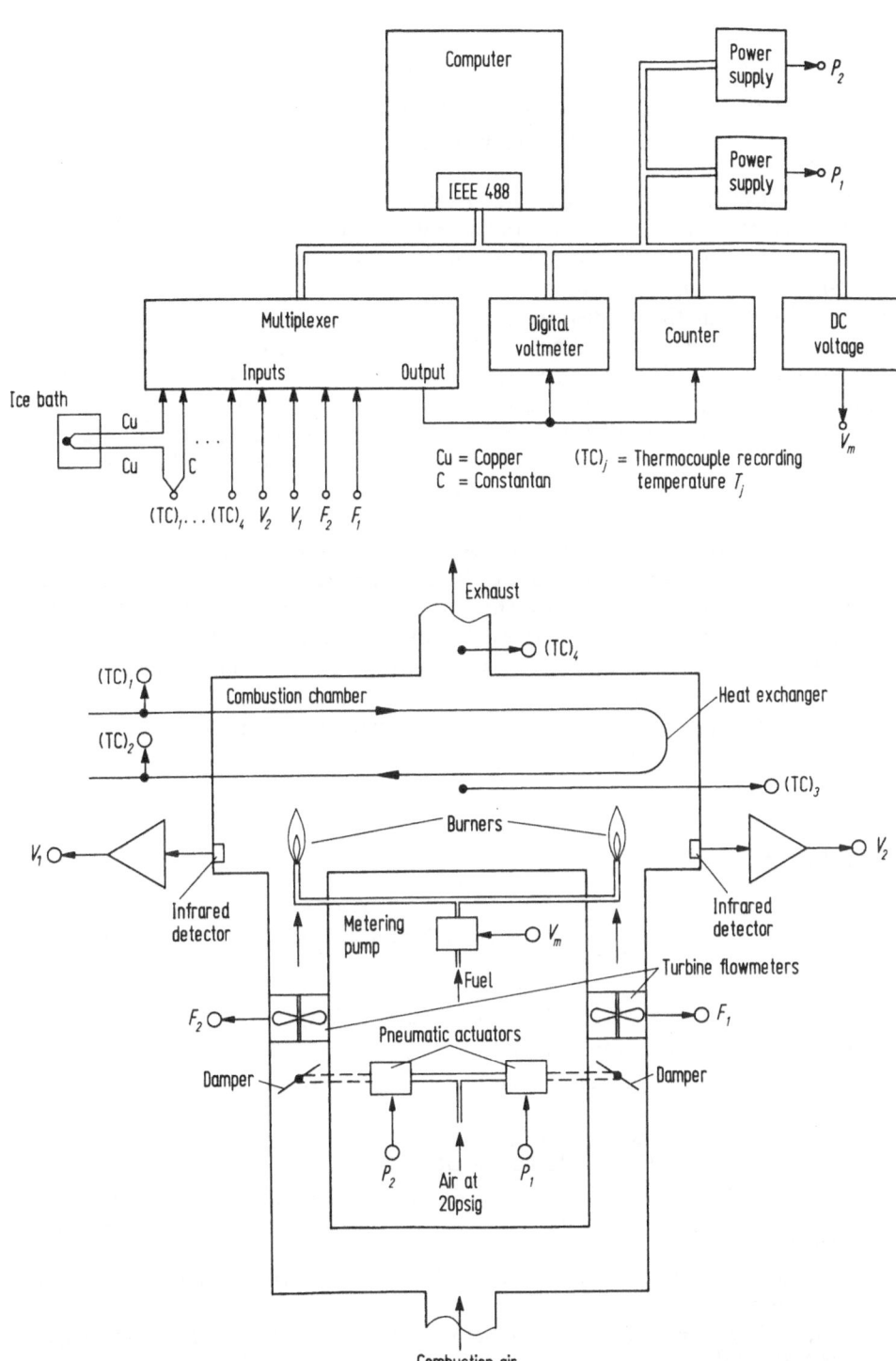

Figure 8.51 Experimental setup and instrumentation for the optimization of a two-burner combustion process

The heat exchange process was monitored by recording the inlet and outlet temperatures T_1 and T_2, respectively, in the heat exchanger and the temperatures T_3 and T_4 in the combustion chamber and flue gases, respectively. The actual combustion process of each burner was individually monitored by a broadband (non-filtered) infrared detector. The spatial distribution of the flame's infrared spectrum was such that each detector responded only to the one closest to it.

At the start of the test the two burners were ignited with a pilot light at some arbitrary air-to-fuel ratio for which a stable flame could be maintained. The intensity of each infrared detector was then recorded. The amount of air flow was increased 10% and the intensity of each detector compared to its previous reading. As discussed above, depending on whether the flame's intensity increased or decreased the air-to-fuel ratio was either decreased or increased by instructing the damper in each duct to either reduce the air flow passage or increase it. The process was repeated until either the maximum intensity was reached or exceeded. If the maximum was exceeded then by using appropriately smaller and smaller changes to the air-to-fuel ratio the maximum air-to-fuel ratio was attained by iteration. After having reached the conditions of maximum flame intensity the four temperatures were monitored every 5 s until steady-state conditions were reached. At steady-state all sensor outputs were recorded. The air-to-fuel ratio was then varied slightly to verify that for this combustion chamber, burner geometry and fuel flow rate the maximum flame intensity corresponded to a maximum heat absorption by the heat exchanger. The fuel flow rate was then varied, while still maintaining the desired air-to-fuel ratio and the same flow rate in the heat exchanger, to obtain an empirical relationship between the fuel flow rate and the temperature rise in the heat exchanger.

References

1. Benedict, R. P., *Fundamentals of Temperature, Pressure, and Flow Measurements,* 3rd Ed., John Wiley and Sons, New York, 1984.

2. Bentley, J. P., *Principles of Measurement Systems,* 2nd Ed., Longman Scientific and Technical, England (Co-published with John Wiley and Sons, New York), 1988.

3. Beranek, L. L., *Acoustics,* McGraw-Hill Book Co., New York, 1954.

4. Boldea, I., and Nasar, S. A., *Linear Motion Electromagnetic Systems,* John Wiley and Sons, New York, 1985.

5. Bouche, R. R., *Calibration of Shock and Vibration Measuring Transducers,* The Shock and Vibration Information Center, Washington, DC, 1979.

6. Bouche, R. R., "Accelerometers for Shock and Vibration Measurements," in *Vibration Testing-Instrumentation and Data Analysis,* E. B. Magrab and O. A. Shinaishin, Eds., ASME Publication AMD Vol. 12, 1975.

7. Buzdugan, G., Mihailescu, E., and Rades, M., *Vibration Measurement,* Martinus Nijhoff, Dordrecht, Holland, 1986.

8. Dally, J. W., and Riley, W. F., *Experimental Stress Analysis,* 2nd Ed., McGraw-Hill Book Co., New York, 1978.

9. Dally, J. W., Riley, W. F., and McConnell, K. G., *Instrumentation for Engineering Measurements*, John Wiley and Sons, New York, 1984.

10. Dereniak, E. L., and Crowe, D. G., *Optical Radiation Detectors*, John Wiley and Sons New York, 1984.

11. Doebelin, E. O., *Measurement Systems: Applications and Design*, 4th Ed., McGraw-Hill Book Co., New York, 1990.

12. Estler, W. T, and Magrab, E. B., "Validation Metrology of the Large Optics Diamond Turning Machine", National Bureau of Standards Report NBSIR 85-3182(R), Gaithersburg, Maryland, June, 1985.

13. Fingerson, L. M., and Freymuth, P., "Thermal Anemometers", in *Fluid Mechanics Measurements*, R. J. Goldstein, Ed., Hemisphere Publishing Corp., Washington, DC, 1983.

14. Herceg, E. E., *Handbook of Measurement and Control*, Shaevitz Engineering, Pennsauken, New Jersey, 1983.

15. Holman, J. P., *Experimental Methods for Engineers*, 5th Ed., McGraw-Hill Book Co., New York, 1989.

16. Hordeski, M. E., *Transducers for Automation*, Van Nostrand Reinhold Co., New York, 1987.

17. Johnson, C. D., *Process Control Instrumentation Technology*, 3rd Ed., John Wiley and Sons New York, 1988.

18. Karditsas, P., "Blow Down Tests for the Calibration of Nozzles, Orifices and Short Pipes", MS Thesis, Department of Mechanical Engineering, University of Maryland, College Park, Maryland, 1982.

19. Kenjo, T., *Stepping Motors and Their Microprocessor Controls*, Clarendon Press, Oxford, 1984.

20. Keyes, R. J., Ed., *Optical Infrared Detectors*, Springer-Verlag, Berlin. 1977.

21. Krause, P. C., and Wasynczrik, O., *Electromechanical Motion Devices*, McGraw-Hill Book Co., New York, 1989.

22. Krohn, D. A., *Fiber Optic Sensors: Fundamentals and Applications*, Instrument Society of America, Research Triangle Park, North Carolina, 1988.

23. Loxton, R., and Pope, P., Eds., *Instrumentation: A Reader*, Open University Press, Milton Keynes, England, 1986.

24. Luxmoore, A. R., *Optical Transducers and Techniques in Engineering Measurement*, Applied Science Publishers, London, 1983.

25. Lyons, J. L., *The Designer's Handbook of Pressure Sensing Devices*, Van Nostrand Reinhold Co., New York, 1980.

26. Magrab, E. B., *Environmental Noise Control*, John Wiley and Sons, New York, 1975.

27. Malmstadt, H. V., and Enke, C. G., *Digital Electronics for Scientists*, W. A. Benjamin, Inc., New York, 1969.

28. McGee, T. D., *Principles and Methods of Temperature Measurement*, John Wiley and Sons, New York, 1988.

29. Mylroi, M. G., and Calvert, G., *Measurement and Instrument for Control*, Peter Peregrines Ltd., London, 1984.

30. Neubert, H. K. P., *Instrument Transducers*, Clarendon Press, Oxford, 1975.

31. Norton, H. N., *Handbook of Transducers for Electric Measuring Systems*, Prentice-Hall, Englewood Cliffs, New Jersey, 1969.

32. Nunley, W., and Bechtel, J. S., *Infrared Optoelectronics*, Marcel Dekker, New York, 1987.

33. Oliver, F. J., *Practical Instrumentation Transducers*, Hayden Book Co., New York, 1971.

34. Ozisik, M. N., *Heat Conduction*, John Wiley and Sons, New York, 1980.

35. Penzias, G. J., "Spectroscopic Measurements of Flame Radiation for Improved Combustion Control", 9th National Power Instrumentation Symposium, Detroit, Michigan, May 16-18, 1966.

36. Ratz, A. G., and Bartlett, F. R., Vibration Simulation Using Electrodynamic Exciters", in *Vibration Testing-Instrumentation and Data Analysis*, E. B. Magrab and O. A. Shinaishin, Eds., ASME Publication AMD Vol. 12, 1975.

37. Robinson, D. C., Serbyn, M. R., and Payne, B. F., "A Description of NBS Calibration Services in Mechanical Vibration and Shock", NBS TN 1232, National Bureau of Standards, Gaithersburg, Maryland, February, 1987.

38. Schmidt, V. A., Edelman, S., Smith, E. R., and Jones, E., "Optical Calibration of Vibration Pickups at Small Amplitude," *Journal Acoustical Society of America*, 33, 6, pp. 748-751, 1961.

39. Schmidt, V. A., Edelman, S., Smith, E. R., and Pierce, E. T., "Modulated Photoelectric Measurement of Vibration," *Journal Acoustical Society of America*, 34, 4, pp. 455-458, 1962.

40. Schooley, J. F., *Thermometry*, CRC Press, Boca Raton, Florida, 1986.

41. Seippel, R. G., *Transducers, Sensors and Detectors*, Reston Publishing Co., Reston, Virginia, 1983.

42. Seippel, R. G., *Transducer Interfacing: Signal Conditioning for Process Control*, Prentice-Hall, Englewood Cliffs, New Jersey, 1988.

43. Sheingold, D. H., Ed., *Transducer Interfacing Handbook*, Analog Devices Inc., Norwood, Massachusetts, 1980.

44. Trietley, H. L., *Transducers in Mechanical and Electronic Design*, Marcel Dekker, New York, 1986.

45. Usher, M. J., *Sensors and Transducers*, Macmillan, London, 1985.

46. Wobschall, D., *Circuit Design for Electronic Instrumentation*, 2nd Ed., McGraw-Hill Book co., New York, 1987.

47. Woolvet, G. A., *Transducers in Digital Systems*, Peter Pereguinus Ltd, Southgate House, Stevenage, Herts., England, 1977.

48. Catalog #8000, "Programmable Motion Control", Compumotor Division, Parker Hannifin Corporation, Petaluma, California, 1988.

Exercises

1. The capacitance gage shown in Figure 8.4 has a sensing area of 0.44 in^2, an offset of 0.01 inches and an excitation voltage of 15 V_{dc}. If the frequency linearity is to be within ±10%, what is the (a) minimum input impedance of the connecting amplifier if the lowest frequency of interest is 10 Hz and (b) sensitivity of the system in volts/0.001 inch.

2. A capacitor microphone is to be used to record a transient sound. It has a sensitivity at 1 kHz of 50 mV/Pa. The microphone is connected to a unity gain charge amplifier whose output goes to a variable gain voltage amplifier and anti-aliasing filter. The output of the anti-aliasing filter is connected to an A/D converter whose output in turn is stored in a computer. The A/D converter's full scale input is ± 10 V.

   (a) If the lower -3 dB bandwidth of the microphone is to be 0.1 Hz, what should be the minimum time constant of the charge amplifier.

(b) If the expected peak value of the sound is 72 dB *re* 20 μPa, what should be the gain of the voltage amplifier.

(c) What is the minimum number of bits, including sign, that the A/D converter must have for its dynamic range to be at least 64 dB.

(d) If the anti-aliasing filter's selectivity is 80 dB/octave and its cutoff frequency is set to 8000 Hz, what is the minimum sampling frequency for the n-bit A/D converter determined in c).

3. An accelerometer-amplifier system is to have a frequency linearity of ±2%. The accelerometer weighs 10 grams and is to be mounted to an object that weighs 80 grams. Assume that the accelerometer's internal mass is half of its total mass. The accelerometer has an unmounted resonance of 45,000 Hz and a capacitance of 1200 pF, including its cable. The input impedance of the connecting amplifier is 800 MΩ What is the frequency range of this system.

4. An accelerometer has a charge sensitivity of 2.2 pC/m/s^2 and a capacitance of 1090 pF. It is connected to a voltage amplifier by a cable whose capacitance is 110 pF. What is the accelerometer's voltage sensitivity.

5. The sensing element of a transducer is to be a cantilever beam that undergoes a displacement at its free end. As close to the fixed end of the beam as possible are placed four strain gages. This beam transducer is to be calibrated statically using a micrometer.

(a) Draw a circuit diagram for the bridge incorporating the strain gages and indicate the gage locations and orientation on the beam to maximize the signal's output.

(b) State several important electrical characteristics that the bridge amplifier should have so that the resulting transducer has the broadest possible operating range.

(c) If the micrometer induces a small displacement d_o at the free end of the beam, then derive an expression for the sensitivity of the transducer in terms of the properties of the strain gages, the cantilever beam and the gain of the amplifier. Define all terms in the final result.

9 Case Studies

9.1 Introduction

In this chapter four case studies are presented. These cases have been selected to illustrate different aspects of computer integrated experimentation. They emphasize the need for one to understand the relationship between the physical process and the instrumentation, the level of detail required to insure that the instruments and devices selected have the capabilities to perform the functions needed and the importance in obtaining an estimate of the error (uncertainty) of the measurement and their sources.

Each case study provides a brief introduction that states the purpose and objectives for the experiment. This is followed, where appropriate, by a brief summary of the theory governing the process. Next a description of the physical apparatus, transducers and controllable devices are given along with their important characteristics as they pertain to the experiment. Then the instrumentation setup and the experiment's protocol are discussed. Finally typical experimental results are presented.

The particular aspect of computer integrated experimentation that each case study illustrates is listed below.

Case Study	Aspect Illustrated
9.2	A method for removing the effects of the transfer functions of the electronics between the transducers and the digital conversion instruments.
9.3	Procedures used to perform high accuracy static and dynamic measurements.
9.4	Implementation of a hierarchical control system for concurrent asynchronous operation of a manufacturing cell.
9.5	Signal processing techniques used in the simultaneous digital conversion of dynamic force and acceleration signals.

9.2 Determination of the Viscoelastic Shear Modulus From Torsional Vibrations[1]

Introduction
In order to use certain materials effectively for sound and vibration isolation and for the development of new materials, the dynamic material properties have to be known. There is, in particular, a class of materials known as linear viscoelastic materials for which the elastic material constants can be replaced by a frequency-dependent function when the material is subjected to harmonic excitation. In the case of a material subjected to a shear stress only, the material constant is called the complex shear modulus $G^*(f) = G'(f) + jG''(f)$, where $G'(f)$ is the shear storage modulus and $G''(f)$ is the shear loss modulus. The objective is to determine $G^*(f)$ over a range of frequencies and temperatures.

Theory
One method used to determine $G^*(f)$ is the forced torsional vibrations of a cylindrical specimen in which a mass with mass moment of inertia J and a spring of constant k is attached to one end. The other end of the cylinder is subjected to a harmonically varying torque. The experimental setup is shown in Figure 9.1 and

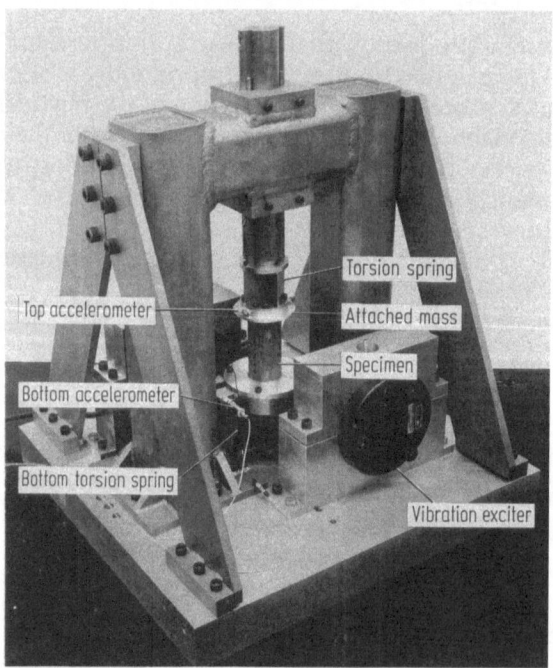

Figure 9.1   Experimental apparatus for the torsional vibrations of a cylindrical specimen

1 Magrab [1984].

the details of the upper torsion spring and specimen are shown in Figure 9.2. The expression for the angular acceleration response of the top plane of the test specimen to its bottom plane is given by

$$A_o = \frac{r_t}{r_b}[C_1 \Omega^* \sin \Omega^* + \cos \Omega^*]^{-1} \qquad (9.1)$$

where

$$C_1 = \frac{2J[(f_m/f)^2 - 1]}{\pi \rho h (b^4 - a^4)}, \qquad f_m = \frac{1}{2\pi}\sqrt{\frac{k}{J}}$$

$$\Omega^* = x - jy, \qquad p = \sqrt{G'^2 + G''^2}, \qquad \theta = \tan^{-1} G''/G'$$

$$x = 2\pi f h \sqrt{\frac{\rho}{p}} \cos(\theta/2), \qquad y = 2\pi f h \sqrt{\frac{\rho}{p}} \sin(\theta/2)$$

and ρ is the density of the cylinder, h its length, b its outer radius, a its inner radius, and G''/G' the loss tangent. The quantity f_m is the natural frequency of the attached spring-mass system with the specimen removed. The quantity r_t is the distance from the center of the axis of the cylinder to the center of the top accelerometer, and r_b is the distance from the axis to the point of application of the applied force, which coincides with the location of the bottom accelerometer.

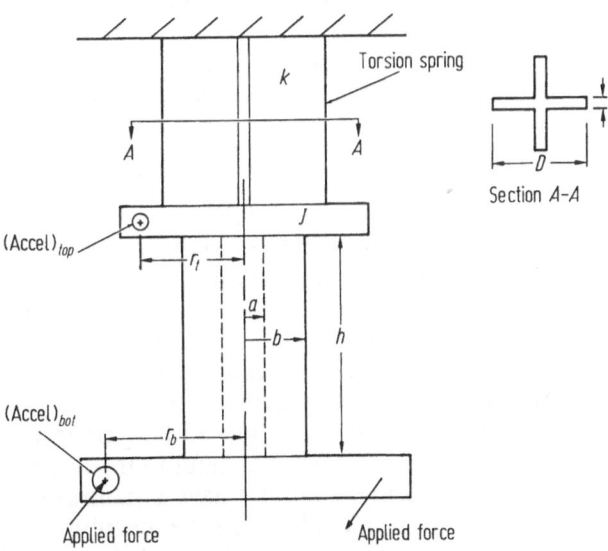

Figure 9.2 Geometric description of the torsion specimen and the attached spring and mass

Equation (9.1) can also be expressed as

$$A_o = R e^{j\phi}$$

where

$$R = \frac{r_t}{r_b}\sqrt{A^2 + B^2}, \qquad \phi = \tan^{-1} B / A \qquad (9.2)$$

and

$$A = [C_1 x \sin x + \cos x]\cosh y - C_1 y \cos x \sinh y \qquad (9.3a)$$

$$B = [C_1 x \cos x - \sin x]\sinh y + C_1 y \sin x \cosh y. \qquad (9.3b)$$

Thus, if R and ϕ are the measured amplitude ratio and phase angle, respectively, and all the physical and geometric parameters of the specimen are determined by other means, then G' and G'' can be found using Eqs.(9.2) and (9.3). However, because of the complexity of these equations G' and G'' cannot be solved for explicitly. A numerical procedure has to be used (Magrab [1984]).

The errors associated with the measurement of the amplitude ratio and phase angle cannot be easily related to the resulting errors in the determination of the shear modulus because of the highly nonlinear nature of Eqs.(9.2) and (9.3). However, to obtain an estimate of the sensitivity of the determination of G' and G'' to errors in the measurement of R and ϕ the following procedure was used. All the parameters appearing in Eqs.(9.2) and (9.3) were given typical values. Additionally, values for G' and G'' were assumed over the frequency range of interest. This resulted in a set of amplitude ratios and phase angles as a function of frequency. This set of R and ϕ were then altered by the accuracy limits of the voltmeter and phasemeter used in the experiment and the inverse problem was considered. Using this procedure it was found that for an uncertainty of ± 0.2 dB in the amplitude ratio and $\pm 0.3°$ in the phase angle the largest errors, which were $\pm 15\%$, occurred in the loss tangent.

Experimental Apparatus
According to the theory used to determine the shear modulus from the experimentally determined data, the specimen had to be subjected to torsional motion only. In addition, the fixture itself had to be free from structural resonances over a broad frequency range. To minimize the possibility of bending the top and bottom torsional springs had the cross-section shown in Figure 9.2, which for this particular apparatus produced a bending stiffness that was a thousand times greater than its torsional stiffness. To further decrease the possibility of bending two vibration exciters were used. The bottom torsion spring played no part in the experiment other than to provide an almost purely torsional input to the bottom of the specimen. This bottom spring-crank-arm system did have a natural frequency, but was of no consequence in the experiment because the magnitude of the input acceleration was kept relatively constant throughout the usable frequency range.

Instrumentation and Experimental Procedure

The computer controlled instrumentation system used to measure the accelerometers' amplitude ratio and phase angle is shown in Figure 9.3. The two electrodynamic vibration exciters were connected in parallel and received their input from an amplifier with the capacity to provide 15 A_{rms} into 1 ohm. The input to the amplifier was connected to an oscillator. The exciters themselves had an impedance of approximately 2 ohms and required 5 A_{rms} to obtain their rated output force. The oscillator's output voltage amplitude and frequency were under computer control.

The accelerometers had a sensitivity of nominally 10 mV/g and had a unity gain preamplifier (impedance converter) built into them. The output signals from the accelerometers were then amplified 30 dB and passed through two tracking filters, each with a 2 Hz bandwidth. The tracking frequency was provided by a second output signal from the oscillator and remained constant at 1 V. This 1 V signal was amplified to meet the tracking filters' requirement of 3.5 V. The 30 dB gain given to each accelerometer signal provided better use of the dynamic range of the tracking filters and used the digital phasemeter in a voltage range in which its accuracy was better, namely at levels above 50 mV_{rms}. To maintain control over the voltages throughout the system, the output voltage of the oscillator was con-

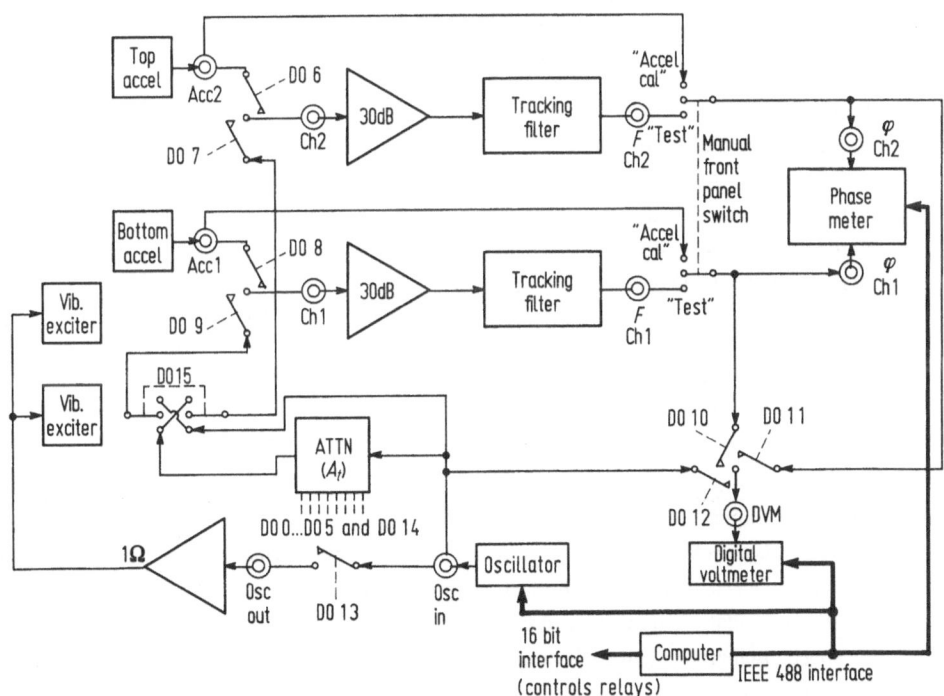

Figure 9.3 Equipment block diagram for the torsional vibration measurements

tinually adjusted so that the output voltage of the bottom accelerometer remained approximately constant at 2.5 mV$_{rms}$ over the entire test frequency range.

Signals from the output of the tracking filters were sent directly to the digital phasemeter and the digital voltmeter, both of which were under computer control. However, the signals to the voltmeter were read sequentially with the aid of a computer-controlled array of sixteen switches. The functions of these switches are given in Table 9.1. The phasemeter had an autocalibration feature that was employed every sixth measurement to ensure that the phase measurements were as accurate as possible. The digital voltmeter was used in its autoranging mode, since the output voltage from the top accelerometer differed widely over the frequency range. The purpose of the attenuator will be discussed subsequently.

To accurately measure the acceleration ratio of the top accelerometer to the bottom one and the phase angle between them, the influences of electronics inserted between the signals at the output of the accelerometers and the voltmeter and the phasemeter had to be removed. The ultimate accuracy of the measurement, however, depended on the accuracy of these two instruments. The voltmeter had an accuracy of approximately ±0.2 dB for values above 100 mV$_{rms}$ and ±0.3 dB for voltages less than 100 mV and greater than 100 μV over the frequency range

Table 9.1 Computer controlled switch functions shown in Figure 9.3

Line designation	Function
D00	Sets attenuator to 5 dB
D01	Sets attenuator to 10 dB
D02	Sets attenuator to 15 dB
D03	Sets attenuator to 20 dB
D04	Sets attenuator to 25 dB
D05	Sets attenuator to 30 dB
D06	Connects accelerometer #2 to fixed gain amplifier of channel #2 when the external switch is in the "test" position, otherwise it connects directly to the phasemeter and the digital voltmeter
D07	Connects oscillator to fixed gain amplifier of channel #2 or the output of the attenuator to the fixed gain amplifier of channel #2
D08	Connects accelerometer #1 to fixed gain amplifier of channel #1 when the external switch is in the "test" position, otherwise it connects directly to the phasemeter and the digital voltmeter
D09	Connects oscillator to fixed gain amplifier of channel #1 or the output of the attenuator to the fixed gain amplifier of channel #1
D010	Connects the accelerometer #1 or the tracking filter output of channel #1 to the digital voltmeter
D011	Connects the accelerometer #2 or the tracking filter output of channel #2 to the digital voltmeter
D012	Connects oscillator output to the digital voltmeter
D013	Connects oscillator output to the amplifier
D014	Sets attenuator to 0 dB
D015	Reverses D07 and D09: connects D07 to the attenuator output and D09 to the oscillator output

1 to 2000 Hz. The phasemeter had an accuracy of $\pm 0.1°$ for voltages greater than 50 mV$_{rms}$ and $\pm 0.2°$ for voltages as low as 1 mV$_{rms}$ from 50 to 50,000 Hz.

The effects of the intervening electronics were removed in the following manner. Consider the simplified equipment diagram given in Figure 9.4. Let $H_j(f)$ be the transfer function of the accelerometers' built-in amplifiers, $H_{Gj}(f)$ the transfer function of the 30 dB gain fixed gain amplifiers and $H_{Fj}(f)$ the transfer function of the tracking filters, where $j = 1$ refers to the bottom accelerometer and $j = 2$ the top one. The true output signals from the accelerometers V_j are related to the signals appearing at the inputs to the digital voltmeter and phasemeter $V_{Rj}(f)$ as follows (recall Eq.(3.16)):

$$V_{Rj} = V_j H_j H_{Gj} H_{Fj} \qquad j = 1,2.$$

Solving these two equations for the acceleration ratio $A_o = V_2/V_1$ yields

$$\frac{V_2}{V_1} = \left(\frac{V_{R2}}{V_{R1}}\right)\left(\frac{1}{C}\right)$$

where C is the complex quantity

$$C = \left(\frac{H_2}{H_1}\right)\left(\frac{H_{G2}H_{F2}}{H_{G1}H_{F1}}\right).$$

Thus, the amplitude ratio and phase angle are, respectively,

$$\left|\frac{V_2}{V_1}\right| = \left|\frac{V_{R2}}{V_{R1}}\right|\left|\frac{1}{C}\right|, \qquad \phi_{21} = \phi_{R21} - \phi_c \qquad (9.4)$$

where $|V_{R2}/V_{R1}|$ is the ratio determined from the digital voltmeter readings and ϕ_{R21} is the value obtained from the phasemeter. The quantities $|C^{-1}|$ and ϕ_c are the corrections that were isolated and measured in order to remove the influence of the electronic components on the amplitude and phase angle measurements, respectively. To remove the effects of C at a given temperature H_{Gj} and H_{Fj} were first removed from the measurement chain. Then a back-to-back calibration of

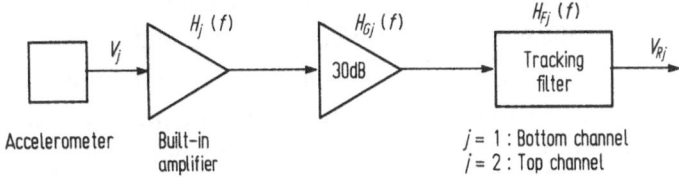

Figure 9.4 A portion of the instrumentation system of Figure 9.3

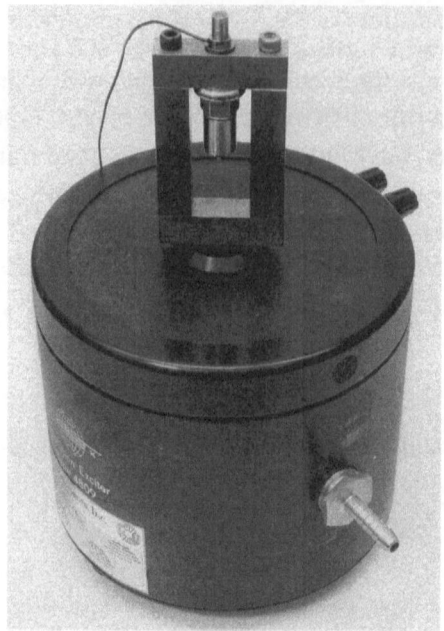

Figure 9.5 Back-to-back calibration of accelerometers

the two accelerometers as shown in Figure 9.5 was performed. Since one accelerometer was upside down with respect to the other

$$V_2 = V_1 e^{j\pi}$$

and

$$\left| \frac{H_2}{H_1} \right| = \left| \frac{V_{R2}}{V_{R1}} \right|_{BB}, \qquad \phi_{21} = \phi_{BB} + \pi$$

where $|V_{R2}/V_{R1}|_{BB}$ was determined directly from the digital voltmeter readings and ϕ_{BB} directly from the phasemeter.

The accelerometers were then disconnected from the measurement chain and replaced with two signals that were at the same frequency and phase with respect to each other, but with different amplitudes. Thus $V_1 H_1 = V_I$ and $V_2 H_2 = A_t V_I$, where V_I was the amplitude of the input voltage to both channels and $0 < A_t < 1$. The attenuation A_t was introduced because of the high degree of amplitude and phase nonlinearity of the tracking filters. The introduction of the attenuator allowed the input signals to each channel to be approximately equal (within ±2.5 dB) to each accelerometer's output signal at each frequency. In the actual test procedure the attenuation A_t could be introduced to either channel depending on whether or not the accelerometer's output voltage was greater than or less than one. If the channels were switched then A_t was replaced by $1 / A_t$ in the subsequent results. Equation (9.4) then yields

$$\left| \frac{H_{G2} H_{F2}}{H_{G1} H_1} \right| = A_t \left| \frac{V_{R2}}{V_{R1}} \right|_I , \qquad \phi_{21} = \phi_I$$

where $| V_{R2} / V_{R1} |_I$ was determined directly from the digital voltmeter readings and ϕ_I directly from the phasemeter. Thus Eqs.(9.4) become, respectively

$$\left| \frac{V_2}{V_1} \right| = \left| \frac{V_{R2}}{V_{R1}} \right| \left[A_t \left| \frac{V_{R2}}{V_{R1}} \right|_{BB} \left| \frac{V_{R2}}{V_{R1}} \right|_I \right]^{-1} \tag{9.5a}$$

$$\phi_{21} = \phi_{R21} - \phi_{BB} - \pi - \phi_I . \tag{9.5b}$$

In the actual testing of the accelerometer's back-to-back response it was found that at each test temperature the accelerometer's relative amplitudes were constant over the frequency range of interest and the phase was essentially zero. Thus

$$\left| \frac{V_{R2}}{V_{R1}} \right|_{BB} = C_3$$

and $\phi_{BB} = 0$. The constant C_3 was then determined at each temperature at a nominal frequency of 200 Hz.

Experimental Results
Typical amplitude ratios and phase angles at 20 °C are shown in Figure 9.6. The computed storage modulus and loss tangent are shown in Figure 9.7 for this and other temperature values.

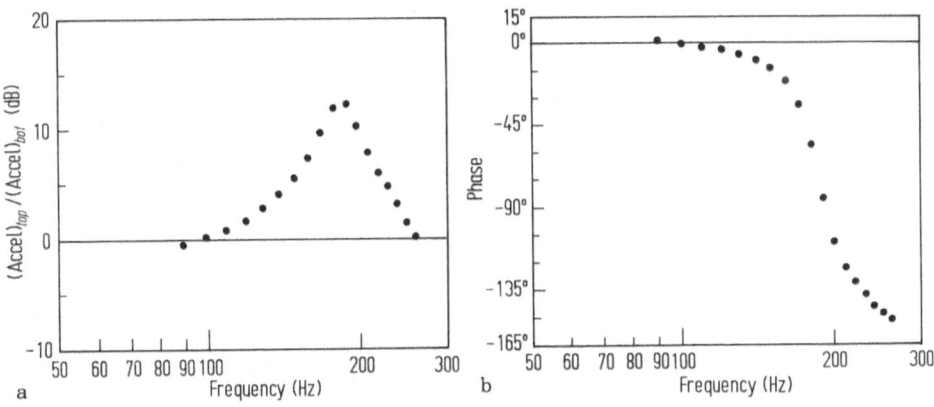

Figure 9.6 Typical results at 20 °C for   (a) amplitude ratio   (b) phase angle

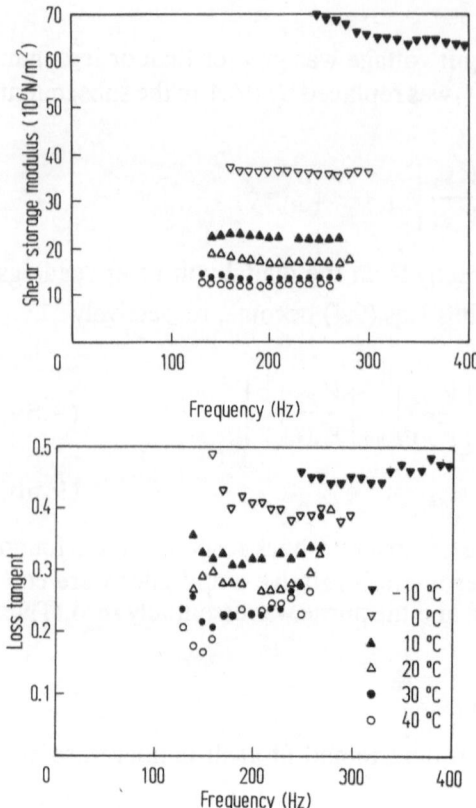

Figure 9.7 Some results for the storage and loss modulii at various temperatures

9.3 Testing of a Fast-Tool Servo for Diamond Turning[2]

Introduction

The fast tool servo (FTS) is a self-contained and independently operated device that was designed to act as a tool post for a diamond cutting tool on a large vertical lathe. Its main function was to move its diamond cutting tool in response to signals from the lathe's controller in such a manner that real-time error corrections could be made to compensate for any errors in the relative position between the tool and the work piece. The FTS was basically a piezoelectric vibration exciter whose displacement was continually monitored and controlled through feedback. A section view of the device is shown in Figure 9.8. The displacements of the piezoelectric-driven tool post were monitored by a very sensitive capacitance gage that was an integral part of the FTS. The desired characteristics of the FTS were to have a displacement range of $\pm 50\,\mu$in over a frequency range from dc to 200 Hz with an error of $0.1\,\mu$in (2.5 nm) or less. Described below are the test procedure and method used to determine its static and dynamic characteristics.

---

2 Patterson and Magrab [1985].

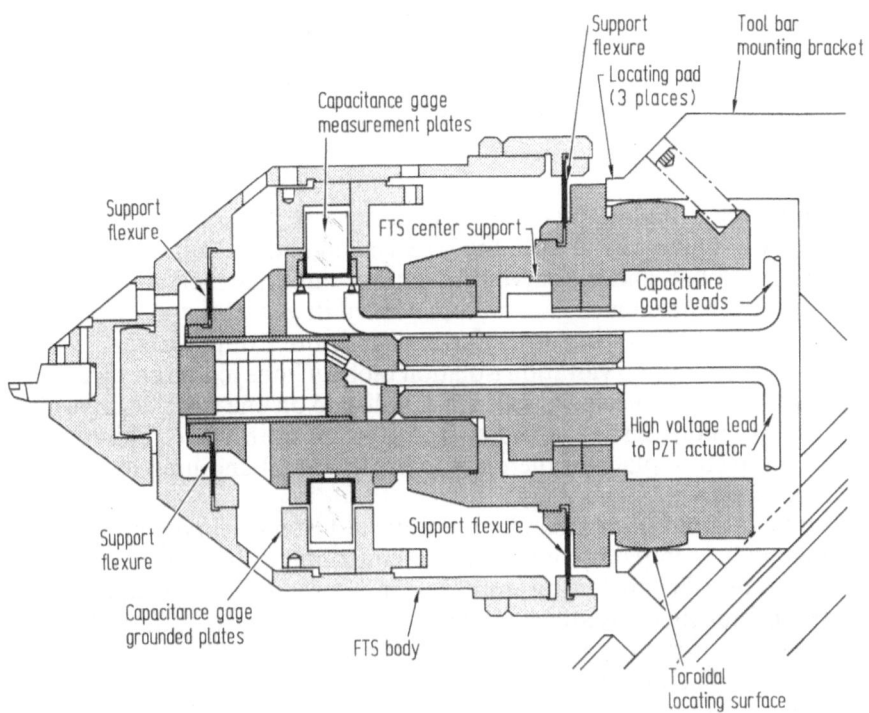

Support flexure
Support flexure
Tool bar mounting bracket
Locating pad (3 places)
Capacitance gage measurement plates
FTS center support
Capacitance gage leads
High voltage lead to PZT actuator
Support flexure
Support flexure
Capacitance gage grounded plates
FTS body
Toroidal locating surface

Figure 9.8 Cross-section of the fast tool servo

Instrumentation

The transducer used to measure the displacement of the FTS tool post was a specially designed and built capacitance gage that had a circular sensing area with a 0.125 inch diameter and a sensitivity of 3.7 mV/μin at an offset of approximately 2200 μin. Its built-in circuits gave an output voltage that was proportional to both a static (dc) and dynamic displacement. The capacitance gage was calibrated on a coordinate measuring machine that moved in 0.5 μin increments and which was monitored by an ac laser interferometer displacement measuring system described in Section 8.2.1. After repeated testing of this capacitance gage it was found that its sensitivity varied less than 0.5%, which corresponded to an uncertainty of approximately 0.3 μin over 50 μin.

It is recalled from Eq.(8.14) that the response of a capacitance gage is inversely proportional to the displacement d. Therefore the results of the capacitance gage testing were used to determine an accurate fit to the data using the following relation:

$$d + d_{off} = \sum_{n=1}^{4} C_n V^{-n} \tag{9.6}$$

260

where d_{off} was the gage's offset from the measured surface and V was the gage's output voltage. The data were fit over a displacement range of $\pm 100\,\mu$in using a general-purpose nonlinear solution algorithm to determine C_n.

Static Displacement Tests

Prior to running the dynamic tests a series of static tests were run to determine the repeatability and linearity of both the FTS system and the capacitance gage, its associated electronics and the curve-fitting procedure indicated above.

The FTS and the capacitance gage were mounted in a steel fixture that duplicated the mounting configuration on the lathe. The entire fixture was then placed on a large, thick steel-topped vibration isolation table of honeycomb construction. The target for the capacitance gage was a disk polished on one surface and which had a concentric cylindrical rod protruding from its opposite surface. This rod was inserted into the FTS in place of the diamond cutting tool and held in place by means of a set screw.

The instrumentation configuration for the static test is shown in Figure 9.9. The entire test was under computer control, with all instruments interfaced using the IEEE 488 protocol. The air temperature was monitored with an independently calibrated thermistor that was attached to a digital ohmmeter (recall Eq.(8.62)). The output voltage of the capacitance gage was read with a 6½ digit dc digital voltmeter (DVM). The output voltage from the barometric pressure sensor and the voltages from the various stages of the FTS electronics were monitored with

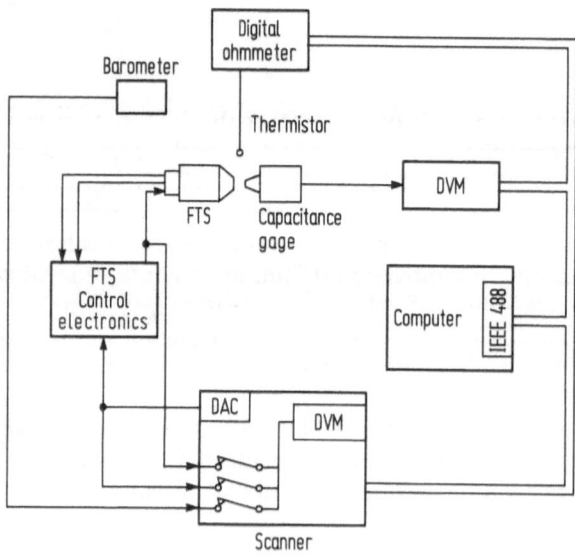

Figure 9.9 Instrument block diagram for the fast tool servo static tests

a scanner unit consisting of a low speed multiplexer, a DVM and digital-to-analog convertor (DAC). The DAC was used to provide the dc voltage to drive the FTS.

The following test procedure was used. The output of the DAC was set to 0.000 V and the output of the capacitance gage read. Then the DAC was set to V_o and the capacitance gage read again. At each of the voltages the corresponding values of the displacements were determined from Eq.(9.6). The difference between these two displacement values was the displacement of the FTS for the input voltage V_o. This method eliminated any effects of drift on the measurements; that is, the effects of drift were avoided and their magnitude remained undetected. The values of V_o ranged from ±10 V in increments of 0.5 V in the following manner: 0.0 V to 10 V to -10 V to 0.0 V.

Typical results are shown in Figure 9.10. The variation of the individual displacement values were within ±0.03 μin for the most part with a few differences as large as ±0.05 μin. These highly reproducible results indicated that all the instrumentation and devices were properly working and the dynamic tests could now be run.

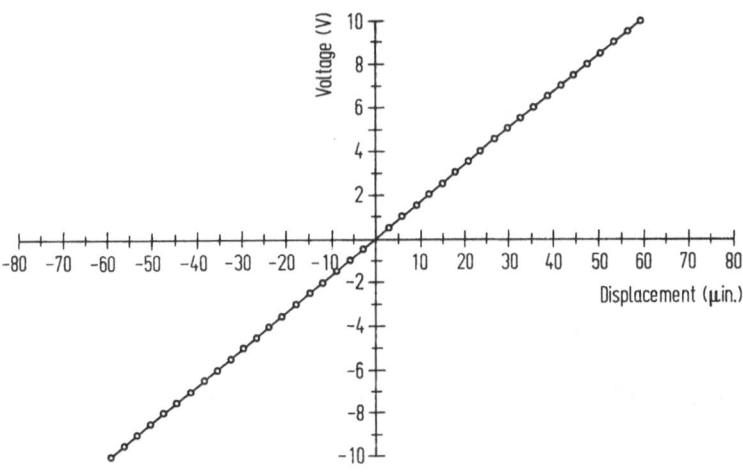

Figure 9.10 Typical static displacement linearity as a function of the input voltage to the fast tool servo

Dynamic Displacement Tests
The mechanical test setup for the dynamic tests was the same as that for the static tests except for the fact that the test fixture was now clamped at three points to the steel-topped table.

The instrumentation system for the dynamic tests are shown in Figure 9.11. The differences between this and the static setup is that the DAC was replaced with a frequency synthesizer and the output of the capacitance gage was recorded with

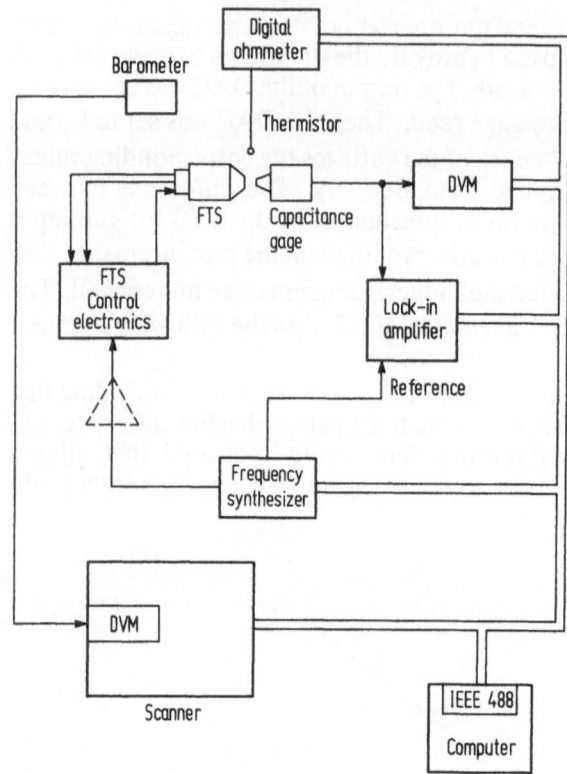

Figure 9.11 Instrument block diagram for the dynamic tests of the fast tool servo

a lock-in amplifier (recall Section 6.1.3). The lock-in amplifier was used for two reasons: its low frequency response (0.2 Hz) and its narrow-band filtering capabilities with the correspondingly long averaging times. The output recorded by the lock-in amplifier was an rms voltage.

The maximum voltage from the frequency synthesizer was 10 V peak-to-peak. In order to obtain the required output voltage of 20 V peak-to-peak an amplifier (shown in Figure 9.11 with dotted lines) was used to obtain the FTS displacements above 30 μin. This amplifier's frequency linearity was virtually perfect from 1 to 200 Hz with a slight attenuation of 0.02 dB (0.2%). This corresponded to 0.1 μin over 50 μin and was well within the experimental error; therefore, it was considered negligible. However, the value of the output of the amplifier at dc was approximately -0.1 dB relative to its output between 1 and 200 Hz. This difference was taken into account by increasing the value of the nominal static displacements by 1.0%. Lastly, the frequency synthesizer also had the capability of producing a dc voltage over the range of ±10 V.

The dynamic tests were conducted at frequencies from 1 to 200 Hz with the following frequency resolution: from 1-10 Hz, every 1 Hz; from 10-100 Hz, every

5 Hz, and from 100-200 Hz, every 20 Hz. Each of these frequency sweeps was conducted at the following seven nominal static displacement amplitudes: 4.6, 9.2, 18.4, 27.9, 36.9, 46.1, 55.2 μin. When the frequency was greater than 10 Hz the averaging time of the lock-in amplifier was reduced. The following test protocol was used at each nominal static displacement and at each frequency and virtually eliminated any effects of drift. The output voltage of the frequency synthesizer was set to $0.000 \, V_{dc}$ and the capacitance gage's output voltage read with the DVM exactly as was done in the static tests. Using Eq.(9.6) this voltage corresponded to a displacement d_b. The frequency synthesizer then put out a dc voltage of magnitude V_o and the capacitance gage's output voltage was again read by the DVM. This voltage corresponded to a displacement d_s. The difference $d_o = d_s - d_b$ was the static displacement corresponding to a dc voltage V_o. Lastly, the frequency synthesizer put out a sine wave of frequency f and amplitude V_o, the same peak amplitude as the dc voltage. The output of the capacitance gage was then read by the lock-in amplifier. Since the output voltage from the lock-in amplifier V_L was in volts-rms, its peak voltage was simply $V_p = V_L\sqrt{2}$ [recall Eq.(6.13)]. Converting V_p to the displacement d_p, the difference d_d due to the FTS motion at frequency f was $d_d = d_p - d_b$. Ideally the displacement corresponding to the output voltage at frequency f should have been the same as that obtained from the static test. Therefore, the error was $d_d - d_o$. Values for the error are shown in Figure 9.12.

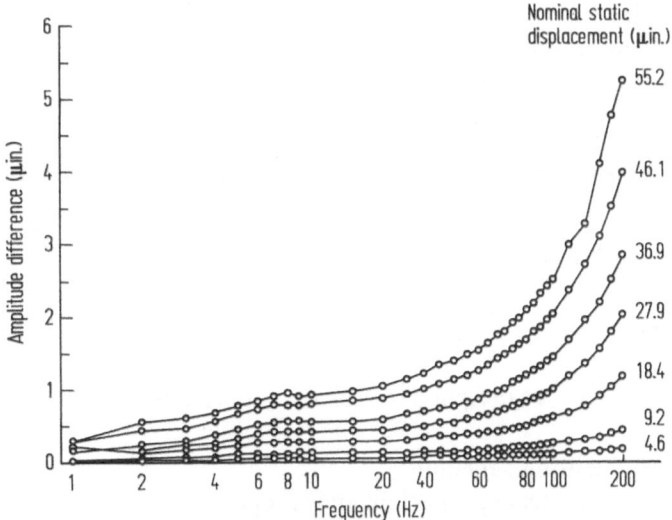

Figure 9.12 Difference between peak dynamic displacement and the static displacement of the fast tool servo as a function of frequency for several nominal static displacements

264

9.4 Integration of an Automated Manufacturing Machining Cell[3]

Introduction
The Automated Manufacturing Research Facility was a cooperative research effort between the National Bureau of Standards (NBS; now called the National Institute of Standards and Technology) and the US Navy's Manufacturing Technology Program. It included three machining workstations, a cleaning and deburring workstation, an inspection workstation and a materials handling workstation. These workstations were linked together using a computer architecture developed at NBS. One of its principal research goals was to obtain an automated prototype factory system that had a high degree of flexibility, integration and sophistication. This case study details the manner in which the vertical machining workstation (VMW) attempted to attain these goals. It will concentrate on the equipment control aspects by describing the types of operations and functions of each piece on equipment, the type and placement of the sensors used to insure proper execution of the commands and the equipment and the equipment controller's software structure and operation. The intelligence to run the equipment controller resided in the workstation controller in the form of process plans, work elements and agreed-upon procedures for the safe operation of the VMW.

The major features and capabilities of the VMW shown in Figure 9.13 were:

Robot
*Programs did not have to reside in robot controller's memory
*Access to locations within its various work volumes required only one taught point per work volume
*Wide position tolerance (±6 mm) for all tasks

Equipment Controller
*Multiprocessor computer executed multiple commands simultaneously
*Software interlocks prevented unsafe operations
*Continuously updated equipment and executed command status always available; robot position available on request
*Automatic error recovery from equipment and communications failures
*Variety of sensors used to verify each command's execution
*Equipment controller's command set easily modified to accommodate changes in workstation's functions

Robot Gripper
*Fixed fingers contained rotary indexer to rotate grasped object either 0°, 90°, 180° or 270°
*Interchangeable fingers were designed to grasp small prefixtured pallets (30 cm × 30 cm)

Vertical Spindle CNC Milling Machine
*Touch probe was used to locate part origin and its operation was an integral part of the NC code

---

3 Magrab [1986].

Figure 9.13   Major components comprising the vertical machining workstation

*Using the rotary feature of the robot's grippers it could perform multiple surface machining of a part
*Automatic tool length offset determination
*Part blanks did not have to be premachined

Vise
*Accepted parts ranging in height from 19 to 75 mm, in length from 95 to 152 mm and in width from 65 to 125 mm.

Chip Removal
*Robot manipulated vacuum hose

In addition the workstation controller provided NC code generation from a feature-based design system and a NC code verifier to prevent unsafe, impossible and inefficient machining operations.

VMW Mechanical Components and Transducers
The mechanical components comprising the VMW and their associated sensors are now described.

Robot Gripper   The gripper assembly, shown in Figure 9.14, used a split rail, rack and pinion mechanism to obtain nonbinding, parallel movement of its fingers. The fingers for the VMW were designed to rotate a grasped part either 0°, 90°, 180° or

270°, and to permit the fingers to be changed so that the gripper could pick up small prefixtured parts on a pallet. The finger changing mechanism was passive, in that after being placed by the robot in the finger changing assembly it only required the coordinated opening and closing of the gripper's fixed fingers. The removable fingers would then either be snapped into, or unsnapped from, the fixed fingers.

The rotary indexing capability of the fixed fingers was accomplished with two pneumatically actuated rotary indexers mounted in tandem (see Figure 9.14). The indexer closest to the finger frame rotated either 0° or 180°. Attached to its rotating disc was another indexer, which rotated either 0° or 90°. By rotating the indexers either singly or together the four positions 0°, 90°, 180° and 270° were obtained. The rotation of each rotary actuator was monitored by a contact proximity switch. The opposing finger contained only an idler disk. Both of these disks had rubber pads glued to them to provide a high-friction grasping surface.

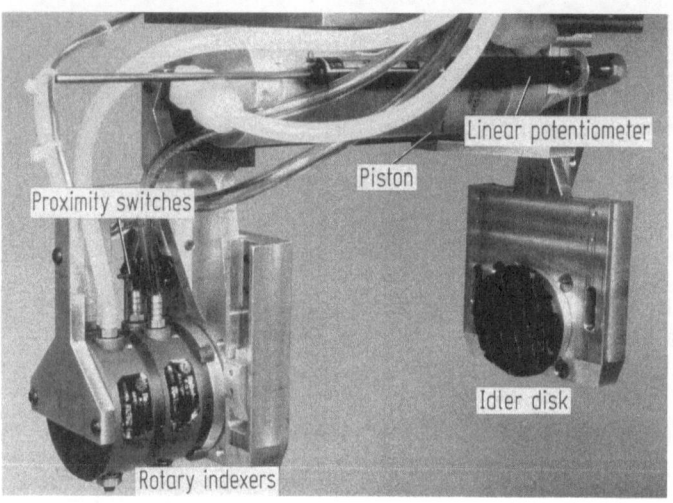

Figure 9.14   Robot gripper

The gripper's fingers were opened and closed with a pneumatic piston that had a 89 mm stroke. Because of the opposing rack and pinion design the maximum opening between finger's disks was 178 mm. The combination of the piston's diameter and air supply pressure resulted in a force of approximately 310 N. The size of the finger's opening was monitored by a linear potentiometer. The linear potentiometer was used to both verify that the gripper had opened and, when directed to close, the amount it had closed. The amount of closure was used either to verify a dimension of the part as grasped or to determine if the part was dropped.

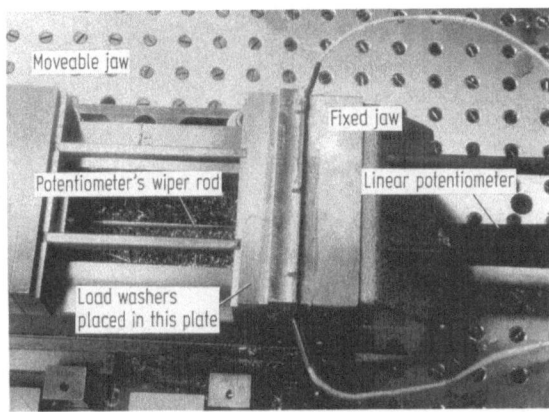

Figure 9.15   Pneumatically controlled vise

Vise   The vise, shown in Figure 9.15, had its moveable jaw controlled by a pneumatically propelled turbine that rotated its screw drive mechanism. When the motion of the moveable jaw was resisted by material placed between it and the fixed jaw, the jaw stopped moving; however, the screw rotation continued until a cam-like mechanism mechanically locked the part in place. This locking force was approximately 20 kN. After the mechanical locking action took place any additional rotation of the turbine had no effect. To insure that the vise had mechanically locked the part in place, the fixed jaw of the vise was modified so that this 20 kN locking force could be monitored with two strain-gauge type load washers. These load washers were placed between the fixed support of the vise and the interchangeable plate that attached to the vise's jaw.

The size of the vise's jaw opening was monitored with a linear potentiometer having a maximum stroke of 178 mm. The potentiometer was oil-filled and placed behind a shroud to eliminate any effects from the coolant.

Vertical Machining Center   The vertical machining center (VMC), partially visible in Figure 9.13, was a CNC vertical milling machine having accuracy and repeatability of better than 0.13 mm, a work area of 1 m × 0.5 m, a 7.5 kW variable speed spindle motor capable of speeds to 5200 rpm and a 40 tool automatic tool changer. Two important accessories to this system were a tool length indicator (TLI) station and a touch probe with telemetry, which is shown in Figure 9.16. Both of these accessories were supported by the machine's controller and by vendor-supplied software. The probe was used by the NC programs and the workstation controller either to determine the origin of a blank part or to locate the coordinates of a feature of a partly machined part prior to the start of additional machining. The TLI station was used after each set of tool exchanges or after the machine's W-axis had been moved. W-axis movement was the repositioning of the entire vertical axis assembly, which included the spindle, motor and tool changer.

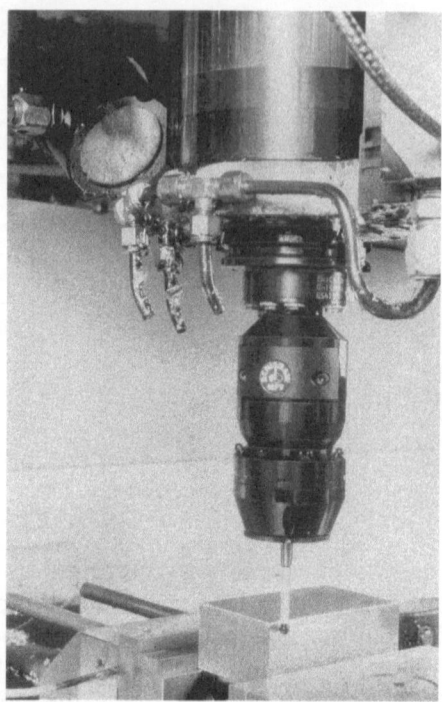

Figure 9.16 Touch probe with telemetry

The last important feature of the VMC was its direct-numerical-control (DNC) interface, which permitted both the transfer of NC programs and subroutines to and from the machine tool controller's memory and the toggling of some of its front panel switches with commands from the equipment controller. This interface communicated with the workstation controller using a RS 232 line at 4800 baud. (The functions that were controlled with this interface are given in Figure 9.19.) Unfortunately the DNC interface did not provide programmatic access to the cycle start switch, which had to be used to start the execution of all NC programs. Consequently this front panel button was wired in parallel to an independently programmable switch, which when activated produced the same result as if the front panel button were pressed.

The VMC was supplemented by two independent systems: one was a hydraulically activated pallet clamping system shown in Figure 9.17 and the other was a robot manipulated vacuum chip removal system. The chip removal system was a 11.3 kW industrial vacuum cleaner capable of generating a minimum of 200 mm Hg vacuum at the end of a 15 m long, 5 cm diameter hose. A vacuum gage was placed at the end of the hose to verify whether the vacuum was on or off. The end of the hose had an adapter attached so that the robot could grasp it as shown in Figure 9.18. This robot manipulated vacuum chip removal system was used either

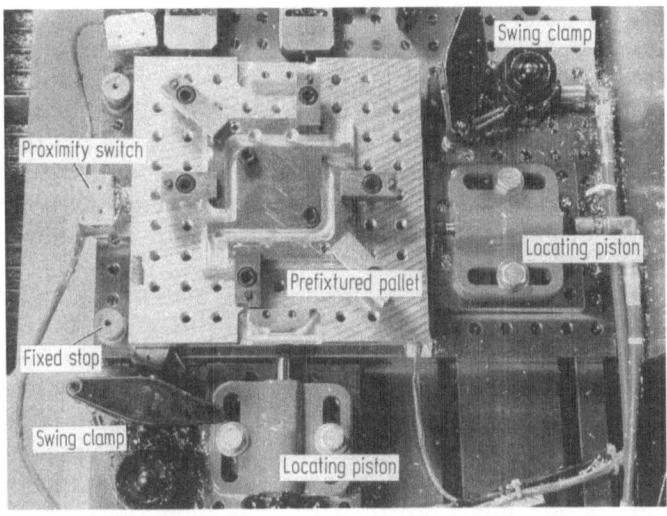

Figure 9.17   Hydraulically operated pallet positioning and clamping system

Figure 9.18   End of vacuum hose with its adapter positioned in its fixture so that the robot gripper could grasp it

at the end of the entire machining cycle or, for those parts subjected to large amounts of material removal, at appropriate times during the machining operations. All vacuuming operations, however, took place when the VMC table was at its extreme forward, right-hand position, which was denoted its "parked" position. This "parked" position was the one that the VMC table always had to be positioned before the robot was able to approach the table for vacuuming, or, for that matter, any function. To determine that the table was in its "parked" position two contact proximity switches were attached to the machine tool's frame, one to detect the x-axis position and the other the y-axis position.

The hydraulic system supplied the pressure necessary to actuate the locating and clamping devices used to register and hold the prefixtured pallets. This hydraulic system was run under the control of a separate computer system that communicated with the equipment controller using a separate IEEE 488 interface. The pallet was locked into its position on the table in the following manner. The robot placed the pallet on top of several 30 mm high posts. After the robot released the pallet two sequentially operated, orthogonally positioned pistons pushed the pallet against fixed stops. Two contact proximity switches monitored this positioning. Then two hydraulically actuated swing clamps locked two diagonally opposed corners of the pallet to the table via two posts, one under each corner of the pallet. These posts contained strain-gauge type load washers which monitored the swing clamps' clamping action.

Robot The robot, shown in Figure 9.13, was a hydraulically powered arm having six articulations (joints), three for the arm itself and three for the wrist. Robot control was accomplished using a straightforward English-like language entered from either the terminal or from an executed program. The communication capabilities of the microprocessor based operating system permitted the equipment controller to completely supervise the operation of the system, using a very rigorous communication protocol over a 9600 baud RS 232 link. Essentially the equipment controller replaced the robot's terminal and, without the use of memory resident robot programs, directly implemented all the necessary commands to move the robot through its desired motions.

The command-by-command control of the robot was simplified by partitioning its overall work volume into a set of N local work volumes. Each local volume was characterized by one "taught" reference point called VW_j, $j = 1, 2, \ldots, N$. The orientation of the robot at each VW_j was such that its tool (local) coordinate system was aligned perpendicular to each surface of interest within each local work volume. Using standard commands all useful points and orientations within each VW_j were accessible.

Trays The material handling workstation (MHS), on request from the workstation controller, directed the flow of the raw materials and machined parts to and from the workstation. The loading and unloading of the trays from the roller tables at each workstation onto the automatically guided carts were under the control of the MHS. The location of the materials placed on each tray were available to the workstation controller and were related to a particular process plan. For parts destined for the vise a maximum of three parts could be placed on a tray. The location of the placement of the nominal geometric center of each part was indicated by a grid permanently painted on each tray.

In order to insure coordinated and proper operation of the MHS with the VMW, the equipment controller used a software interlock system with the material handling controller in the manner described in the subsequent description of the station's operational safety. This software interlock prevented the MHS controller from moving trays when the robot was attempting to access them and, conversely, it prevented the robot from attempting to approach the trays when they were in, or about to be in, motion. In addition to the software interlock a contact proximity switch was placed at the end of each tray's forward travel onto the roller table. Before the robot could approach either tray, the tray's switch had to have been activated.

Equipment Controller

The equipment controller was a 32-bit desktop computer with 2.25 Mbytes of memory and three central processing units (CPU). The communication ports to the equipment controller were comprised of two IEEE 488 and three separate RS 232 interfaces. The programming language was an extended BASIC, which, in addition to other features permitted one to take full advantage of its multiprocessor environment. The three CPU system permitted the simultaneous running of three different programs, each of which has been assigned its own memory partition. If more than three programs were being run (each program still simultaneously residing in its own partition) the operating system arbitrated, allocating time intervals to each of the programs. Consequently some programs may have ended up running in a time-shared environment.

The equipment controller's program was really a collection of independent programs, each assigned to its own memory partition and each intended for a specific task. This type of structure permitted one to easily make programmatic changes to accommodate any changes in the functional operation of each task and to include new tasks. It also effectively uncoupled one set of operations associated with a physical device from another set, resulting in the controller's ability to continue functioning when there was a mechanical or communications failure with any of these devices, including the workstation controller itself. This configuration divided the workstation's equipment control into the following eight separate programs (tasks):

Task-1. Communicated with the workstation controller and parsed and distributed its commands to the appropriate programs within the equipment controller.

Task-2 Periodically read all the sensors and operated all the subsystems' electrically actuated valves and switches.

Task-3 Transferred commands to the robot controller for their execution.

Task-4 Communicated with the machine tool controller and loaded, selected and executed NC programs.

Task-5 Operated the gripper.

Task-6 Operated the vise.

272

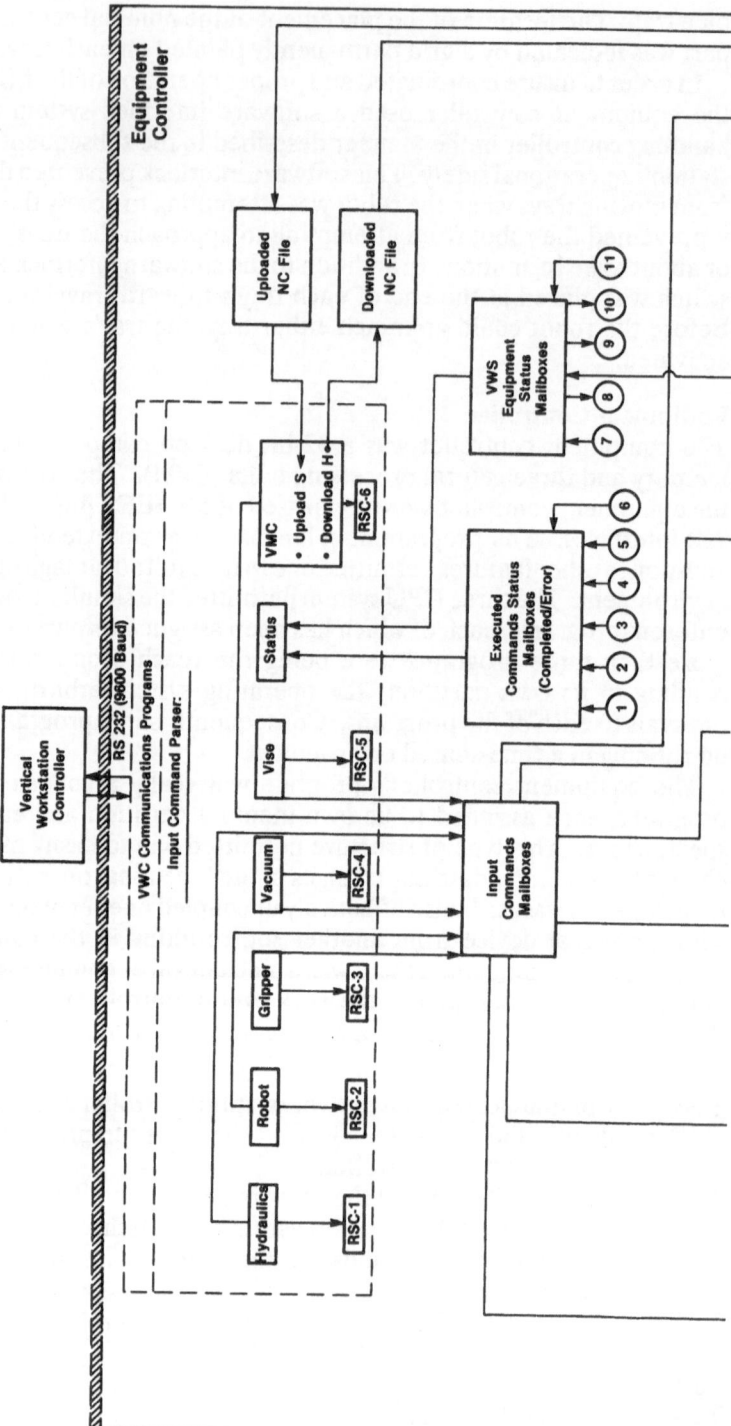

Figure 9.19 Multiprocessor equipment controller program structure, communication links and sensor and actuator functions. The programs within the dashed boxes reside in their own memory partition. All files and mailboxes are memory resident. A = Actuator; S = Sensor; RSC-n = Request for Service of Command from set n; VMC = Vertical Machining Center; VWS = Vertical Workstation; and VWC = Vertical Workstation Controller

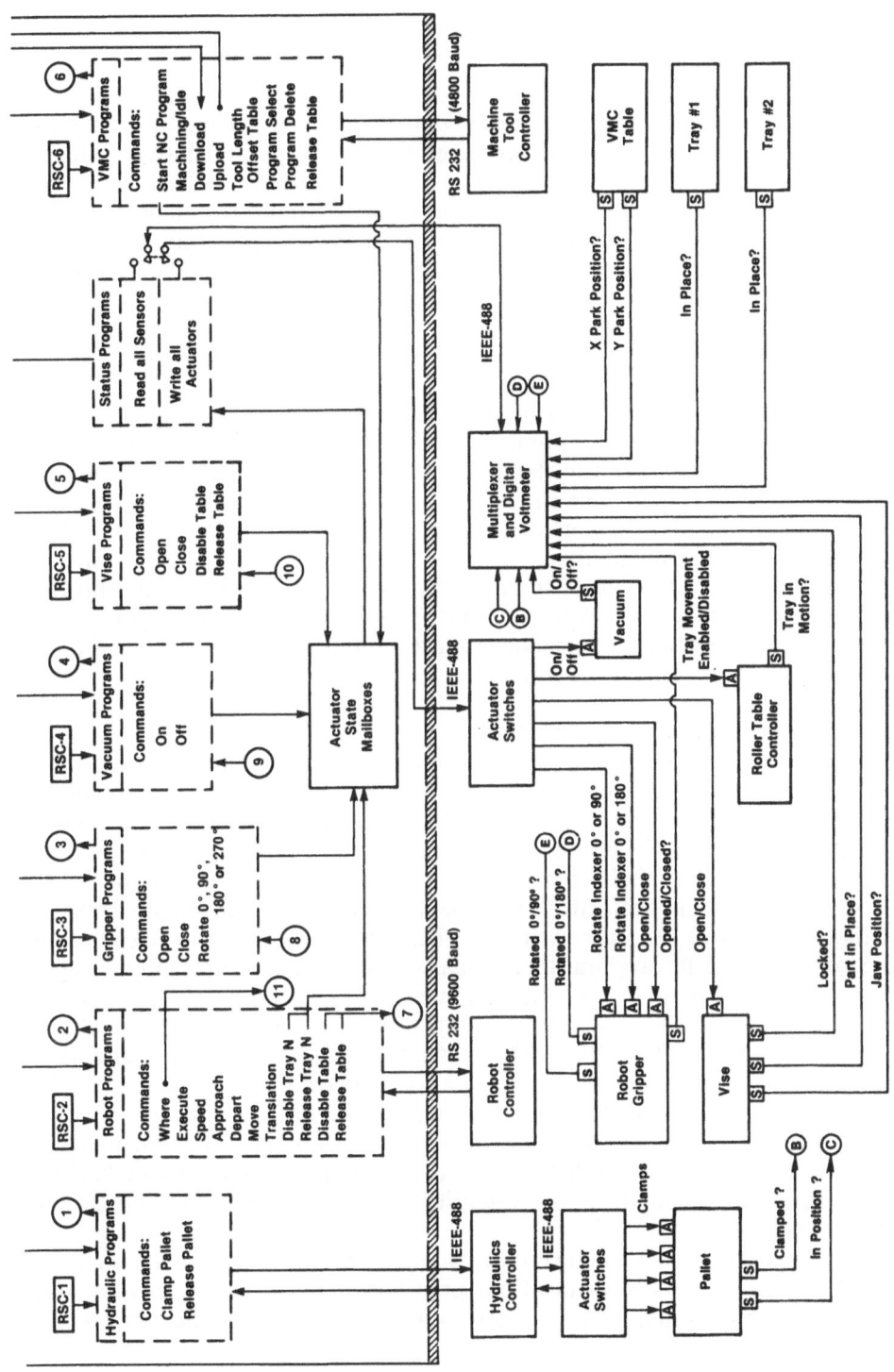

Task-7   Operated the vacuum chip removal system.
Task-8   Operated the hydraulically actuated pallet clamping devices.

Each of these programs controlling a subsystem responded to its own set of commands, and could only execute one command from its command set at a time. In order for these programs to perform their individual operations asynchronously, yet in an orderly, safe and controlled manner, the following simple rule governed all of the equipment controller's operations. Prior to transferring a new command from a given command set to the equipment controller, the workstation controller must have first examined the status of the previously transferred command from that set. Only after determining that the previous command was executed without error did it send a new command for that subsystem. The only exception to this rule was the transferring of status commands (see task 2 below), which could be done anytime.

As stated previously each of the eight tasks had its own command set. The details of these command sets and the operations they cause are not presented. However, a reasonably self-explanatory list of them appear in their appropriate contexts in Figure 9.19. The individual commands governing the operation of equipment of the vertical workstation had no functional meaning *per se*. It was the ordered collection of these commands by which functional meaning was attained. These collections of commands resided in the workstation controller's set of equipment work elements, which were designed to perform different sequences of operations for specific purposes: for example, to get a part from the vise and place it in one of the trays.

Task-1: Communications.   The eight programs comprising the equipment controller are shown in Figure 9.19 and interacted in the following manner. First, all communications with the workstation controller were handled by the communications program. The incoming command string contained, in general, multiple commands. This command string was then parsed and the individual commands written to their appropriate locations in a memory resident, random access file, which is henceforth called a mailbox. (In this case the "input command" mailbox in Figure 9.19.) After writing the commands to their respective mailboxes the communications program then signaled the appropriate programs to go to their designated mailbox, get the command and execute it. After a program executed its command a message was written to another mailbox, denoted the "status mailbox" in Figure 9.19. This message stated whether the command was successfully executed or whether an error had occurred.

The equipment controller did not initiate communications with the workstation controller. It only transferred information to the workstation controller when requested. The only exception was to let the workstation controller know the outcome of the communications integrity check that was performed for each transmission received.

In addition to parsing the incoming command strings and signaling the appropriate programs the communications program also had access to both status mailboxes, which contained the executed command status, and the interpretation of each sensor reading. This permitted the workstation controller to request and receive equipment and command status information at anytime.

Lastly the communications program initialized and created all the mailboxes.

Task-2: Equipment and Command Status Information. The status programs were a set of continuously running programs that were independent of the work-station controller. These routines were programmed to read all the sensors and to operate all the electrically actuated valves and switches approximately every 650 ms. The regularly updated sensory information and the regularly operated valves uncoupled the command's execution from its actual implementation, thus eliminating the inevitable contention problems that arise in resource-sharing asynchronous systems. Consequently, to change the state of a valve one wrote the desired state (open/close) to the appropriate mailbox. In typically less than 650 ms the valve began to change state. If the valve was already in that state nothing happened. Immediately before operating the pneumatic and hydraulic valves all the workstation's sensors were queried and the interpretations of their values were stored in their respective mailboxes. During the time between the next sensor readings and valve operations the mailboxes could be read by the other programs.

The sensor readings and valve operations were made with a multipurpose data acquisition and control unit that was externally addressable using the IEEE-488 protocol. This device had been configured to both read all the sensors and to operate all the valves and switches. The sensors were read using a 20-channel sequentially scanned multiplexer that fed the selected channel's voltage into its built-in digital voltmeter (DVM). The digital output of the DVM for each channel selected was transferred to the equipment controller. Depending on the placement of the sensor corresponding to that channel the equipment controller translated the value into an English language message, which was then stored in the appro-priate mailbox. The actuators were controlled by a 16-channel programmable array of relays, each rated at 2 A. The opening and closing of each relay connected and disconnected a 12 V dc voltage to each pneumatic and hydraulic valve, causing it either to close or open as the case may have been.

Task-3: Communications with the Robot Controller. The equipment controller's robot program communicated, using a very complex and rigorous protocol, with the robot controller's operating system through its special supervisory port. As stated previously this port permitted the robot to respond to commands from the equipment controller as if it had received instructions from its own terminal. Since virtually all programming commands could also be implemented from the robot's terminal, memory resident robot programs were not necessary.

Task-4: Communications with the Machine Tool Controller. Communications with the machine tool controller took place through a special interface board in the machine tool controller called the direct numerical control (DNC) interface. This board had a relatively simple communication protocol that provided the capability of downloading and uploading NC programs and subroutines, selecting programs for execution and deleting programs from its memory. It also provided the important status information of whether the machine was active (executing a program) or whether it was idle. When the machine tool was active, only active/idle information could be obtained from the machine tool controller; no other access was possible. It was for this reason that the downloading and uploading of programs and subroutines were performed in two steps. A program to be downloaded to the machine tool controller was first downloaded to a memory resident file in the equipment controller. Then at a future time a different command caused this program to be downloaded from the equipment controller's memory file to the machine tool controller. A reverse two-stage procedure was used for the uploading of a NC program or subroutine from the machine tool controller to the workstation controller.

Tasks-5 to 8: Operation of the Gripper, Vise, Vacuum and Hydraulics. The operations governing the gripper, vise and vacuum were straightforward, requiring only the opening or closing of the appropriate valves. The hydraulic system also only opened and closed valves; however, in this case another computer system directly controlled these actions, although the verification of the execution of these commands was determined by the equipment controller's status program.

Operational Safety: Software Interlocks Certain operations of the workstation controller were subject to software interlocks to prevent unsafe conditions. Although numerous combinations of operations were potentially unsafe it was decided to remove only the most obviously dangerous ones. The rest of the burden for safe operations resided in the implementation of the workstation controller's equipment work elements. From the equipment control perspective this provided for reasonably safe operation while still maintaining sufficient flexibility.

The software interlock procedure worked in the following manner. Consider two mailbox locations M1 and M2 and two independent operations A1 and A2 that could not occur simultaneously. When operation A1 was to be performed a message, F1, was written to mailbox M1. Similarly when operation A2 was to be performed a message, F2, was written to mailbox M2. Consider now operation A1. After setting F1, mailbox M2 was read to determine the state of F2. If F2 indicated that A2 was not in operation then operation A1 started. If F2 indicated that operation A2 was in progress then operation A1 did not start. Instead it withdrew message F1 and, after a short pause, again wrote F1 into M1 and read M2 and evaluated F2. This process was repeated until either F2 indicated that operation A1 could start or the request to perform A1 had timed out. The same procedure took place for operation A2. Consequently operations A1 and A2 could never occur at the same time.

The following operations were governed by the software interlock procedure:

1a. The robot could not approach the machine tool table unless the machine tool was idle and the table was in its "parked" position.

1b. A machine tool NC program could not start unless the robot had released the machine tool table.

2a. The robot could not approach a parts tray unless the material handling system had released that tray.

2b. The material handling system could not access (move) a parts tray unless the robot had released that tray.

3a. The vise could not be opened unless the machine tool was idle and its table was in its "parked" position.

3b. A machine tool NC program could not started unless the vise had released the machine tool table.

9.5 Identification of a Machine Tool's Dynamic Structural Response[4]

Introduction
A major source of error in machining operations is the relative dynamic motion between the cutting tool and the work piece. There are two types of dynamic phenomena that cause such motions: (a) externally excited oscillations, and (b) machine tool chatter. The first category includes both transient vibrations induced by a shock load and vibrations induced by periodic excitations. In contrast with these forced, damped vibratory motions, chatter is a "self-exited" oscillation. Under certain conditions energy from the cutting process excites the machine tool structure in one or more of its modes. If not enough energy is dissipated by either the damping of the structure or the friction of the cutting process, then the relative motion between the tool and the work piece at one of the system's natural frequencies grows beyond acceptable limits. This phenomenon is termed chatter and leads to poor surface finish, reduced dimensional accuracy, increased tool wear, and even tool fracture and damage to the machine tool itself.

In order to predict the onset of chatter both the cutting process parameters and the dynamic structural properties of the machine tool have to be known. This case study will describe a method that can be used to determine a particular machine tool's dynamic structural response properties during actual machining operations.

Theory
In continuous machining processes the structure is often excited by abrupt changes in the cutting force, such as those generated by a hard spot in the work piece, material clearing from a built-up edge on the cutting tool or a chip breaking. The structure responds to the change in the applied force $\vec{F}$ with a relative displacement $\vec{x}$ between the tool and the work piece. This displacement affects the values of the parameters of the cutting process, such as chip thickness, effective cutting angle, etc., which in turn results in a further variation of the cutting force.

The interaction between the machine tool and the cutting process can be described by a closed loop system that incorporates two fundamental blocks as shown in Figure 9.20. The machine tool block describes the relationship between

---

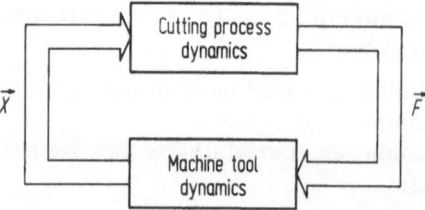

Figure 9.20   Representation of metal cutting dynamics

the dynamic cutting force $\vec{F}$, which is applied on the tool, and the response of the structure x. The effect of the relative displacement x on the variation of the cutting force $\vec{F}$, which can be described by "cutting equations," is represented by the cutting process block.

Although the machine tool is a highly complex dynamical system, for chatter analysis it is sufficient to describe the structure's dynamics only at the cutting point. The dynamic cutting force, which is applied on the cutting point, is decomposed into three components along the directions of a cartesian coordinate system fixed on the structure. If F_x, F_y and F_z are the Fourier transforms of these force components, and x, y and z are the Fourier transforms of the components of the relative displacement between the tool and the work piece, then the dynamics at the cutting point are described by the matrix equation

$$\vec{x} = G \vec{F}$$

or

$$\begin{bmatrix} x \\ y \\ z \end{bmatrix} = \begin{bmatrix} G_{xx} & G_{xy} & G_{xz} \\ G_{yx} & G_{yy} & G_{yz} \\ G_{zx} & G_{zy} & G_{zz} \end{bmatrix} \begin{bmatrix} F_x \\ F_y \\ F_z \end{bmatrix} \tag{9.7}$$

where G is the symmetric transfer matrix; that is, $G_{ab} = G_{ba}$. This matrix consists of six transfer function elements $G_{ab}(\omega)$, which provide the relative response of the structure in the ath direction due to a force acting in the bth direction when the other two force components are zero.

The six transfer functions in Eq.(9.7) can be determined by direct measurements using any of a wide variety of experimental techniques. The parameters of the dynamic system are then extracted from these experimental results. If the modes are well separated the system can be treated as a single-degree-of-freedom system in the vicinity of the resonance (recall Eq.(3.21)). The desired system parameters are then found by curve-fitting this model to the results at each mode.

Although the dynamic cutting force provides the ideal excitation of the structure for identification experiments, such an approach can be used only if the following conditions are met: (a) the dynamics of the machine tool structure are uncoupled from the influence of the cutting process, and (b) the cutting force excites all the modes of the structure in the frequency band of interest. Additional difficulties are presented by the unknown direction of the cutting force.

Instrumentation

The identification of the machine tool's transfer matrix relies on the simultaneous measurement of the various force and displacement components given in Eq.(9.7). The machine tool used was a computer numerically controlled lathe with a turret assembly that supported up to eight cutting tools. The measurement of the three cutting force components at the tool tip was performed by a specially designed force dynamometer. The use of a commercially available device was not possible because of the particular geometry of the lathe's turret. The force dynamometer consisted of three, three-axis piezoelectric force transducers that were sandwiched under high preload between a thick steel base plate and a thick steel cover plate. The base plate was placed on five ground blocks that were inserted into what was normally the tool holder channels of the lathe's turret and were fastened by screws. The cover plate supported the cutting tool and its shape was an exact copy of the turret's design.

The use of three force transducers was necessary in order to balance the moments that were transmitted by the cutting forces to the cover plate. The triangular arrangement of the three force transducers is shown in Figure 9.21. The load that the transducers experienced depended on the geometry of this arrangement and the relative position of the application point of the cutting force. A static analysis of the forces on these transducer's showed that the maximum

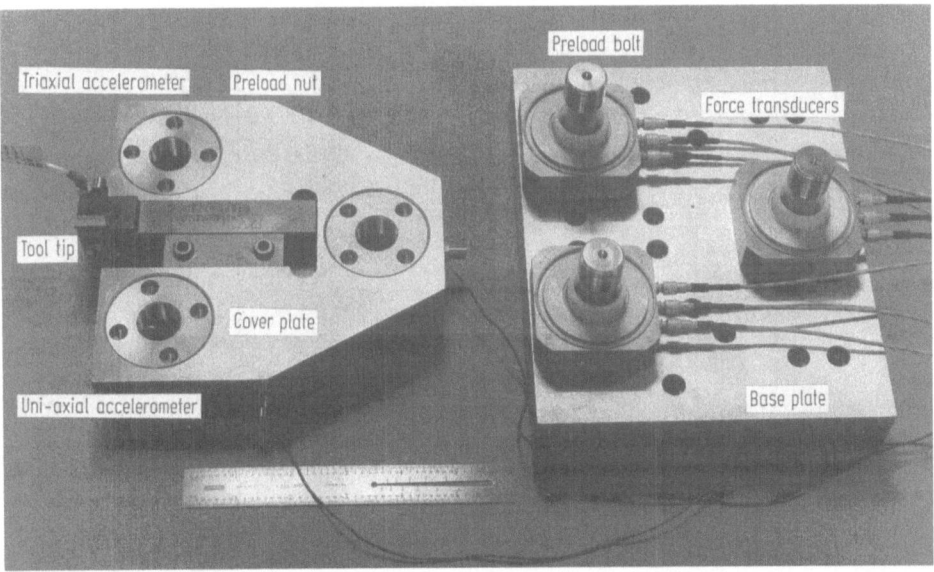

Figure 9.21 Components comprising the force dynamometer (Assembled configuration is shown in Figure 9.25a)

280

force on the dynamometer was limited to 13 kN. The magnitude of the cutting forces during the experiments were considerably below this limiting value.

Each cutting force component was balanced by the sum of the reactions of the transducers in their respective directions. Hence the measurement of each force component required the summation of the three electrical charges generated by the piezoelectric transducers for a particular direction. The transducers were selected such that they each had the same sensitivity in each of their respective axes. Therefore each sum was obtained by simply connecting the three outputs to a summing junction. Three charge amplifiers were then used to convert the summed electrical charges into three output voltages that were proportional to the magnitudes of the respective cutting force components. The sensitivities of the charge amplifiers were selected so that a full scale output of 10 V for each channel corresponded to 10 kN. A block diagram of the dynamometer's electrical operation is shown in Figure 9.22.

A high preload of 160 kN was required for each transducer in the dynamometer. For this purpose the normal force (F_z) of each transducer was monitored separately by a digital voltmeter and a preload of 10 kN was applied sequentially to each transducer. This static 10 kN preload was able to be recorded because of the very long time constant of the transducer/charge amplifier combination (recall Eq.(8.38)). Then each amplifier was reset (discharged) and the process repeated 16 times until the preload reached the final value of 160 kN.

The extremely high stiffness of the transducers combined with a relatively low mass of the cover plate resulted in a dynamometer with a first natural frequency

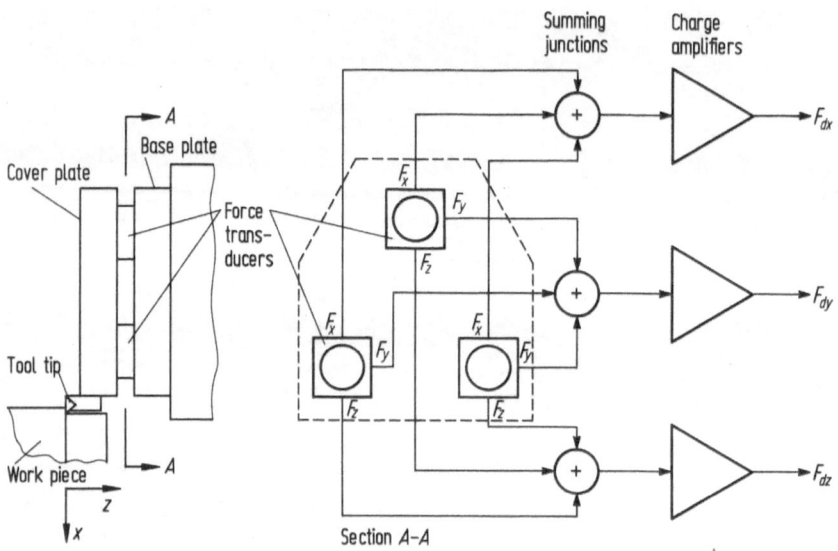

Figure 9.22   Interconnections of the force gages' output signals (All transducers have the same sensitivity)

of 6000 Hz. The actual frequency response was ±1.5 dB from 30 to 1000 Hz with a corresponding phase response of ±2.5°. If the measured force in the q-direction is $F_{dq}, q = x, y, z$ then the value of each applied force component F_q is given by (recall Eq.(8.54))

$$F_q = F_{dq} + m\ddot{q} \qquad q = x, y, z \qquad (9.8)$$

where $\ddot{q}$ is the translational acceleration of the center of mass of the cover plate in the q-direction. The mass of the cover plate was 6.7 kg. As can be seen in Figure 9.23 there is a large difference between the measured force F_{dq} and the compensated force F_q. The data shown in this figure are the normalized transfer function of the response of the force gages in the z-direction to an impulsive force in the same direction. It is seen from Eq.(9.8) that the mass compensation therefore required the measurement of the three acceleration components of the center of mass. These three acceleration components were measured with uniaxial accelerometers mounted along mutually orthogonal axes that passed through the center of mass of the cover plate as shown in Figure 9.21 (and in Figure 9.25a). Each force component F_q was then obtained using the simple operational amplifier configuration shown in Figure 9.24.

The relative displacement between the tool and the work piece was derived from the simultaneous measurement of the acceleration components of both the tool and the work piece. The Fourier transform of the acceleration signals were

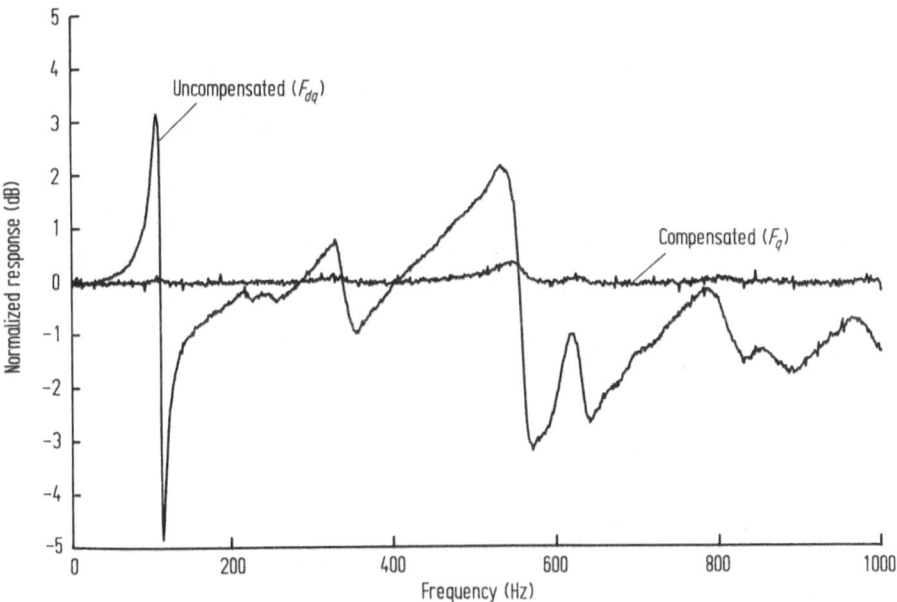

Figure 9.23 Dynamic mass compensation of the force dynamometer according to Eq.(8.54)

282

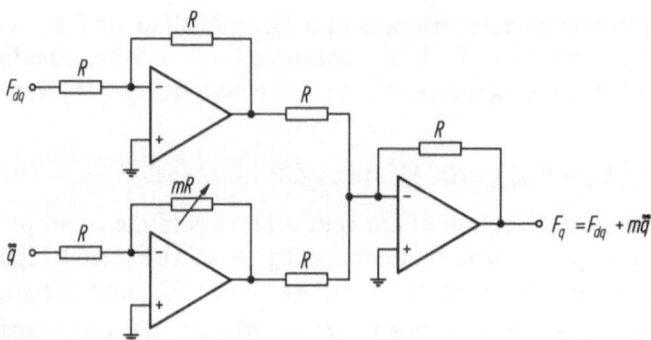

Figure 9.24 Operational amplifier configuration used to attain the mass compensation shown in Figure 9.23

subtracted from each other and the results integrated twice to yield the displacement. The three acceleration components of the tool's cutting edge were measured by a triaxial accelerometer mounted on the cutting tool as shown in Figures 9.21 and 9.25a. The work piece acceleration was measured by another triaxial accelerometer that was mounted on the cutting center of the lathe, which was pressed against the work piece with a constant load of 100 N. The diameter-to-length ratio of all the work pieces used was greater than 1 and, therefore, could be considered a rigid body within the bandwidth of interest. Consequently its displacement at the cutting point was directly determined by measuring the displacement at the cutting center. This assumed negligible rotation of the work piece.

The data acquisition system was a nine-channel simultaneously sampled A/D conversion system described in Section 5.6 and shown in Figure 5.17. Each of the three force and six acceleration signals was amplified and then passed through an anti-aliasing filter. The outputs of the filters were connected to sample-and-hold amplifiers, whose outputs were then connected to a multiplexer. The signals were sampled simultaneously and the multiplexer's channels were sequentially scanned and each output digitized by the A/D converter. The digital outputs were directed to an array processor that was programmed to receive the digitized time signals, compute their Fourier transforms, and store the results to a disk. This process was performed in real time; that is, with such speed that there were no gaps in the sampled data between the end of one time slice and the start of the next one, up to a sampling rate of about 10 kHz. The system worked in the following manner: While one time record, containing nine channels, each of 1024 samples, was being acquired the previous record was being transformed by a FFT routine. Simultaneously a third record, captured two time slices previously, was being written to the disk. In addition to this continuous data acquisition mode the system provided for a synchronous acquisition mode, which permitted the recording of a group of one or more time slices of data each time it received an external trigger signal. The system could store 62,548 1024-point transforms. The acquisition-transformation-storage process was controlled by a host computer and directed

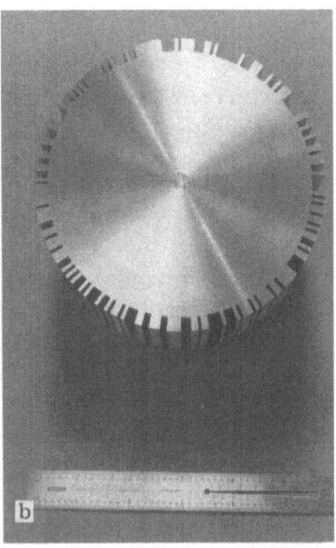

Figure 9.25 (a) Experimental setup (b) pseudorandom work piece profile used to generate interrupted cutting

by special programs running in the array processor. The actual analysis of the Fourier transformed data was done after the completion of the acquisition process by a very robust data analysis software system specially written to compute the various quantities of interest, display them and fit selected models to the results. The method of calibration of this system has been described in Section 5.7.

Experimental Procedure and Results

The broadband cutting force signal was generated by the interrupted cutting of specially machined aluminum work pieces shown in Figure 9.25b. This special work piece decoupled the measurement of the dynamic structural response from the cutting process; that is, the closed loop process shown in Figure 9.20 became open loop. The surfaces of these work pieces had been modulated with a precise pattern of channels in the axial direction to create a pseudorandom profile, i.e., a discrete-interval binary sequence of teeth (1) and channels (0) that changed state only at specified integer multiples of an elementary length Δl, where Δl was the circumference of the work piece divided by the total number of binary lengths. The total number of binary lengths selected was 255. The outer diameter of the work piece was 140 mm and, therefore, the elementary length was 1.72 mm. The height of each tooth was 6.35 mm.

Since the cutting of a single tooth of the work piece generated a pulse-like cutting force, the interrupted cutting of this work piece generated a pseudorandom force signal. If T was the period of revolution of the work piece, then the force components were pseudorandom periodic signals of period T. Hence their spectra were discrete, with amplitude information appearing at the harmonics

$$f_k = \frac{k}{T} = k \Delta f \qquad k=0,1,2,\ldots,$$

where Δf was the spacing of the spectral components (recall Eq.(2.8)). Furthermore the bandwidth of the first spectral lobe is the inverse of the duration of the elementary length; that is (recall Figure 2.4 and Example 2 of Chapter 2),

$$BW = \frac{255}{T}$$

where, as previously stated, 255 was the number of elementary lengths. Both the frequency resolution and the bandwidth of the force spectrum are functions of T, or equivalently, of the rotational speed N. Hence the choice of the rotational speed was a compromise between the resolution of the excitation and its bandwidth.

The frequency band where the identification of the transfer matrix was valid depended on the selection of N. The measured pseudorandom periodic signals were comprised of the impulse responses of the various vibration modes of the structure. Since the data were to be analyzed with the FFT implementation of the DFT it was necessary to minimize the magnitude of the response corresponding to the mode with the lowest natural frequency. According to the discussion in Section 2.5.2 this required that $T > \tau$, where τ was the settling time of the lowest mode. Assuming a lightly damped single-degree-of-freedom system with natural frequency f_n and damping ratio c, the time to settle to within 5% of its final value is, from Eq.(3.25), given by

$$\tau = \frac{3}{2\pi c f_n}.$$

For a damping ratio $c = 0.1$ (a reasonable value for the structure) the mode could be identified only if its natural frequency satisfied the inequality

$$f_n \geq \frac{15}{\pi \tau} \geq \frac{15}{\pi T} = f_l$$

where f_l denotes the lower limit of the method.

The upper limit of the frequency band was chosen to be equal to one half of the bandwidth of the first spectral lobe; that is, $f_h = BW/2$. Thus, within the band

$$f_l = \frac{15}{\pi} \Delta f \leq f \leq \frac{BW}{2} = \frac{255}{2} \Delta f = f_h \qquad (9.9)$$

the cutting force had enough energy to excite the machine tool structure. This is shown in Figure 9.26, which gives the measured components of the cutting force as a function of frequency.

The intrinsic periodicity of the pseudorandom cutting method imposed specific signal processing requirements for the correct Fourier analysis of both the excitation and the response signals. Since an FFT algorithm was used for this purpose,

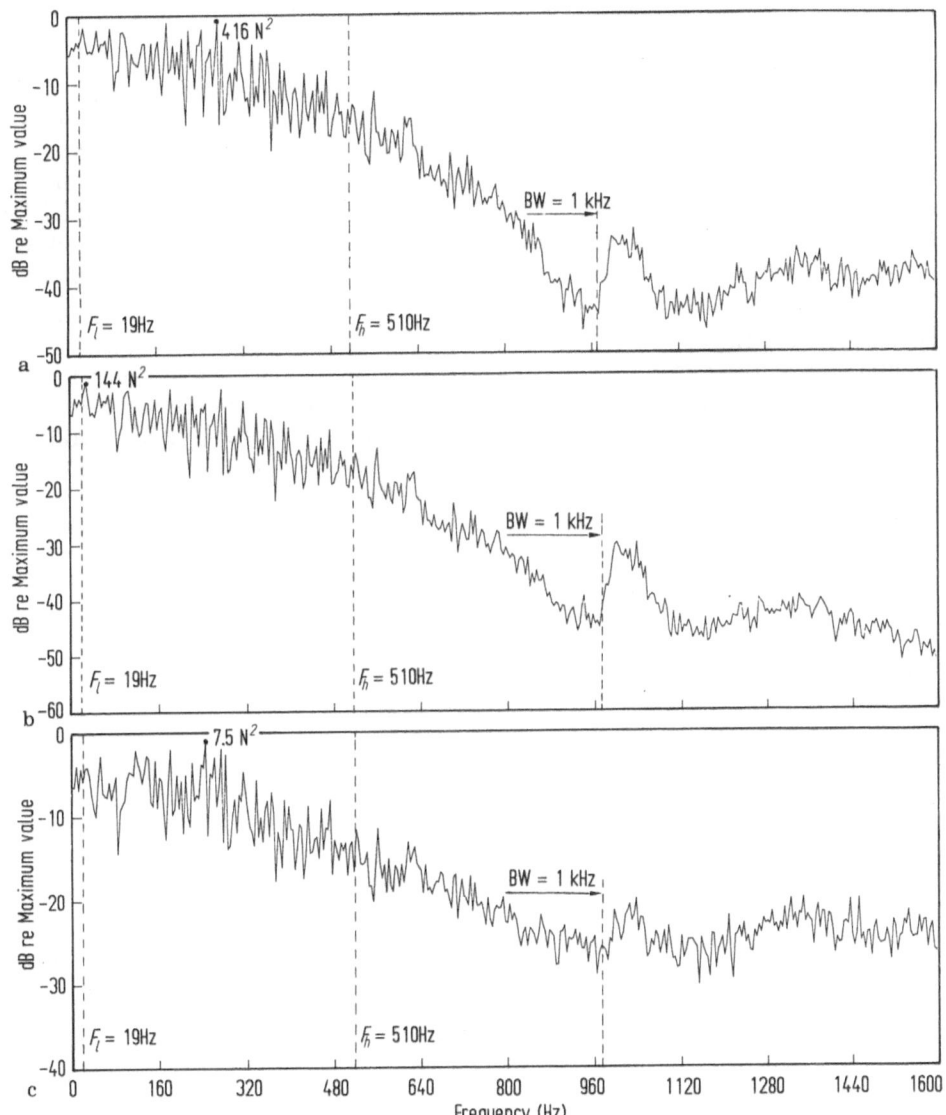

Figure 9.26 Power spectra of the cutting force components: (a) F_x (b) F_y (c) F_z

the time slice length had to be synchronized with the signal period T. If this were not the case leakage would occur (recall Section 2.5.2). In this setup the sampling rate f_s samples/s was constant, independent of the spindle's rotational speed. Thus $f_s T = N_s$, where N_s was the chosen number of samples per time slice, 1024 in this case. The cutting experiments were performed at 240 rpm, which corresponded to a period of revolution of $T = 0.25$ s; therefore, the sampling rate of the A/D converter was set to 4096 samples/s ($= 1024/.25$). Since f_s was chosen to be $2.56 f_c$, where f_c was the filter's cutoff frequency, this required an anti-aliasing filter with $f_c = 1600$ Hz. The selected spindle speed was the lowest value that gave a speed deviation of less than $\pm 1\%$ from the nominal speed. Thus the variation of the period of the measured signals was less than 1%. In order to minimize further the slight leakage resulting from the period's variation, a trigger signal was used to initiate the sampling of each time slice from a specific point on the circumference of the work piece. Possible overlapping of a small portion of the consecutive time slices was avoided by recording the data every other revolution.

Shown in Figure 9.27 are the transfer functions G_{zz} and G_{zx} generated by interrupted, orthogonal facing of the pseudorandom work piece at a rotational speed of 240 rpm. The resolution of the spectra at this speed was 4 Hz ($= 1/.25$) and the bandwidth of the first lobe was 1020 Hz ($= 255/.25$). Then from Eq.(9.9) the lower and upper limits of the usable identification bandwidth were approximately 19 and 510 Hz, respectively. Fitting a single-degree-of-freedom model to the data shown in Figure 9.27 yielded a damping coefficient for G_{zz} of 0.127 and for G_{zx} a value of 0.14.

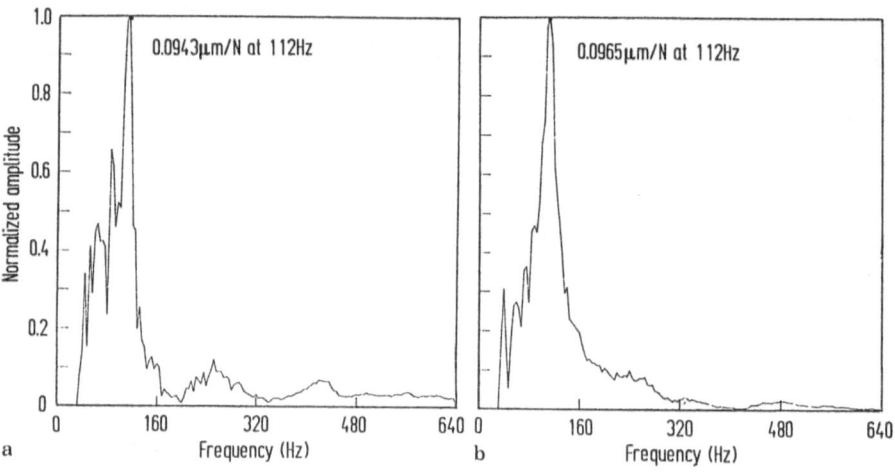

Figure 9.27 Normalized magnitudes of the transfer functions at 240 rpm: (a) G_{zz} (b) G_{zx}

References

1. Magrab, E. B., "Determination of the Viscoelastic Shear Modulus Using Forced Torsional Vibrations," *Journal Research National Bureau of Standards*, Vol. 89, No. 2, pp. 193-207, March-April 1984.

2. Magrab, E. B., "Vertical Machining Workstation of the AMRF: Equipment Integration", in *Integrated and Intelligent Manufacturing*, C. R. Liu and T. C. Chang, Eds., ASME Publication PED-Vol. 21, pp. 83-100, 1986.

3. Minis, I. E., Magrab, E. B., and Pandelidis, I. O., "Improved Methods for the Prediction of Chatter in Turning, Part I: Determination of Structural Response Parameters", *Journal of Engineering for Industry*, 112, 1, pp.12-20, February 1990.

4. Patterson, S. R., and Magrab, E. B., "Design and Testing of a Fast Tool Servo for Diamond Turning," *Precision Engineering*, Vol. 7, No. 3, pp. 123-128, July 1985.

Copyright Acknowledgements

Figure 1.1 and Table 1.1 from H. W. Coleman and W. G. Steele, *Experimentation and Uncertainty Analysis for Engineers*. Copyright (c)1989 John Wiley & Sons, Inc. Reprinted by permission of John Wiley & Sons, Inc. Figures 2.11 and 2.14 from E. O. Brigham, *The Fast Fourier Transform*. Copyright (c) 1974, pp. 81, 95. Reprinted by permission of Prentice Hall, Inc., Englewood Cliffs, New Jersey. Figure 2.15 from M. J. Miller, "Discrete Signals and Frequency Spectra", in *Handbook of Measurement Science, Vol. 1, Theoretical Fundamentals*, P. H. Sydenham, Ed. Copyright (c) 1982 John Wiley & Sons, Ltd. Reprinted by permission of John Wiley & Sons, Inc. Figure 2.16 and Table 6.1 from E. B. Magrab and D. S. Blomquist, *The Measurement of Time-Varying Phenomena: Fundamentals and Applications*. Copyright (c) 1971 John Wiley & Sons, Inc. Reprinted by permission of John Wiley & Sons, Inc. Figures 3.2 and 3.4 from J. Javid and E. Brenner, *Analysis, Transmission, and Filtering of Signals*. Copyright (c) 1963 McGraw-Hill, Inc. Reprinted by permission of McGraw-Hill, Inc. Figures 5.1 to 5.3 from D. H. Sheingold, Ed., *Analog-Digital Conversion Handbook*, 3rd Ed. Copyright (c) 1986, pp. 317, 319, 320. Reprinted by permission of Prentice Hall, Inc., Englewood Cliffs, New Jersey. Figures 5.4, 5.15 and 5.17 from G. B. Clayton, *Data Converters*. Copyright (c) 1982 Macmillan Press, Ltd. Reprinted by permission of Macmillan Press, Ltd. Figure 5.8 from Patrick H. Garrett, *Analog I/O Design: Acquisition, Conversion, Recovery*. Copyright (c) 1981, p.180. Reprinted by permission of Prentice Hall, Inc., Englewood Cliffs, New Jersey. Figures 5.9 and 5.19 from Richard J. Higgins, *Electronics with Digital and Analog Integrated Circuits*. Copyright (c) 1983, pp. 262, 275. Reprinted by permission of Prentice Hall, Inc., Englewood Cliffs, New Jersey. Figures 5.10 and 5.11 from A. Kinstler, "Are All High Resolution DMMs Created Equal?", *Test & Measurement World*, June, 1987. Copyright (c) 1987 Cahners Publishing Company. Reprinted by permission of Cahners Publishing Company. Figures 6.1 and 6.2 from "A Lock-in Primer". Copyright (c) 1986 EG&G Princeton Applied Research. Reprinted by permission of EG&G Princeton Applied Research. Table 6.2 from *DC Power Supply Catalog*. Copyright (c) 1989 Hewlett-Packard Company. Reprinted courtesy of Hewlett-Packard Company. Figures 6.5, 8.28 and 8.30 to 8.33 from J. W. Dally, W. F. Riley and K. G. McConnell, *Instrumentation for Engineering Measurements*. Copyright (c) 1984 John Wiley & Sons. Reprinted by permission of John Wiley & Sons. Figure 7.3 and Section 7.4.1 from G. F. Lang, "Bits, Bytes, Baud, Bell, and Bull", *S)V Sound and Vibration*, 21, 9. Copyright (c) 1987 Acoustical Publications, Inc. Reprinted by permission of Acoustical Publications, Inc. Figures 8.3 to 8.6 and 8.18 from E. O. Doebelin, *Measurement Systems: Applications and Design*, 4th Ed. Copyright (c) 1990 McGraw-Hill, Inc. Reprinted by permission of McGraw-Hill, Inc. Figure 8.8 from D. A. Krohn, *Fiber Optic Sensors: Fundamentals and Applications*. Copyright (c) 1988 Instrument Society of America. Reprinted by permission of Instrument Society of America. Figure 8.10 from H. V. Malmstadt and C. G. Enke, *Digital Electronics for Scientists*. Copyright (c) 1969, p. 282. Reprinted by permission of The Benjamin/Cummings Publishing Company. Figures 8.11 to 8.13 and Section 8.2.9 from *Compumotor Programmable Motion Control Catalog*. Copyright (c) 1988 Compumotor Division, Parker-Hannifin Corporation. Reprinted by permission of Parker-Hannifin Corporation. Figure 8.21 and part of Section 8.3.1 from E. B. Magrab, *Environmental Noise Control*. Copyright (c) 1975 John Wiley & Sons, Inc. Reprinted by permission of John Wiley & Sons, Inc. Figures 8.42 and 8.43 from G. Buzdugan, E. Mihailescu and M. Rades, *Vibration Measurement*. Copyright (c) 1986 Kluwer Academic Publishers. Reprinted by permission of Kluwer Academic Publishers. Figure 8.47 from L. M. Fingerson and P. Freymuth, "Thermal Anemometers", in *Fluid Mechanics Measurements*, R. J. Goldstein, Ed. Copyright (c) 1983 Taylor & Francis Group. Reprinted by permission of the Taylor & Francis Group. Figures 9.25 to 9.27 and Section 9.5 from I. E. Minis, E. B. Magrab and I. O. Pandelidis, "Improved Methods for the Prediction of Chatter in Turning, Part I: Determination of Structural Response Parameters", *Journal of Engineering for Industry*, 112, 1. Copyright (c) 1990 American Society of Mechanical Engineers. Reprinted by permission of American Society of Mechanical Engineers.

Copyright Acknowledgments

Subject Index